Hydrothermal Processing of Biomass for Hydrogen and Bio-oil Production

Yang Guo · Donghai Xu · Shuzhong Wang

Hydrothermal Processing of Biomass for Hydrogen and Bio-oil Production

西安交通大學出版社
XI'AN JIAOTONG UNIVERSITY PRESS

Springer

Yang Guo
School of Energy and Power Engineering
Xi'an Jiaotong University
Xi'an, Shaanxi, China

Donghai Xu
School of Energy and Power Engineering
Xi'an Jiaotong University
Xi'an, Shaanxi, China

Shuzhong Wang
School of Energy and Power Engineering
Xi'an Jiaotong University
Xi'an, Shaanxi, China

ISBN 978-981-96-6335-4 ISBN 978-981-96-6336-1 (eBook)
https://doi.org/10.1007/978-981-96-6336-1

Jointly published with Xi'an Jiaotong University Press
The print edition is not for sale in China (Mainland). Customers from China (Mainland) please order the print book from: Xi'an Jiaotong University Press.

This Springer imprint is published by the registered company Springer Nature Singapore Pte Ltd.
The registered company address is: 152 Beach Road, #21-01/04 Gateway East, Singapore 189721, Singapore

Contents

Introduction

Biomass offers numerous advantages, such as low pollution, abundant availability, and renewability. It ranks as the fourth-largest global energy source by total consumption, following coal, oil, and natural gas, with an estimated resource equivalent to about 85 billion tons of standard coal [1]. As the only sustainable organic carbon source currently available, biomass is considered the most promising alternative to fossil fuels for energy and chemical production. Utilizing biomass not only decreases dependence on fossil fuels and lowers greenhouse gas emissions but also fosters sustainable development in agriculture and industry, leading to job creation. However, the biomass sector faces challenges like technological efficiency, supply of raw materials, and production costs. Nevertheless, advancements in technology and supportive policies are paving the way for broader commercial applications in biomass energy and chemicals. This sector has the potential to make significant contributions to global energy transition and sustainable development efforts.

Biomass consists of organic materials sourced from plants, animals, and microorganisms. As a renewable resource, it encompasses various forms, including wood, crop residues, animal manure, algae, and organic waste. These materials can be categorized into different types based on their sources and chemical compositions, such as woody biomass, agricultural biomass, animal biomass, microbial biomass, and urban organic waste. Biomass development technologies are divided into conventional methods, including anaerobic digestion [2] and sanitary landfill [3]. Thermochemical methods [4], such as carbonization [5], liquefaction, pyrolysis/gasification, incineration, and hydrothermal treatment [6]. Hydrothermal treatments are gaining attention for their high efficiency, cost-effectiveness, and environmental benefits. This technology is particularly effective for processing biomass with high moisture content, as it reduces the energy needed for drying and dehydration. Among hydrothermal treatment methods, hydrothermal liquefaction (HTL) and supercritical water gasification (SCWG) are noteworthy for their ability to convert biomass directly into liquid and gas fuels [7]. HTL typically operates at temperatures between 250 and 370 °C, producing biocrude (a type of bio-oil) as the primary output. In

Y. Guo et al., *Hydrothermal Processing of Biomass for Hydrogen and Bio-oil Production*, https://doi.org/10.1007/978-981-96-6336-1_1

contrast, SCWG operates at temperatures and pressures exceeding the critical point of water ($\geq$ 374 °C and 22.1 MPa), resulting in the generation of hydrogen-rich gases as the main product [8].

1 Hydrogen Production by Supercritical Water Gasification

Supercritical water gasification is an innovative biomass conversion technology that utilizes the unique properties of supercritical water (SCW) to convert biomass into valuable products. When water is subjected to supercritical conditions, its physicochemical properties change dramatically compared to those in gaseous or liquid states. This transformation allows SCW to act not only as a reactant but also as a catalyst, influencing the reaction pathways. In the SCWG process, biomass is rapidly heated in supercritical water, which possesses gas-like diffusivity and liquid-like solvating power. This enables SCW to penetrate biomass efficiently, facilitating chemical reactions. As a result, biomass undergoes swift decomposition and hydrolysis, leading to the formation of smaller organic molecules and gases, including hydrogen and methane. The entire process can be completed in just seconds and does not require oxygen, which helps prevent biomass oxidation and reduces energy losses [9, 10].

The distinct physicochemical properties of SCW are what primarily differentiate this reaction from other thermochemical processes. A key factor in this distinction is hydrogen bonding, which highlights the differences between water's macroscopic and microscopic characteristics. As temperature increases, the chemical structure of water, typically stable at room temperature, can be altered or disrupted. Under standard temperature and pressure conditions, water has a high dielectric constant of around 80, largely due to the strong hydrogen bonding. However, as temperature and pressure rise, this dielectric constant gradually decreases, usually falling within the range of 20 to 30 in supercritical conditions. This change in the dielectric constant affects water's solubility properties. In supercritical conditions, the dielectric constant of water approximates that of many common organic compounds, allowing SCW to act as an effective solvent for nonpolar organic substances. Conversely, the solubility of highly polar inorganic compounds significantly decreases, resulting in the separation of dissolved organic compounds from water.

An increase in temperature significantly affects the viscosity of water. At the critical point, the viscosity of water (2.98×10^{-5} Pa s) becomes comparable to that of vapor (2×10^{-5} Pa s). This reduced viscosity enhances the diffusion coefficient of water, creating favorable conditions for rapid reactions. As illustrated in Fig. 1, near the critical point, both the density and ion product of water decrease compared to their normal state. The density transitions from that of a liquid (1 g/cm^3) to that of low-pressure vapor (<0.0011 g/cm^3) with variations in temperature and pressure. The ion product of water is influenced by both temperature and density, with density

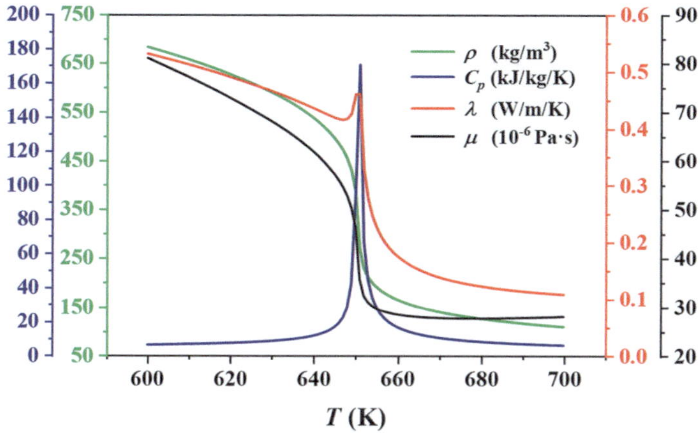

Fig. 1 Physical properties of supercritical water vary with temperature for 23 MPa [12]

having a more pronounced effect. Below the critical point, higher density corresponds to an increased ion product, promoting ionic reaction mechanisms. Conversely, as water density decreases beyond the critical point, the ion product declines, indicating a shift toward free radical mechanisms under these conditions [11].

The gasification of biomass in SCW presents several advantages, including high efficiency, a significant molar fraction of hydrogen, and the elimination of drying requirements for wet biomass, distinguishing it from other biomass conversion methods [9]. SCWG is particularly effective for biomass with high moisture content, as it enhances energy conversion efficiency by avoiding the drying step. This streamlined process produces hydrogen-rich gas with minimal carbon monoxide and suppresses the formation of tar and char [13]. Despite these benefits, SCWG faces challenges such as material compatibility issues and the necessity for high temperatures and pressures. The distribution of products in the SCWG process is highly dependent on operational conditions. Research indicates that high hydrogen yields can only be achieved at temperatures exceeding 600 °C [14, 15]. At temperatures below 450 °C, methane predominates. As water density rises, methane yield increases, but hydrogen yield decreases. Unlike many exothermic reactions, SCWG cannot sustain the minimum required temperature autonomously, making economic efficiency a significant barrier to its advancement [16]. To address the challenges of high reaction conditions and reduce capital and operational costs, one viable strategy is to lower activation energy by using appropriate hydrothermal catalysts. Thus, exploring catalysts that function effectively in hydrothermal environments is crucial for the development of this technology.

2 Bio-oil Production via Hydrothermal Liquefaction in Subcritical Water

Three primary methods are used for biocrude production: transesterification, pyrolysis, and HTL [17, 18]. Transesterification entails the reaction of triglycerides with methanol at 60 °C and atmospheric pressure, leading to the formation of diglycerides and monoglycerides, and ultimately producing glycerol and fatty acid methyl esters with the aid of an alkali catalyst [19]. Pyrolysis involves the thermal decomposition of biomass in an inert atmosphere at temperatures ranging from 300 to 900 °C, resulting in the generation of combustible gases or bio-oil [20]. HTL is a method that converts biomass into gases, solids, biocrude, and aqueous phases using hot compressed water at temperatures between 300 and 350 °C and pressures ranging from 5 to 20 MPa. This process can be performed with or without a catalyst. Afterward, an organic solvent is added to extract and separate the oil fraction [21]. The primary chemical reactions involved in biomass HTL can be categorized into three groups based on the main organic components of biomass: proteins, carbohydrates, and lipids. These reactions include hydrolysis, degradation, decarboxylation, deamination, repolymerization, and various other chemical processes [22]. HTL presents several advantages over other methods [23, 24]. First, HTL is an energy-efficient process that avoids the drying of biomass, significantly lowering energy consumption, which usually represents over 75% of the total energy required. Secondly, the resulting biocrude has a low water content and a high heating value. Third, heat loss is minimized since the reaction medium maintains a constant state. Finally, HTL achieves high biomass utilization rates by not only extracting lipids but also converting other biomass components into biocrude. Thus, HTL has considerable potential for producing biocrude from biomass.

Figure 2 depicts the key conversion pathways for the three main components—lipids, proteins, and carbohydrates during HTL. Lipids play a crucial role in biocrude production, breaking down into glycerol and fatty acids through hydrolysis. The glycerol then degrades and condenses, leading to the formation of water-soluble substances and gaseous products. Some fatty acids undergo decarboxylation and hydrogenation to generate long-chain alkanes, while others are converted to alcohols through deoxygenation, which can subsequently react with fatty acids to produce fatty acid esters. Additionally, specific fatty acids may react with NH_3 produced from amino acids to form fatty acid amides [25]. Importantly, lipids do not produce solid by-products. Proteins are the main source of nitrogenous compounds in HTL products. They undergo hydrolysis to release amino acids, which then further decompose into carboxylic acids, ammonia, amides, and carbon dioxide through concurrent deamination and decarboxylation reactions [26].

Following hydrolysis, non-cellulosic carbohydrates yield glucose and fructose, with glucose also able to isomerize into fructose, albeit at a much lower rate. Consequently, glucose primarily degrades into phenols and furans through ring-opening and cyclization processes. The dehydration of fructose produces 5-hydroxymethyl furfural, while glucose participates in Maillard reactions with amino acids resulting

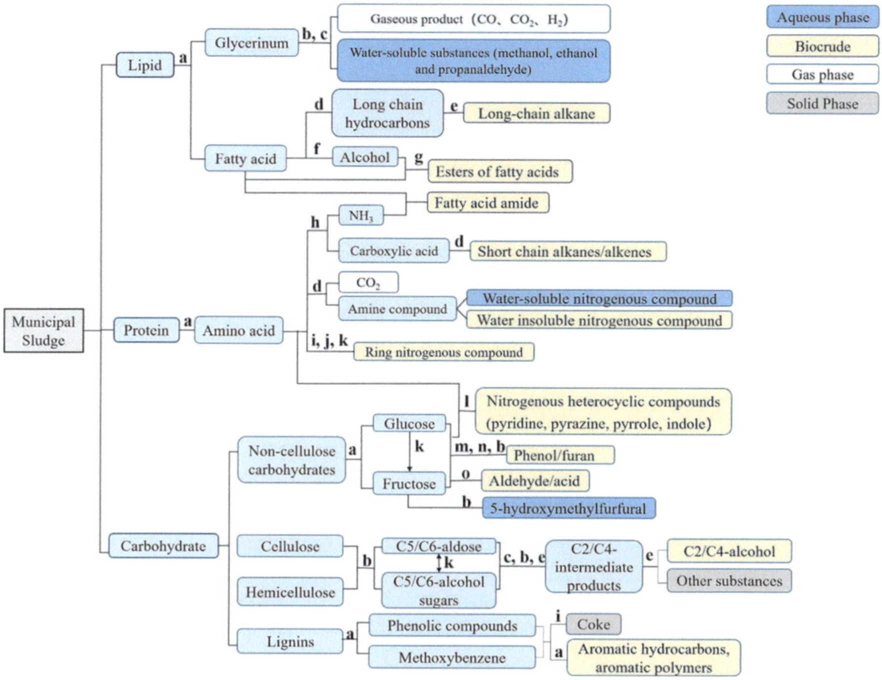

Fig. 2 Simplified conversion pathways of major components during HTL of municipal sludge. **a** hydrolysis, **b** degradation, **c** condensation, **d** decarboxylation, **e** hydrogenation, **f** deoxygenation, **g** esterification, **h** deamination, **i** polymerization, **j** dehydration, **k** isomerization, **l** Maillard reactions, **m** cyclization, **n** ring opening, **o** C–C cleavage

from protein hydrolysis, leading to the formation of nitrogenous heterocyclic compounds such as pyridine, pyrazine, pyrrole, and indole [26, 27]. Lignocelluloses consist of cellulose, hemicellulose, and lignin, each following unique reaction pathways during HTL. Initially, cellulose and hemicellulose decompose into glucose, pentoses, and hexoses, which subsequently break down into aldehydes, ketones, and other compounds [28]. Lignins contain a variety of bond types, including β-O-4, α-O-4, 4-O-5, and β-5 [29]. The cleavage of these bonds leads to the formation of various phenolic compounds and methylbenzenes. Subsequently, some of these compounds undergo hydrolysis to form aromatic hydrocarbons and aromatic polymers, while others polymerize into coke through an ionic mechanism [30].

3 Catalytic Hydrothermal Upgrading of Biocrude

Biocrude typically has a dark brown color, a tar-like viscosity, and a smoky odor [31]. It contains over 400 organic compounds, mainly phenols, hydrocarbons, fatty acids, carbonyl compounds, nitrogenous substances, as well as various gases and solids [32]. Based on its functional groups, biocrude includes monoaromatic hydrocarbons, monoheterocyclic compounds, aliphatic compounds, and oxygen-containing compounds (e.g., 4-chloro-2-nitrobenzyl alcohol, and 3-methylbutanol ether), nitrogen-containing compounds (e.g., 2,5-piperazine dione and 3-benzyl-6-isopropyl), and polyaromatic compounds (e.g., 7-ethyl-4,6-heptanedione and 2-(m-torrazole) naphthalene) [33]. Biocrude holds potential as a gas turbine fuel [34], and its physical and chemical properties primarily hinge factors such as raw material [35], reaction conditions, including temperature [36], residence time [37], solvent type [38], catalyst type and amount [39] and process route [40, 41]. The complex and variable nature of these conditions leads to notable differences in both the yield and quality of biocrude [42]. Table 1 shows the physical and chemical properties of biocrude produced from biomass via HTL, highlighting the significant variation in characteristics based on the type of biomass used. For example, the nitrogen content in biocrude from microalgae, cow manure, or sludge is significantly higher than that from bark and straw, which is indicative of the protein content in the biomass. While the oxygen content does not show a clear relationship with the type of biomass, its average is around 16.8 wt%, which is much higher than that of standard diesel fuel. The carbon and hydrogen contents of biocrude vary from 45.7 to 80.2 wt% and 6.0 to 11.6 wt%, respectively, both of which are lower than those in diesel. Additionally, factors such as water content, higher heating value (HHV), density, and other properties of biocrude derived from biomass still differ considerably from conventional transportation fuels.

The composition and characteristics of biocrude produced through biomass HTL can closely mimic those of fossil fuels when suitable HTL conditions and catalysts are employed. However, several challenges still impede the widespread adoption of biocrude [32, 42, 57]. The high viscosity of biocrude complicates transportation logistics. Moreover, it has higher oxygen and carbon contents, as well as a greater HHV compared to conventional fuels, which limits its efficiency for power generation. Additionally, the presence of nitrogen and sulfur compounds leads to the formation of nitrogen oxides and sulfur oxides during combustion. Additionally, with a total acid value reaching 256 mg of KOH/g, biocrude can lead to considerable corrosion issues. These challenges emphasize the necessity of upgrading biocrude after biomass HTL.

Figure 3 illustrates various typical biocrude upgrading technologies. Among these, hydrogenation stands out as an effective method for improving biocrude quality. This process involves treating biocrude with hydrogen under specific conditions, which aids in converting oxygenated compounds into water and hydrocarbons through the hydrodeoxygenation (HDO) reaction, representing the simplest form of this transformation. Simultaneously, the nitrogen in biocrude is removed as ammonia [42].

Table 1 Physical and chemical properties of biocrude derived from biomass HTL

Biocrude	Element composition (wt%)					Density (kg/m^3)	HHV (MJ/kg)	References
Biomass	C	H	O	N	S			
Microalgae (Spirulina platensis)	45.73	5.96	47.04	1.27	–[a]	1380	15.60	[18]
Microalgae (D. tertiolecta)	68.53	8.65	19.75	2.89	0.18	–	31.97	[43]
Microalgae (Chlorella)	74.70	10.40	5.50	7.70	0.76	–	39.80	[44]
Microalgae (C. pyrenoidosa)	56.14	6.49	32.40	4.97	–	1100–1250	26.62	[45]
Microalgae (Dunaliella tertiolecta)	78.60	10.40	5.30	4.20	0.50	960	–	[46]
Microalgae	66.01	11.56	12.43	9.54	0.46	1020	36.76	[41]
Microalgae (Spirulina)	70.31	8.72	11.27	9.70	–	970	34.33	[47]
Microalgae and lemon peel	69.18	8.17	12.92	9.73	–	–	33.00	[48]
Swine manure	72.58	9.76	13.19	4.47	–	–	36.05	[49]
Swine manure	73.66	11.61	11.48	3.25	–	–	33.98	[50]
Cow manure	73.72	8.13	16.76	1.39	–	–	35.53	[51]
Willow	78.15	8.50	12.40	0.95	–	–	38.57	[52]
Aspen	75.20	8.20	15.80	0.50	0.3	–	34.30	[53]
Bark	80.21	9.42	9.95	0.42	–	–	38.87	[54]
Beech wood	65.95	7.04	27.01	–	–	–	27.53	[55]
Straw	67.89	6.42	23.18	0.75	0.56	–	27.29	[51]
Sludge	75.13	9.41	8.97	5.31	1.17	–	37.35	[56]
Diesel[b]	86.00	13.20	0.80	–	–	817	47.70	[18]

[a] "–" means that the value is not reported
[b] It refers to general diesel

Moreover, specific hydrotreatment processes can effectively eliminate sulfur from biocrude. The removal of heteroatoms can lead to a notable increase of 14–28% in the HHV of biocrude, along with a decrease in total acidity, bringing it more in line with refinery specifications [58]. Catalytic hydrotreatment of biocrudes from municipal solid waste HTL and food waste HTL demonstrates that the steady-state densities of

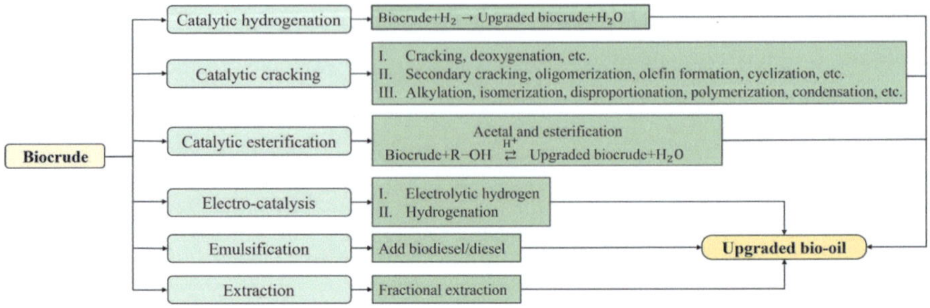

Fig. 3 Upgrading technologies for biocrude

these biocrudes reach up to 0.79 and 0.81 g/mL after treatment, respectively. Additionally, the increase in alkane (cetane) content makes the biocrude more similar to diesel-like biofuel.

In the hydrogenation process, temperatures above 350 °C can lead to the conversion of biocrude into coke, which reduces the yield of upgraded bio-oil. The formation of coke on the catalyst surface results in catalyst deactivation [59]. Nevertheless, high temperatures can improve the physicochemical properties of biocrude and facilitate deoxygenation. Increased hydrogen pressure, generally between 8 and 12 MPa, improves hydrodeoxygenation efficiency but raises operational costs. Additionally, methanol and ethanol can be used as in situ hydrogen sources for biocrude hydrogenation. For instance, using ethanol as a hydrogen source can achieve a hydrocarbon yield of 82.8 wt% at 360 °C [60]. Rapid feed rates can lead to shorter residence times, resulting in a potential 10–30% decrease in deoxygenation conversion. Hydrogenation commonly utilizes catalysts such as sulfides, oxides, noble metals, phosphides, carbides, and nitrides. To mitigate catalyst deactivation, strategies include blending biocrude feedstock with hydrogen-donating solvents, using activated carbon as catalyst carriers, and lowering reaction temperatures. For example, Kunwar et al. [61] developed a biocatalyst (bio-Pd/C) from low-grade waste through biorefining. This catalyst reduced the O, N, and S content in biocrude by about 72%, 52%, and 93%, respectively, while enhancing the biocrude's HHV from 35 MJ/kg to 41 MJ/kg, and is notable for its significantly lower cost.

4 Significance of This Book

This book holds significant importance in the realm of biomass energy utilization for several reasons. Firstly, it offers a meticulous exploration into the intricate technology behind harnessing energy from biomass sources. By delving into specific technological processes and dissecting reaction mechanisms, it provides readers with a comprehensive understanding of the underlying principles. One of the book's standout features is its detailed examination of key topics such as hydrogen production through SCWG of biomass, bio-oil production via hydrothermal liquefaction, and the hydrotreating process for refining bio-oil. Through thorough investigation and analysis, it elucidates not only the catalysts utilized in these processes but also the intricate reaction pathways involved in biomass gasification, liquefaction, and oil refining.

Moreover, the book's systematic approach to elucidating the reaction mechanisms associated with biomass energization sets it apart. This in-depth treatment not only enhances readers' comprehension but also serves as a valuable resource for researchers, engineers, and graduate students seeking to delve deeper into the realm of bio-oil and hydrogen production from biomass. By catering to a diverse audience interested in new energy science, energy and power engineering, and green energy development, this book serves as a pivotal tool for advancing knowledge and innovation in sustainable energy technologies. Its wealth of insights and comprehensive coverage make it an indispensable resource for those at the forefront of shaping the future of renewable energy.

References

1. L. Dessbesell, M. Paleologou, M. Leitch, R. Pulkki, C. Xu, Global lignin supply overview and kraft lignin potential as an alternative for petroleum-based polymers. Renew. Sustain. Energy Rev. **123**, 109786 (2020)
2. B.H. Yu, X.M. Xiao, J.W. Wang, M. Hong, C. Deng, Y.Y. Li, J.Y. Liu, Enhancing phosphorus recovery from sewage sludge using anaerobic-based processes: current status and perspectives. Biores. Technol.. Technol. **341**, 125899 (2021)
3. W. Wu, Z. Zhou, J.Z. Yang, G. Chen, J. Yao, C.Q. Tu, X.D. Zhao, Z. Qiu, Z.C. Wu, Insights into conditioning of landfill sludge by $FeCl_3$ and lime. Water Res. **160**, 167–177 (2019)
4. G.Y. Jiang, D.H. Xu, B.T. Hao, L. Liu, S.Z. Wang, Z.Q. Wu, Thermochemical methods for the treatment of municipal sludge. J. Clean. Prod. **311**, 127811 (2021)
5. V.S. Varma, S. Nashine, C.V. Sastri, A.S. Kalamdhad, Influence of carbide sludge on microbial diversity and degradation of lignocellulose during in-vessel composting of agricultural waste. Ecol. Eng. **101**, 155–161 (2017)
6. G. Boniardi, A. Turolla, L. Fiameni, E. Gelmi, F. Malpei, E. Bontempi, R. Canziani, Assessment of a simple and replicable procedure for selective phosphorus recovery from sewage sludge ashes by wet chemical extraction and precipitation. Chemosphere **285**, 131476 (2021)
7. Z.J. Xu, J. Zhou, Y.D. Liu, L.F. Gu, X.J. Wu, X.Y. Zhang, Effect of hydrothermal carbonization on dewatering performance of dyeing sludge. RSC Adv. **8**, 38574–38581 (2018)
8. S.S. Toor, L. Rosendahl, A. Rudolf, Hydrothermal liquefaction of biomass: a review of subcritical water technologies. Energy **36**, 2328–2342 (2011)
9. P.E. Savage, Organic chemical reactions in supercritical water. Chem. Rev. **99**, 603–621 (1999)
10. A. Kruse, E. Dinjus, Hot compressed water as reaction medium and reactant: properties and synthesis reactions. J. Supercrit. FluidsSupercrit. Fluids **39**, 362–380 (2007)
11. J.T. Henrikson, Z. Chen, P.E. Savage, Inhibition and acceleration of phenol oxidation by supercritical water. Ind. Eng. Chem. Res. **42**, 6303–6309 (2003)
12. X. Li, H. Wang, Y. Li, H. Jin, Effects of physical properties of supercritical water on coarse graining of particle cluster. Particuology **82**, 166–178 (2023)
13. L.J. Guo, Y.J. Lu, X.M. Zhang, C.M. Ji, Y. Guan, A.X. Pei, Hydrogen production by biomass gasification in supercritical water: a systematic experimental and analytical study. Catal. Today. Today **129**, 275–286 (2007)

14. H.Q. Tang, K. Kitagawa, Supercritical water gasification of biomass: thermodynamic analysis with direct Gibbs free energy minimization. Chem. Eng. J. **106**, 261–267 (2005)
15. J. Yanik, S. Ebale, A. Kruse, M. Saglam, M. Yüksel, Biomass gasification in supercritical water: II. Effect of catalyst. Int. J. Hydrogen Energy **33**, 4520–4526 (2008)
16. Y. Calzavara, C. Joussot-Dubien, G. Boissonnet, S. Sarrade, Evaluation of biomass gasification in supercritical water process for hydrogen production. Energy Convers. Manage. **46**, 615–631 (2005)
17. C.Y. Yang, R. Li, C. Cui, S.P. Liu, Q. Qiu, Y.G. Ding, Y.X. Wu, B. Zhang, Catalytic hydroprocessing of microalgae-derived biofuels: a review. Green Chem. **18**, 3684–3699 (2016)
18. Y.F. Xu, Y.H. Hu, Y.B. Peng, L.L. Yao, Y.H. Dong, B.X. Yang, R.H. Song, Catalytic pyrolysis and liquefaction behavior of microalgae for bio-oil production. Biores. Technol.. Technol. **300**, 122665 (2020)
19. Y. Chisti, Biodiesel from microalgae. Biotechnol. Adv.. Adv. **25**, 294–306 (2007)
20. P. Manara, A. Zabaniotou, Towards sewage sludge based biofuels via thermochemical conversion—a review. Renew. Sustain. Energy Rev. **16**, 2566–2582 (2012)
21. Z. He, D.H. Xu, S.Z. Wang, H.F. Zhang, Z.F. Jing, Catalytic upgrading of water-soluble biocrude from hydrothermal liquefaction of chlorella. Energy Fuels **32**, 1893–1899 (2018)
22. L.J. Leng, W.J. Zhang, H.Y. Peng, H.L. Li, S.J. Jiang, H.J. Huang, Nitrogen in bio-oil produced from hydrothermal liquefaction of biomass: a review. Chem. Eng. J. **401**, 126030 (2020)
23. F. Cheng, J.M. Jarvis, J.L. Yu, U. Jena, N. Nirmalakhandan, T.M. Schaub, C.E. Brewer, Biocrude oil from hydrothermal liquefaction of wastewater microalgae in a pilot-scale continuous flow reactor. Biores. Technol.. Technol. **294**, 122184 (2019)
24. D.H. Xu, P.E. Savage, Effect of temperature, water loading, and Ru/C catalyst on water-insoluble and water-soluble biocrude fractions from hydrothermal liquefaction of algae. Biores. Technol.. Technol. **239**, 1–6 (2017)
25. B.T. Hao, D.H. Xu, G.Y. Jiang, T.A. Sabri, Z.F. Jing, Y. Guo, Chemical reactions in the hydrothermal liquefaction of biomass and in the catalytic hydrogenation upgrading of biocrude. Green Chem. **23**, 1562–1583 (2021)
26. Y. Chen, Y. Huang, J. Xie, X. Yin, C. Wu, Hydrothermal reaction of phenylalanine as a model compound of algal protein. J. Fuel Chem. Technol. **42**, 61–67 (2014)
27. M. Watanabe, F. Bayer, A. Kruse, Oil formation from glucose with formic acid and cobalt catalyst in hot-compressed water. Carbohyd. Res.. Res. **341**, 2891–2900 (2006)
28. Y.H. Xu, M.F. Li, Hydrothermal liquefaction of lignocellulose for value-added products: mechanism, parameter and production application. Biores. Technol.. Technol. **342**, 126035 (2021)
29. L. Feng, X.H. Li, Z.Z. Wang, B.Z. Liu, Catalytic hydrothermal liquefaction of lignin for production of aromatic hydrocarbon over metal supported mesoporous catalyst. Biores. Technol.. Technol. **323**, 124569 (2021)
30. B. Ciuffi, M. Loppi, A.M. Rizzo, D. Chiaramonti, L. Rosi, Towards a better understanding of the HTL process of lignin-rich feedstock. Sci. Rep. **11**, 15504 (2021)
31. P. Duan, X. Bai, Y. Xu, A. Zhang, F. Wang, L. Zhang, J. Miao, Catalytic upgrading of crude algal oil using platinum/gamma alumina in supercritical water. Fuel **109**, 225–233 (2013)
32. Z.H. Wang, S. Adhikari, P. Valdez, R. Shakya, C. Laird, Upgrading of hydrothermal liquefaction biocrude from algae grown in municipal wastewater. Fuel Process. Technol. **142**, 147–156 (2016)
33. M. Wadrzyk, R. Janus, M.P. Vos, D.W.F. Brilman, Effect of process conditions on bio-oil obtained through continuous hydrothermal liquefaction of *Scenedesmus* sp. microalgae. J. Anal. Appl. Pyrol.Pyrol. **134**, 415–426 (2018)
34. S. Hansen, A. Mirkouei, L.A. Diaz, A comprehensive state-of-technology review for upgrading bio-oil to renewable or blended hydrocarbon fuels. Renew. Sustain. Energy Rev. **118**, 109548 (2020)
35. Y. Chen, L. Dong, J. Miao, J. Wang, C. Zhu, Y. Xu, G. Chen, J. Liu, Hydrothermal liquefaction of corn straw with mixed catalysts for the production of bio-oil and aromatic compounds. Biores. Technol.. Technol. **294**, 122148 (2019)

36. H.K. Reddy, T. Muppaneni, S. Ponnusamy, N. Sudasinghe, A. Pegallapati, T. Selvaratnam, M. Seger, B. Dungan, N. Nirmalakhandan, T. Schaub, F.O. Holguin, P. Lammers, W. Voorhies, S. Deng, Temperature effect on hydrothermal liquefaction of *Nannochloropsis gaditana* and *Chlorella* sp. Appl. Energy **165**, 943–951 (2016)
37. W.-T. Chen, Y. Zhang, J. Zhang, G. Yu, L.C. Schideman, P. Zhang, M. Minarick, Hydrothermal liquefaction of mixed-culture algal biomass from wastewater treatment system into bio-crude oil. Biores. Technol.. Technol. **152**, 130–139 (2014)
38. X. Peng, X. Ma, Y. Lin, X. Wang, X. Zhang, C. Yang, Effect of process parameters on solvolysis liquefaction of chlorella pyrenoidosa in ethanol-water system and energy evaluation. Energy Convers. Manage. **117**, 43–53 (2016)
39. R. Cherad, J.A. Onwudili, P. Biller, P.T. Williams, A.B. Ross, Hydrogen production from the catalytic supercritical water gasification of process water generated from hydrothermal liquefaction of microalgae. Fuel **166**, 24–28 (2016)
40. D.L. Barreiro, S. Riede, U. Hornung, A. Kruse, W. Prins, Hydrothermal liquefaction of microalgae: effect on the product yields of the addition of an organic solvent to separate the aqueous phase and the biocrude oil. Algal Res.-Biomass Biofuels Bioprod. **12**, 206–212 (2015)
41. D.C. Elliott, T.R. Hart, A.J. Schmidt, G.G. Neuenschwander, L.J. Rotness, M.V. Olarte, A.H. Zacher, K.O. Albrecht, R.T. Hallen, J.E. Holladay, Process development for hydrothermal liquefaction of algae feedstocks in a continuous-flow reactor. Algal Res.-Biomass Biofuels Bioprod. **2**, 445–454 (2013)
42. L. Leng, H. Li, X. Yuan, W. Zhou, H. Huang, Bio-oil upgrading by emulsification/microemulsification: a review. Energy **161**, 214–232 (2018)
43. Q. Lin, Y. Chen, Y. Tang, K. Wu, M. Yang, H. Hu, Y. Wu, Catalytic hydrothermal liquefaction of D. tertiolecta over multifunctional mesoporous silica-based catalysts with high stability. Microporous Mesoporous Mater. **250**, 120–127 (2017)
44. P. Biller, A.B. Ross, Potential yields and properties of oil from the hydrothermal liquefaction of microalgae with different biochemical content. Biores. Technol.. Technol. **102**, 215–225 (2011)
45. Y. Xu, P. Duan, B. Wang, Catalytic upgrading of pretreated algal oil with a two-component catalyst mixture in supercritical water. Algal Res.-Biomass Biofuels Bioprod. **9**, 186–193 (2015)
46. S. Zou, Y. Wu, M. Yang, C. Li, J. Tong, Bio-oil production from sub- and supercritical water liquefaction of microalgae Dunaliella tertiolecta and related properties. Energy Environ. Sci. **3**, 1073–1078 (2010)
47. H.-J. Huang, X.-Z. Yuan, H.-N. Zhu, H. Li, Y. Liu, X.-L. Wang, G.-M. Zeng, Comparative studies of thermochemical liquefaction characteristics of microalgae, lignocellulosic biomass and sewage sludge. Energy **56**, 52–60 (2013)
48. B. Zhang, F. Zou, M. Sun, S. Zhang, Y. Huang, Zhiqi granules decreased lead level but increased iron level in serum of Chinese children with moderately elevated blood lead levels. Biol. Trace Elem. Res. **193**, 1–6 (2020)
49. S. Xiu, A. Shahbazi, V. Shirley, D. Cheng, Hydrothermal pyrolysis of swine manure to bio-oil: effects of operating parameters on products yield and characterization of bio-oil. J. Anal. Appl. Pyrol.Pyrol. **88**, 73–79 (2010)
50. S. Xiu, A. Shahbazi, C.W. Wallace, L. Wang, D. Cheng, Enhanced bio-oil production from swine manure co-liquefaction with crude glycerol. Energy Convers. Manage. **52**, 1004–1009 (2011)
51. Z. Zhu, S.S. Toor, L. Rosendahl, D. Yu, G. Chen, Influence of alkali catalyst on product yield and properties via hydrothermal liquefaction of barley straw. Energy **80**, 284–292 (2015)
52. F. Conti, S.S. Toor, T.H. Pedersen, A.H. Nielsen, L.A. Rosendahl, Biocrude production and nutrients recovery through hydrothermal liquefaction of wastewater irrigated willow. Biomass Bioenerg.Bioenerg. **118**, 24–31 (2018)
53. T.H. Pedersen, I.F. Grigoras, J. Hoffmann, S.S. Toor, I.M. Daraban, C.U. Jensen, S.B. Iversen, R.B. Madsen, M. Glasius, K.R. Arturi, R.P. Nielsen, E.G. Sogaard, L.A. Rosendahl, Continuous

hydrothermal co-liquefaction of aspen wood and glycerol with water phase recirculation. Appl. Energy **162**, 1034–1041 (2016)
54. S. Feng, Z. Yuan, M. Leitch, C.C. Xu, Hydrothermal liquefaction of barks into bio-crude—effects of species and ash content/composition. Fuel **116**, 214–220 (2014)
55. K. Tekin, S. Karagoz, S. Bektas, Hydrothermal liquefaction of beech wood using a natural calcium borate mineral. J. Supercrit. FluidsSupercrit. Fluids **72**, 134–139 (2012)
56. K. Kapusta, Effect of ultrasound pretreatment of municipal sewage sludge on characteristics of bio-oil from hydrothermal liquefaction process. Waste Manage. **78**, 183–190 (2018)
57. R. Shakya, S. Adhikari, R. Mahadevan, E.B. Hassan, T.A. Dempster, Catalytic upgrading of bio-oil produced from hydrothermal liquefaction of *Nannochloropsis* sp. Biores. Technol.. Technol. **252**, 28–36 (2018)
58. C. Yang, S. Wang, J. Yang, D. Xu, Y. Li, J. Li, Y. Zhang, Hydrothermal liquefaction and gasification of biomass and model compounds: a review. Green Chem. **22**, 8210–8232 (2020)
59. M. Zhang, Y. Hu, H. Wang, H. Li, X. Han, Y. Zeng, C.C. Xu, A review of bio-oil upgrading by catalytic hydrotreatment: advances, challenges, and prospects. Mol. Catal. **504**, 111438 (2021)
60. D. Xu, N. Wei, Y. Liang, H. Wang, L. Liu, S. Wang, Biocrude upgrading in different solvents after microalgae hydrothermal liquefaction. Ind. Eng. Chem. Res. **60**, 7966–7974 (2021)
61. B. Kunwar, S.D. Deilami, L.E. Macaskie, J. Wood, P. Biller, B.K. Sharma, Nanoparticles of Pd supported on bacterial biomass for hydroprocessing crude bio-oil. Fuel **209**, 449–456 (2017)

Chemical Reactions of Organic Compounds in Supercritical Water Reactions

In typical circumstances, water functions as a polar solvent, capable of dissolving the majority of inorganic salts. However, it exhibits minimal solubility or insolubility for gases and the majority of organic substances. The density, dielectric constant, ion product, viscosity, thermal conductivity, diffusion coefficient, and dissolution performance of water undergo significant changes at the supercritical state (T > 374 °C, P > 22.1 MPa) [1–5], as illustrated in Fig. 1. SCW exhibits a small number of distinctive properties, including the presence of only a small number of hydrogen bonds, a low dielectric constant comparable to that of a polar organic solvent, a high diffusion coefficient, and a relatively low viscosity [4, 6, 7]. For example, SCW has good solubility, which makes organic matter, oxygen, and SCW mutually soluble, greatly reducing the mass transfer resistance and making it an ideal reaction medium. On the contrary, inorganic substances, especially salts, have very low solubility in SCW and can be easily separated. In addition, the temperature and pressure of the SCW can be varied to control the reaction environment, coordinate the reaction rate and chemical equilibrium, and adjust the selective activity of the catalyst [8]. The two most common applications of SCW are SCWG and SCWO.

Model commenced research into the production of hydrogen by the SCWG of glucose and wood in the 1970s. The term SCWG is typically used to describe a biomass conversion process that produces a hydrogen-rich gas (including H_2, CO_2, CH_4, and CO, among others) in SCW (400–700 °C) [14]. The degradation of organic matter in SCWG can be summarized by Eq. (1), although intermediate reaction processes may involve a number of complex reactions, including hydrolysis, cracking, and polymerization. It is worthy of note that the reactions which significantly affect the distribution of gaseous products are mainly steam reforming, water–gas shift, and methanation reactions (see Eqs. (2)–(6), and minor hydrogenation reaction (see Eq. 7) are also available in SCWG [15–18].

$$\text{Organic compounds} + H_2O \rightarrow H_2 + CO_2 + CO + CH_4 + \text{other small molecules} \tag{1}$$

Y. Guo et al., *Hydrothermal Processing of Biomass for Hydrogen and Bio-oil Production*, https://doi.org/10.1007/978-981-96-6336-1_2

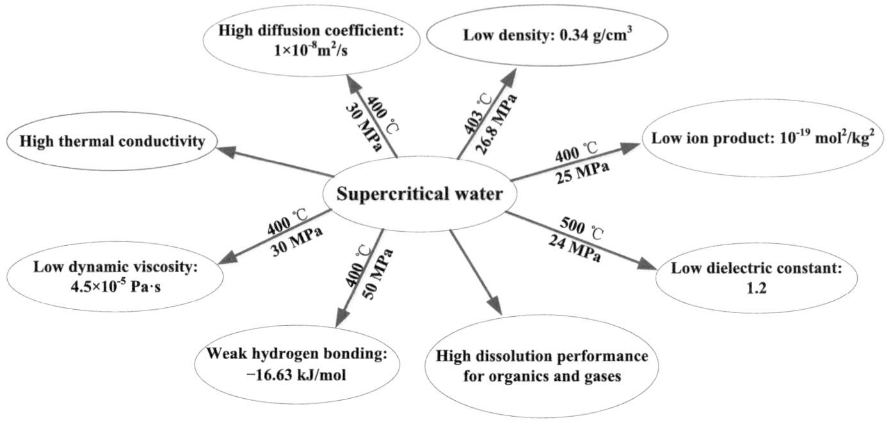

Fig. 1 Main properties of supercritical water [9–13]

$$CH_xO_y + (2-y)H_2O \rightarrow (2-y+x/2)H_2 + CO_2 \tag{2}$$

$$CH_xO_y + (1-y)H_2O \rightarrow (1-y+x/2)H_2 + CO \tag{3}$$

$$H_2O + CO \rightarrow CO_2 + H_2 \tag{4}$$

$$4H_2 + CO_2 \rightarrow CH_4 + 2H_2O \tag{5}$$

$$3H_2 + CO \rightarrow CH_4 + H_2O \tag{6}$$

$$2CO + 4H_2 \rightarrow 2CH_4 + O_2 \tag{7}$$

In SCWG, water serves not only as a reaction medium to accelerate reaction processes, but also participates in the water–gas shift reaction simultaneously. This provides a substantial quantity of H·free radicals and enhances the yield of hydrogen, while reducing coking production [19]. The SCWG has the potential to address the limitations of conventional air-based gasification techniques, such as the high energy consumption associated with drying processes [20]. To date, the SCWG has demonstrated efficacy in the conversion of wet biomass (e.g., organic wastewater) into a hydrogen-rich mixture gas in the context of hydrogen production.

SCWO is a heat-based organic waste treatment process that employs oxidants such as H_2O_2, O_2, and air. In SCWO, the majority of hydrocarbons undergo oxidation into CO_2 and H_2O, with conversion rates exceeding 99.999% within a single minute [21]. The conversion of nitrogen atoms in N-containing compounds into nitrate, ammonium, and other organic nitrogen in liquid products, and ammonia, nitrogen dioxide, and nitrogen monoxide in gaseous products, is a common occurrence in SCWO [22, 23]. Other heteroatoms (e.g., Cl, S, and P) are converted into corresponding inorganic acids (i.e., HCl, H_2SO_4, and H_3PO_4), which can be further neutralized into organic salts [24–26].

At the present time, a considerable number of articles have reported on the effects of specific operating parameters (e.g., oxidation coefficient, catalyst, reaction temperature, residence time, and reactant concentration) on the degradation rate of organic matter in SCWO and the hydrogen production capacity in SCWG. They have made significant contributions to the understanding of the conversion mechanisms of organic matter in SCW. Furthermore, numerous articles have indicated that the reactions of organic compound conversion in SCW encompass hydrolysis, pyrolysis, dehydration, hydrogenation, polymerization, isomerization, and other types of reactions, resulting in the formation of a multitude of diverse intermediate products during the oxidation/gasification process.

In order to gain a deeper understanding of the mechanism of organic matter conversion in SCW, a comprehensive search was conducted for relevant papers published mainly in the past decade. The chemical reactions of different types of organic compounds (e.g., hydrocarbons and heteroatom-containing compounds) and their intermediate products were then summarized in detail, focusing on SCWG and SCWO. The reaction processes of organic compounds with the same functional group were compared, and the bond fracture characteristics of reactants with different functional groups were analyzed. Finally, the distribution characteristics of intermediate products and final products were reviewed. To the best of our knowledge, this is the first documented instance. This information facilitates the elucidation of the reaction mechanism, the establishment of reaction kinetics, and the prediction of the reaction processes of organic compounds in SCW.

1 Free Radical Reactions

The term "free radicals" is used to describe atoms or groups of atoms that have unpaired electrons. These electrons are formed as a result of the cleavage of covalent bonds under specific conditions. They are pivotal species in the initiation of free radical reactions, which are characterized by three principal stages: initiation, propagation, and termination [27, 28]. SCW exhibits characteristics of a weak electrolyte, indicating a preference for free radical reactions and an inhibition of ionic reactions [29]. Additionally, free radical reactions are prone to occur under conditions of elevated temperature [30], thus making SCW an optimal choice for these reactions.

Indeed, the primary reaction mechanisms observed in organic matter during SCWG and SCWO processes can be attributed to free radical reactions [31–34]. Although the concentration of initial free radicals is typically low, the rate constants of free radical reactions are generally high, resulting in a rapid decline in the concentration of the reactants [35]. In the presence of substantial quantities of initial reactants within the reaction system, the production rate of free radicals is markedly higher than their consumption rate. Consequently, the entire reaction process can be conceptualized as a continuous chain, with a circulation that persists until the reactants are almost entirely oxidized and/or decomposed into the final products [28]. In the final analysis, the free radical termination reactions prevail, resulting in the production of stable products.

Table 1 illustrates a number of distinctive free radical reactions that occur in SCW, encompassing a range of reaction types, including decomposition, addition, isomerization, substitution, β-scission, coupling, and disproportionation [27, 30, 36, 37]. It is evident that the most significant free radicals in free radical reactions are H·, HO_2·, and OH·. These are the most significant products in initial reactions and act as promoters of the propagation process. Water serves as a significant source of hydrogen atoms, which contribute to the generation of substantial quantities of H· and OH· (Eq. (9)). Additionally, the minimal hydrogen bond content in SCW facilitates the transfer of hydrogen from H_2O to free radicals. H· is capable of generating H_2 through two distinct mechanisms: the H-abstraction reaction (Eq. (20)) and the coupling reaction (Eq. (29)). Consequently, it plays a pivotal role in SCWG.

In SCWO, the initial reaction of free radicals produced by reactants is closely related to O_2 if it is used as the oxidant (see Eqs. (10)–(11)) [37]. At the preheating stage, the reaction of O_2 and H_2O produces active free radicals, including OH·. The hydroxyl radical (OH·) is a highly reactive species with the ability to oxidize a wide range of organic compounds. Nevertheless, the reaction rate of O_2 and H_2O is relatively slow, which renders O_2 less active than hydrogen peroxide [38]. Additionally, the activation energy associated with the oxidation of oxygen is relatively high, indicating that oxygen is more effective at elevated temperatures. Consequently, an increasing number of researchers employ H_2O_2 as an oxidant in their SCWO experiments. Hydrogen peroxide ($H_2O_2$and the hydroxyl radical (HO_2·) are the primary unstable substances involved in free radical reactions. Despite the lower activity of HO_2 in comparison with OH·, due to its relatively low abundance, HO_2· has become an important chain carrier in propagation reactions. The decomposition of H_2O_2 is the primary source of OH·, and H_2O_2 can be regenerated by the H-abstraction of HO_2·. This process continuously renews OH· from the reservoir compound H_2O_2 [28]. Moreover, the reaction rate of H_2O_2 decompose (Eq. (13)) in SCW is fast, and thus, a large number of ·OH can be accumulated [18]. Therefore, H_2O_2 can greatly accelerate the oxidation rate of organic compounds, improve the reaction rate constant, and become an effective oxidant, although it faces the problem of high cost. In particular, too high a concentration of H_2O_2 may not effectively improve the degradation efficiency of the reactants. This can be explained by the fact that a high concentration of H_2O_2 causes a quenching reaction between OH– and H_2O_2, thus interfering with the chain transfer reaction of OH^- [24]. It is therefore essential to ascertain the optimal oxidant quantity in experimental procedures and to exercise meticulous control over the mixing time of oxidant and reactant.

As listed in Table 1, the propagation process in SCW is the most complex and diverse, with a multitude of free radical types, which ultimately leads to the formation

Table 1 Free radical reactions in SCW

Reaction stage	Reaction type	Equation	Num
Initiation	Decomposition (except O_2)	$R\text{-}X \rightarrow R\cdot + X\cdot$	(8)
		$H_2O \rightarrow H\cdot + OH\cdot$	(9)
		$R\text{–}H + O_2 \rightarrow R\cdot + HO_2\cdot$	(10)
		$O_2 + H_2O \rightarrow OH\cdot + HO_2\cdot$	(11)
		$ROOH \rightarrow RO\cdot + OH\cdot$	(12)
		$H_2O_2 \rightarrow 2OH\cdot$	(13)
Propagation	Addition	$R\cdot + CH_2 = CH_2 \rightarrow$ $R\text{–}CH_2\text{–}CH_2\cdot$	(14)
		$R\cdot + O_2 \rightarrow ROO\cdot$	(15)
	H-abstraction (Special substitution)	$R\cdot + R'\text{-}H \rightarrow R\text{-}H + R'\cdot$	(16)
		$ROO\cdot + R\text{–}H \rightarrow ROOH + R\cdot$	(17)
		$HO\cdot + R\text{–}H \rightarrow R\cdot + H_2O$	(18)
		$HO_2\cdot + R\text{–}H \rightarrow R\cdot + H_2O_2$	(19)
		$H\cdot + R\text{–}H \rightarrow R\cdot + H_2$	(20)
	Substitution	$R\cdot + A\text{–}B \rightarrow R\text{–}A + B\cdot$	(21)
	Rearrangement (Isomerization)	$\cdot CH_2\text{-}C(R)_3 \rightarrow \cdot C(R)_2\text{–}CH_2\text{–}R$	(22)
		$ROO\cdot \rightarrow HOOR'\cdot$ (Intramolecular H-abstraction)	(23)
	β-scission	$R_{a+b}O\cdot \rightarrow R_a = O + R_b\cdot$	(24)
		$R\cdot \rightarrow C = R''H + R'\cdot$	(25)
	Dehydration	$\cdot C(R)OH\text{-}CH_2OH \rightarrow C(R)OH =$ $CH\cdot + H_2O$	(26)
	Disproportionation	$2ROO\cdot \rightarrow O_2 + 2RO\cdot$	(27)
Termination	Coupling	$R\cdot + Y\cdot \rightarrow R\text{–}Y$	(28)
		$R\cdot + R\cdot \rightarrow R\text{–}R$	(29)
	Disproportionation	$2CH_3\text{–}CH_2\cdot \rightarrow CH_3\text{-}CH_3 + CH_2$ $= CH_2$	(30)

of a plethora of liquid-phase intermediates. These intermediates are commonly generated by superposition of multi-step free radical reactions. Accordingly, the formation mechanism of intermediates in SCW can be elucidated. Furthermore, isomerization and substitution of free radicals are more prevalent during the propagation phase [30]; it can be observed that the majority of free radicals generated by the initial reactions are capable of attacking the relatively weak chemical bonds of C–H, leading to the occurrence of H-abstraction reactions (Eqs. (16)–(20)). In accordance with the principles of free radical reactions, H_2 is formed by the combination of H· and H· (Eq. (29)), whereas CO is produced by the decomposition of ·CHO free radicals. Subsequently, the majority of CO is oxidized by OH· to generate CO_2 (Eq. (30)). The peroxide ROOH (Eq. (28)) produced in the termination process is relatively active and readily decomposes into small organic compounds (e.g., formic acid and acetic acid). These compounds are subsequently oxidized into CO_2 and H_2O [37].

It goes without saying that free radical chain reactions exert a considerable influence on the formation of intermediate products, which in turn has a marked effect on the degradation pathways of reactants. Accordingly, in order to gain a comprehensive understanding of the reaction pathways of diverse compounds in SCW, the gasification and oxidation decomposition mechanisms of several representative organic compounds and their intermediates are presented and analyzed below.

2 Hydrocarbons' Conversions

Hydrocarbons can be classified into three principal groups: alkanes, alkenes, and aromatics. Among these three types of hydrocarbons, there is a paucity of studies on the reactions of alkenes in supercritical water (SCW). Alkenes [6] are capable of undergoing isomerization, hydration, and hydrogenation reactions in SCW. Alkenes et al. [6] investigated the conversion of 1-octene in sub- and supercritical water (250–450 °C, 11–33 MPa) catalyzed by TiO_2 and identified the main products as 2-octene (resulting from the double bond isomerization reaction of 1-octene and dehydration reaction of 2-octanol) and 2-octanol (formed by the hydration reaction of 1-octene). Alshammari's research illustrates that the addition of oxidants and an extension of the reaction time can facilitate the conversion of alkenes into alkanes through hydrogenation and partial oxidation in SCW [39].

With regard to alkanes, the majority of research has been conducted on long-chain alkanes (e.g., dodecane [40] and hexadecane [41]), it has been demonstrated that long-chain alkanes degrade at a considerably faster rate than aromatics in SCW, given that aromatic compounds tend to accumulate for a longer period of time [42]. Recent studies have indicated the potential for hydrogen production through the selective catalytic weight reduction of heavy hydrocarbons, with the presentation of disparate key kinetic data. To illustrate, in the n-hexadecane SCWG at 525–605 °C and 22 MPa, the pre-exponential frequency factor and the reaction activation energy are 7.36 × 1014 s^{-1} and 263.68 kJ mol^{-1}, respectively [41]. Picou et al. [43] conducted a non-catalytic SCWG of 1-dodecane at temperatures between 530 and 699 °C and pressures of 24 MPa. Their findings revealed that the reformation reaction exhibited the highest pre-exponential frequency factor and activation energy (1.75 × 108 s^{-1} and 197.3 kJ mol^{-1}), in comparison with the water–gas shift reaction (4.16 × 103 s^{-1} and 107.4 kJ mol^{-1}) and pyrolysis (10.03 s^{-1} and 60.6 kJ mol^{-1}). Furthermore, the pre-exponential frequency factors and activation energies of the reformation reaction (2.20×10^9 s^{-1} and 226.30 kJ mol^{-1}) and pyrolysis (1.12×10^3 s^{-1} and 96.04 kJ mol^{-1}) have been reported for the n-dodecane SCWG at 626–705 °C and 24 MPa. Hexadecane is capable of undergoing cracking into smaller n-alkanes and 1-alkenes, which are hydrocarbon compounds ranging in size from C_{15} to C_1, under conditions of both sub- and supercriticality. The generation of highly reactive long-chain free radicals (e.g., primary and secondary hexadecyl radicals) in this process may potentially involve a series of reactions, primarily including β-scission, hydrogen abstraction, isomerization, and addition. It has been observed that an increase in reaction temperature can facilitate the formation of 1-alkenes via β-scission, while an elevation in pressure can enhance the formation of n-alkanes via H-abstraction [41]. Finally, the reaction process is terminated by the combination of smaller free radicals to form stable products, and thus the whole process is far less complicated than that of aromatic hydrocarbons.

It is therefore pertinent to focus on some aromatic hydrocarbons, in particular polycyclic aromatic hydrocarbons (PAHs) with chemical stability, such as naphthalene, phenanthrene, anthracene, and pyrene. These compounds are typical refractory intermediate products of biomass, coal, oil, and other substances during supercritical water gasification decomposition [19, 44, 45]. In the absence of appropriate treatment, the formation of char and tar can occur, which has the effect of significantly reducing the efficiency of the gasification process. The SCWG process of PAHs identifies the method of ring cracking as a pivotal area of current and future research, given its pivotal role in determining the extent of complete gasification of reactants [46].

The present study demonstrates that SCW is an effective reaction environment for promoting the cleavage of aromatic rings into small molecules. This is achieved by effectively weakening the C–C bond energy in the aromatic rings. To illustrate, the energy barrier of the ring-opening reaction of anthracene can be diminished from 776.7 to 218.5 kJ/mol [19]. In the past decade, many scholars have used PAHs as model compounds to explore the cracking of rings and their reaction mechanisms in SCWG [47–49] In the PAHs SCWG, the initial transformation of polycyclic structures involves the conversion into low ring and small ring structures (e.g., three rings and five rings) through deformation and fracture. This is followed by a process of fragmentation that results in the formation of linear chains. Ultimately, the release of small molecules occurs through bond fracture. Two distinct explanations are provided in greater detail below. The initial process is illustrated in Fig. 2a. The addition of a hydrogen atom in SCW to a C atom on a ring results in the rupture of the C(ring)-C(ring) bond. Concurrently, the OH in SCW forms a bond with another C atom [19]. Then the C (ring)-C (ring) bond on the middle ring of PAHs can also break and bond, and the lower structure can be twisted until the middle benzene ring is completely destroyed. Another explanation is that the cleavage of PAHs is due to the fact that H on the outer ring is easily attacked by highly reactive OH– radicals or H_2O molecule and thus H-abstraction occurs [47, 50]. As indicated in Fig. 2b, when the aromatic ring loses hydrogen, it becomes distorted and unstable and is prone to cracking. It can then be transformed into a macroring or multi-ring structure consisting of three and five rings. These rings are further opened by the breaking of the C–C bond due to the continued attack of OH^- radicals. It is also believed that after these rings are broken, some linear chains are formed and then the C–C bonds are broken to form some small molecules. In addition, OH^- and H^- radicals in SCW play an important role in promoting and breaking the bonds in the whole process.

The SCWO of PAHs in early years has been the subject of study by a number of scholars [51–54], with the initial reaction on rings being oxygen invasion. In the case

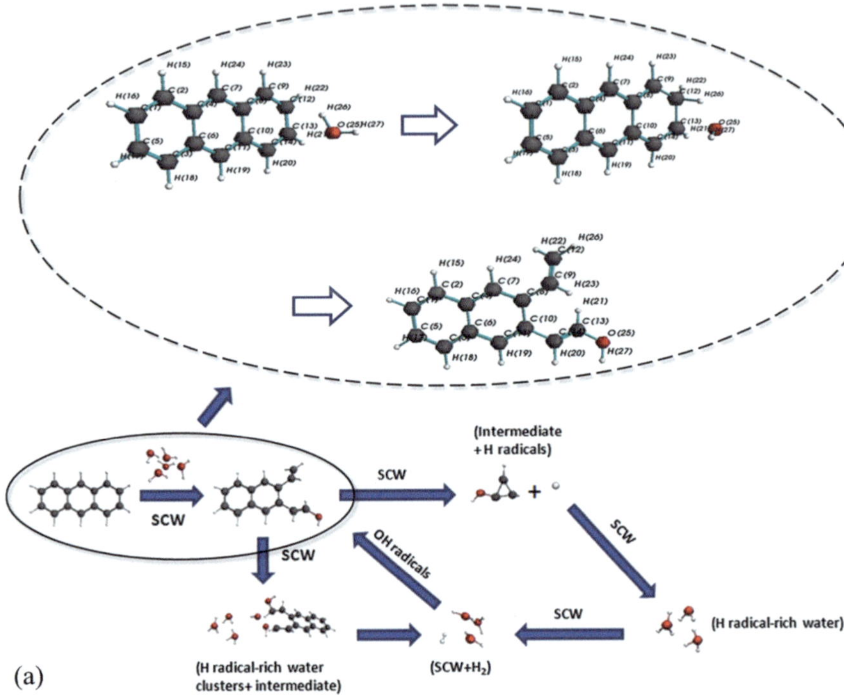

Fig. 2 Different ring-cracking pathways during PAHs SCWG. **a** anthracene SCWG [19], **b** naphthalene SCWG [47]

of naphthalene, the five principal intermediate products for oxygen to invade naphthalene rings are 1,4-naphthoquinone, 1-naphthol, phthalic anhydride, coumarin, and chromone [52]. Subsequently, these compounds undergo a one-ring-cracking process, resulting in the formation of salicylaldehyde, dibenzofuran, fluorenone, and xanthone. They then undergo further fragmentation and oxidation, ultimately yielding acids, water, and carbon dioxide. Furthermore, isomerization and hydrogen addition also occur during the cracking process of PAHs in SCW. Wang et al. [55] obtained hydrogenation products (i.e., tetralin and decalin) by the ring-opening reaction of naphthalene. They also found that Rh and Mo-Ni were active for the hydrogenation reaction, while HY zeolite and Mo–Ni promoted the isomerization and ring-opening reaction. Cheng et al. [56] employed $Fe_2\ (MoO_4)_3$ as a catalyst for the hydrogenation of naphthalene in subcritical water, resulting in the production of tetralin. They also observed the occurrence of a hydrogen/deuterium exchange reaction during the reaction process.

Phenanthrene and anthracene are both members of the PAH family, characterized by the presence of three rings. The three central rings of anthracene are in a straight line, and it is an isomer of phenanthrene. In the anthracene SCWG, Jin et al. [48] demonstrated that the predominant liquid products at 700 °C and 25 MPa

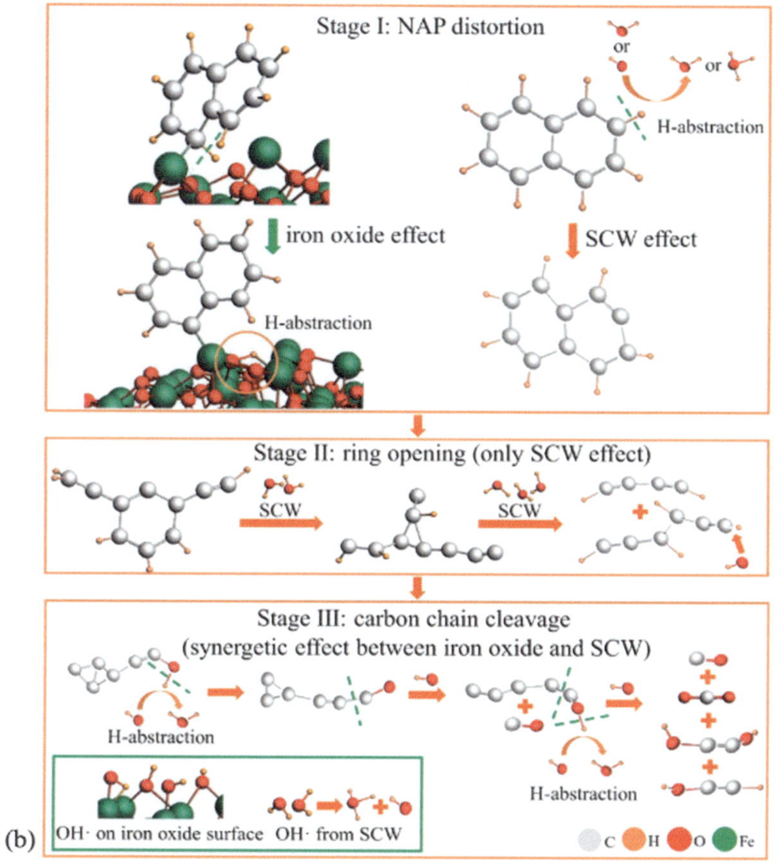

Fig. 2 (continued)

were benzene and naphthalene, followed by fluorene, 9,10-dihydro-anthracene, 2-methyl-naphthalene, and 3-phenyltoluene. The addition of K_2CO_3 was observed to effectively reduce the amount of benzene and naphthalene. The optimal temperatures for benzene and naphthalene SCWGs were identified as 700 °C and 650 °C, respectively. Yeletsky et al. [57] employed a NiMo/SiO_2 catalyst to facilitate the cracking of phenanthrene in SCW at 425 °C and 23 MPa. Their findings indicated that the reactions occurred preferentially in the central ring of phenanthrene, and that the catalyst was capable of promoting the hydrogenation of polycyclic hydrocarbons in SCW. The principal liquid products are dihydrophenanthrene, produced by hydrogenation, and benzophenone and xanthine, formed by oxidation.

Based on intermediate products and reactions mentioned in several literatures [48, 55, 57], potential degradation pathways of phenanthrene, anthracene, and naphthalene in SCW are deduced, as displayed in Fig. 3. Some potential interconnections among the degradation pathways of the three typical PAHs in SCW are postulated.

It is evident that the degradation processes of PAHs in SCW primarily encompass hydrogenation, contraction, ring opening, and dealkylation, which result in a transformation from a polycyclic to a monocyclic to a carbon chain structure. Firstly, for aromatic hydrocarbons with three rings, one possible pathway is the middle ring hydrogenation, which forms dihydrophenanthrene and so forth. Subsequently, the unstable middle ring undergoes contraction, resulting in the production of compounds such as fluorine and ultimately leading to ring opening. An alternative pathway involves the cleavage of the outer C–C bond in PAHs, which gives rise to the formation of double aromatic rings. These subsequently undergo degradation in a manner analogous to that observed in the naphthalene system.

The effectiveness of the addition of K_2CO_3 [48], NiMo [55, 57], iron oxide [47], or oxidant [44] in reducing the content of intermediate products and in greatly improving the gasification effect of aromatic hydrocarbons has been demonstrated. The addition of a small amount of oxidant (with an oxidation coefficient of less than 0.2) has been demonstrated to promote the production of ring-opening products and inhibit the formation of polymers, thereby increasing the yield of hydrogen [58]. The SCWG experiments on the mixture of naphthalene, phenol, and acetic acid indicate that

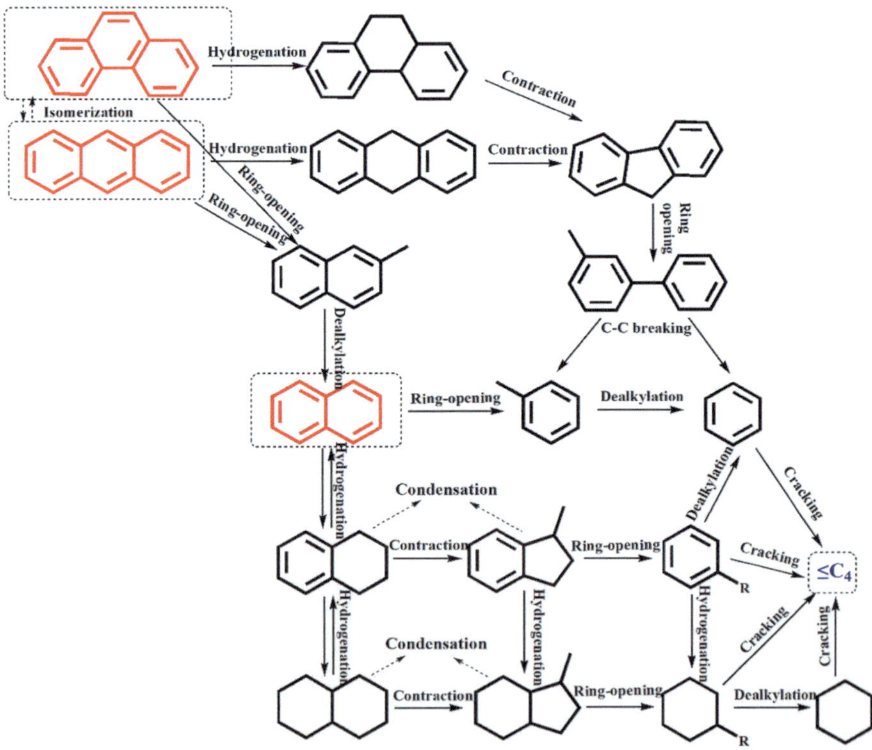

Fig. 3 Potential reaction network of typical PAHs degradation in SCW

the content of phenolic compounds should be sufficiently reduced in biomass/coal SCWG, as these compounds can cause the cross-linking reaction of PAHs [59].

3 Conversions of O-Containing Organic Matters in SCW

3.1 Cellulose, Proteins, and Lignins

Biomass has played an important role in the utilization of global renewable energy resources and is considered to be the fourth-largest energy resource after coal, oil, and natural gas [60]. The technical and economic feasibility of the biomass SCWG for hydrogen production, along with its promising future, has been established [61–63]. SCWG has the potential to directly convert high-moisture biomass without the need for energy-intensive drying processes, which has led to a significant increase in interest in this area in recent years [64]. Biomass is composed of a multitude of complex macromolecular substances, including proteins, celluloses, hemicelluloses, and lignins. The interior of these macromolecules is a complex network comprising covalent bonds, intermolecular bridges, and van der Waals forces. Consequently, an investigation into the degradation pathways of these complex molecules in SCWG is of considerable importance for the optimization of the reaction system and future market applications.

Celluloses represent the most abundant component of biomass. They can be considered to be a glucose polymer, comprising D-glucose subunits linked together by β(1, 4) glycosidic bonds, and containing hydrogen bonds and van der Waals forces. Sandeep et al. [65] reported that celluloses could undergo partial dissolution in subcritical water at 302 °C, with complete dissolution occurring at 330 °C. The addition of acids (e.g., mineral acids or solid acids), enzymes (e.g., cellobiohydrolase) and supported metal catalysts (e.g., Pt/γ-Al_2O_3 and Ru/HUSY (Si/Al = 20)) is conducive to the hydrolysis of celluloses [66–69]. Figure 4 illustrates the potential degradation pathways of celluloses in SCWG. Initially, cellulose undergoes hydrolysis to form oligomers and monomers in SCWG. This process continues until glucose is produced, which can then isomerize and yield fructose. As temperature increases, hydrolysates can undergo a series of reactions, including dehydration, condensation, isomerization, and polymerization, which can result in the conversion of intermediates (e.g., phenols, dimers, and furfurals) and low-molecular weight compounds (e.g., acids, aldehydes, ketones, and alcohols) into gases, primarily consisting of H_2, CH_4, CO, and CO_2 [16, 70, 71]. Glucose is the most significant hydrolysis product of cellulose under high-temperature and high-pressure conditions. The subsequent degradation reaction mechanism is of great interest to researchers, as it offers insights into the complete degradation of cellulose.

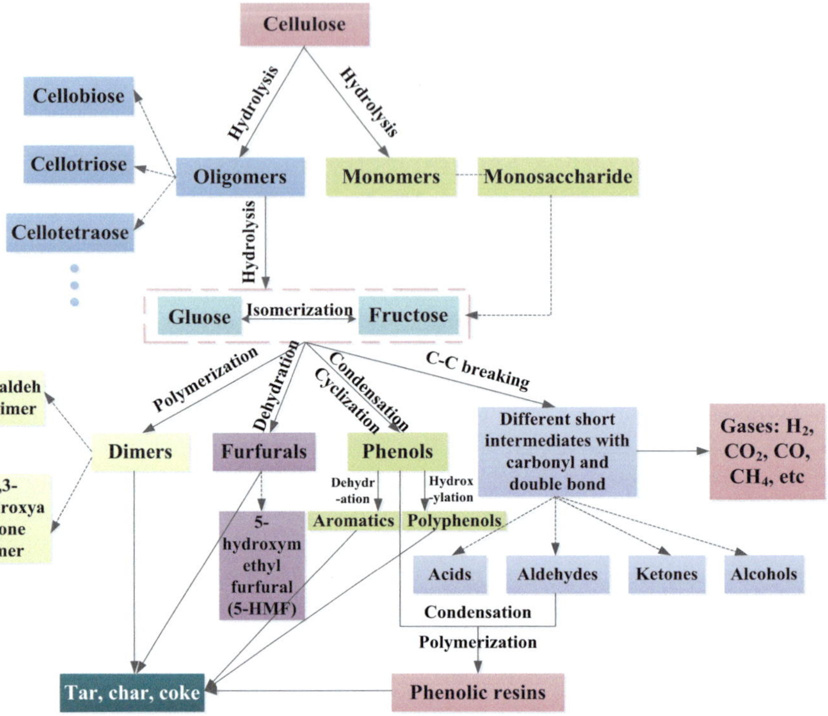

Fig. 4 Potential degradation pathways of cellulose in SCWG. (Dotted line represents typical product distribution, and solid line represents main reaction pathway)

In the absence of a catalyst, glucose is readily dehydrated to form furfural and 5-hydroxymethylfurfural, which then undergo a series of reactions (e.g., decomposition, isomerization and condensation lead to the formation of 7-hydroxy-1-indanone, 2-vinylfuran, 2-methyl-5-hydroxybenzofuran, 3-hydroxy-benzaldehyde, 1-(3-hydroxyphenyl)-ethanone, 2-naphthalenol, 2-methyl-1,4-benzenediol, hydroquinone, and resorcinol, as well as other refractory aromatic liquid products, which inhibit further gasification [72]. In order to reduce tar production and improve gasification efficiency, many researchers add KOH [16], NaOH [73], Ni-based catalysts [71, 74], Pt-based catalysts [70], and a small amount of oxidant [75, 76]. One potential explanation is that the catalyst can facilitate the formation of additional gasification compounds, including hydroxylated ketones, aldehydes, and acids. Conversely, the oxidant can accelerate the decomposition of aldehydes, thereby inhibiting the production of tar, char, and other polymers. The improved products are primarily composed of unsaturated hydrocarbons with low carbon numbers, including cyclic and non-cyclic ketones, aldehydes, carboxylic acids, and their alkylated and hydroxylated derivatives. Notably, furfural formation is absent, significantly enhancing the degradation rate of glucose. Furthermore, Kobiro et al. [77] demonstrated that sugars (e.g., glucose and polysaccharides) serve as effective hydrogen sources, supplying hydrogen free radicals (H·) that facilitate the dehydroxylation reduction of alcohols in SCW and catalyze the hydrogenation reduction of olefins.

Proteins represent the primary constituents of sewage sludge, along with microalgae and select other biomass. Figure 5 illustrates potential degradation pathways in SCWG of proteins. In the SCWG process, proteins are typically hydrolyzed to amino acids and peptides. Subsequently, amino acids undergo deamination, resulting in the production of ammonia and organic acids. Additionally, decarboxylation occurs, leading to the generation of carbonic acid and amines [78–80]. The Maillard reaction also allows amino acids to react with carbonyl compounds, forming N-heterocyclic compounds such as indole, pyrimidine, pyrrole, and pyridine. This results in the production of free radical scavengers, which inhibit gas production in SCWG [34, 81–83]. Peptides can be decomposed into a range of compounds, including aromatic hydrocarbons, aldehydes, diketopiperazines, and aliphatic amines [81]. Consequently, utilizing amino acids as model compounds to investigate their gasification in SCW is beneficial for elucidating the degradation pathways of proteins in SCWG and the impact of nitrogen on biomass SCWG.

The structural diversity of amino acid molecules gives rise to considerable variation in SCWG difficulty and product profiles. The research on glycine and alanine in SCWG [84–86] indicates that the primary products of glycine are ammonia, methylamine, glycolic acid, and formic acid. In contrast, the principal products of alanine are ammonia, ethylamine, propanoic acid, diacetone alcohol, and N-ethyl-propanamide. In the case of amino acids with more complex structures, the hydrothermal reaction products of phenylalanine include [78], styrene, and phenylethanol. These are formed by the decarboxylation and deamination of phenylalanine and the hydration of styrene, respectively. The Samanmulya research group has conducted numerous SCWG experiments on amino acids over the past decade, including studies on glycine, alanine, histidine, aminobutyric acid, and serine, among others [85, 87]. It can be observed that the gasification efficiencies of aminobutyric

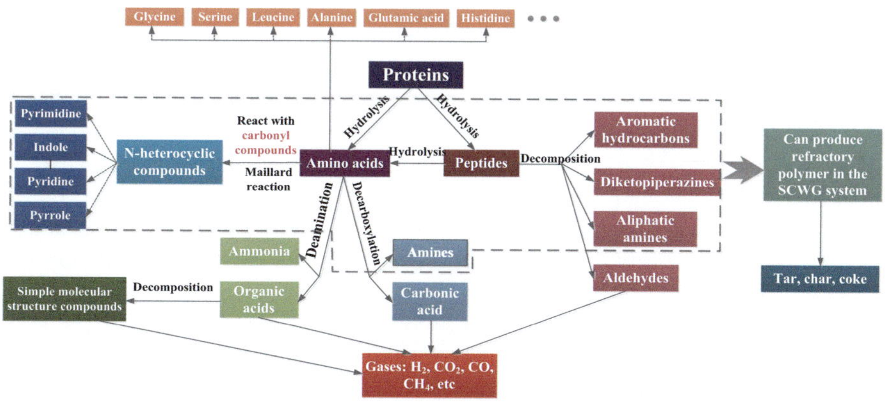

Fig. 5 Potential degradation pathways of proteins in SCWG. (Dotted line represents typical product distribution, and solid line represents main reaction pathway)

acid, glycine, and alanine in SCW are almost identical, with serine exhibiting the highest efficiency. Nevertheless, the gasification efficiency of histidine is relatively low due to the formation of a stable intermediate, 4-methylimidazole. In the study conducted by Caputo et al. [61], the order of gasification resistance was found to be glycine > leucine > glutamic acid. The ease of gasification observed in glutamic acid is attributed to the presence of two carboxyl groups. Furthermore, methane production increases in the order of leucine > glutamic acid > glycine, which can be attributed to the number of methyl groups present in the molecules. This finding is consistent with the explanation of the methane source provided by Samanmulya et al. [85]. However, the concentration of hydrogen follows an inverse order.

Currently, there is a paucity of research examining the degradation behavior of amino acids with highly complex structures under conditions of elevated temperature and pressure. Furthermore, research is being conducted into the interaction between amino acids and other organic compounds, including studies into the effects of amino acids on phenol [86], amino acids, and glucose [61, 82]. The SCWG has identified alkali salts and certain metals as potential catalysts for amino acid research. The addition of salts, specifically sodium nitrate ($NaNO_3$), potassium carbonate (K_2CO_3), and potassium hydrogen phosphate (K_2HPO_4), has been observed to accelerate the decomposition of amino acids [78, 88]. The addition of 0.1wt% Na_2CO_3 [89] or Ni [84, 90] has also been demonstrated to effectively enhance hydrogen production in glycine SCWG. In a separate study, Lee [91] proposed that Ni–Pt/AC and Ni–Pd/AC effectively reformed C^{2+} hydrocarbon intermediates produced in valine to H_2 and CO_2. The research on the degradation of different amino acids with various catalysts demonstrates considerable potential for further development.

The chemical structure of lignin is primarily constituted of repeating units of glyceraldehyde and guaiacol. Figure 6 illustrates the potential degradation pathways of SCWG lignins. Lignins contain a multitude of oxygen-containing functional groups, including the hydroxyl, carboxyl, carbonyl, ether, and ester groups, among others. When lignin decomposes in SCW in the absence of a catalyst, the ether bond and ester bond are susceptible to hydrolysis, resulting in the formation of phenolic compounds and aldehydes (e.g., formaldehyde) [92]. To provide further detail, the phenolic components that lignins are first degraded in SCWG are primarily catechols, guaiacols, and syringols. The intermediate products (e.g., alcohols, aromatics, and aldehydes) are derived from the denaturation of these phenols and subsequently converted into simple gases, including hydrogen, carbon dioxide, and methane, among others. Nevertheless, within the context of the SCWG process, the potential exists for phenolic compounds and aldehydes to undergo condensation or cross-linking, resulting in the formation of high-molecular-weight condensation products. To illustrate, the condensation of phenol and formaldehyde results in the formation of phenolic resin. The condensation products, in conjunction with other refractory phenolic derivatives and polycyclic aromatic hydrocarbons, give rise to the formation of char and tar [75, 93]. The primary reaction involved in the conversion of phenol is isomerization, with other notable reactions including cracking, alkylation and condensation. The cracking of alkyl groups attached to phenol molecules results in the release of gas-phase hydrocarbons and alkoxyphenol. Phenol is frequently

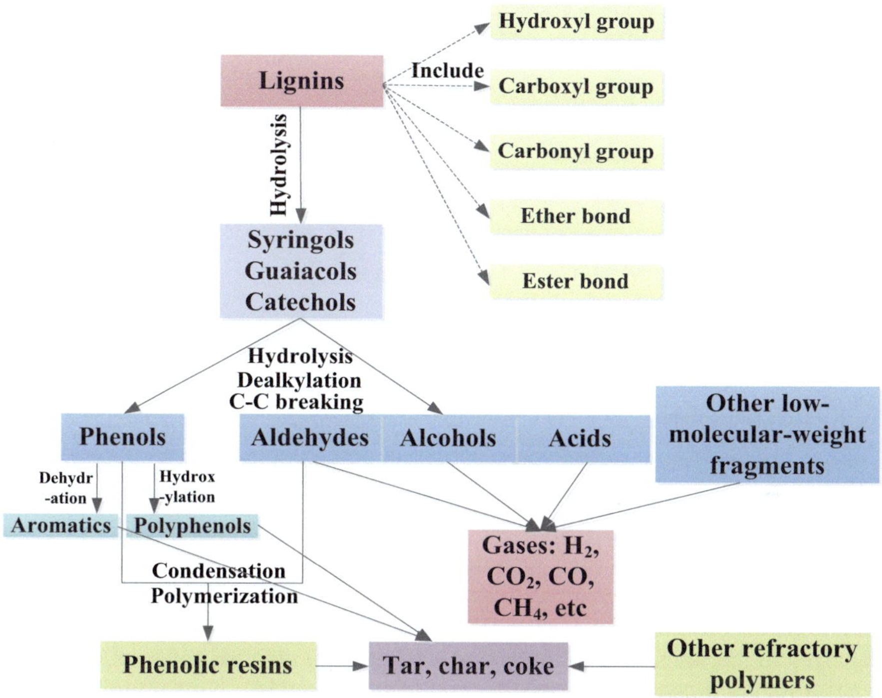

Fig. 6 Potential degradation pathways of lignins in SCWG

employed by researchers as a model compound of lignin for the analysis of its SCWG process. A comprehensive mechanistic explanation can be found in the following section. In recent years, there has been a growing interest in the potential synergistic effects between phenol and other compounds, including glucose and phenol [94], phenol and acetic acid [59, 95], phenol and naphthalene [7], and phenol and cyclohexanol [15].

Based on the above analyses, Fig. 7 depicts the potential reaction processes of polymers in SCWG. Initially, polymers undergo hydrolysis, resulting in the formation of oligomers and monomers. These monomers then decompose in SCW through a series of processes, including isomerization, dehydration, condensation, and hydrolysis, to produce intermediate products. Moreover, the intermediate products have the potential to polymerize with certain substances that have ceased to decompose, thereby forming char and tar. Ultimately, the intermediate products can undergo gas production reactions, including steam reforming, water–gas shift, and methanation, to yield gases (e.g., CO, CO_2, H_2, and CH_4). Alternatively, they can undergo decomposition, pyrolysis and hydrothermal conversion into other intermediate products, which can then be generated as gases through the steam reforming reaction. Additionally, various metal catalysts, including Ru, Ru/γ-Al_2O_3, Pt, Pt/γ-Al_2O_3, Raney Ni, Ni/α-Al_2O_3, Ru/C, Ni/silica–alumina, Ni/MgO, Ni/CeO_2/Al_2O_3, Ni/Al_2O_3, Ni/

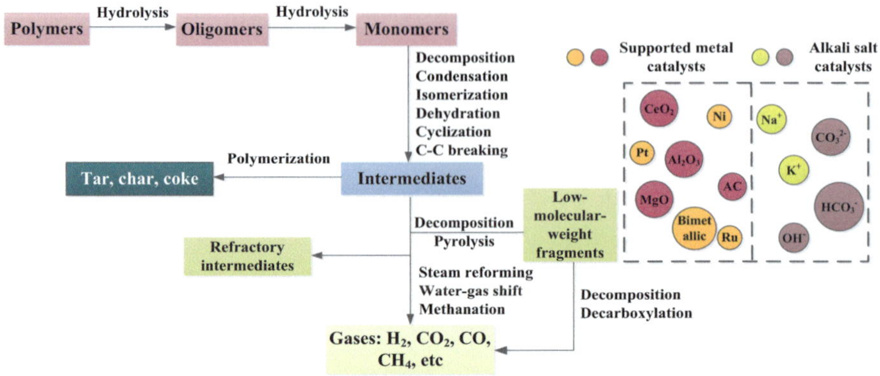

Fig. 7 Potential reaction pathways of polymers in SCWG

AC, NiLa/Al_2O_3, and NiRu/Al_2O_3, are frequently employed in monomer degradation processes [1, 70, 90, 96, 97]. The catalysts most commonly used for the methanation reaction are those containing nickel, rhodium, ruthenium, palladium, or platinum, as well as active carbon. The order of catalytic performance for methane selectivity is as follows: Ru > Rh > Pt > Pd > Ni [90, 98]. Moreover, the catalysts employed in the water–gas shift reaction are primarily alkaline catalysts, including nickel, ruthenium, sodium hydroxide, potassium hydroxide, potassium carbonate, potassium bicarbonate, sodium bicarbonate, and natural alkali, among others [73, 91, 99–101].

3.2 *Phenols*

Phenolic compounds are a type of difficult degradation compound that are widely used in a number of industries, including papermaking, textiles, pharmaceuticals, and petrochemicals [102]. Additionally, they are considered intermediates in the synthesis gas to liquid gas conversion of biomass, coal, and municipal sludge [33, 81, 103, 104]. Phenols can be produced by two principal pathways in SCW [83]. One pathway involves the cleavage of the aryl ether bond in the decomposition of lignin in SCW, while the other entails the condensation or cyclization of sugar produced by cellulose hydrolysis. The degradation behavior of phenol in SCW has attracted considerable attention from the research community, with numerous studies investigating the use of catalysts, including activated carbon [105], CuO/Al_2O_3 [106], MnO_2/TiO_2 [107], CARULITE 150 [108] and MnO_2 [109] to enhance the degradation efficiency of phenol. In SCWO, the primary attackers of phenol are $HO_2\cdot$, $HO\cdot$, and O_2; these generate $C_6H_5O\cdot$, a phenoxy free radical. The free radical then stimulates further reactions, leading to the ring-opening reaction of the phenoxy free radical. The principal free radical reactions of phenol in the initiation process can be expressed by the following equations Eqs. (31)–(34).

$$C_6H_5OH \rightarrow C_6H_5O \cdot +H\cdot \quad (31)$$

$$C_6H_5OH + O_2 \rightarrow C_6H_5O \cdot +HO_2\cdot \quad (32)$$

$$C_6H_5OH + HO_2\cdot \rightarrow C_6H_5O \cdot +H_2O_2 \quad (33)$$

$$C_6H_5OH + OH\cdot \rightarrow C_6H_5O \cdot +H_2O \quad (34)$$

The isomerism exhibited by phenoxy has a significant impact on the chemical reactions that occur. Phenoxy exists in three distinct isomeric forms. In SCW, three potential reaction pathways have been identified for phenoxy free radicals, including coupling to form a dimer, hydroxylation to form phenylglycol, and ring opening [110]. Nevertheless, the principal pathways of phenol degradation are dimer formation and ring-opening reactions. Dimerization is the consequence of the recombination of free radicals and the tautomerization of free radical chains. A high concentration of phenol has been observed to enhance dimerization [111]. The ring-opening process can be elucidated by the addition of $HO_2\cdot$ to phenoxy free radicals, which then undergo a series of chemical reactions leading to the formation of hydroperoxides. The subsequent breakage of the relatively weak O–O bond in these hydroperoxides culminates in β-scission [110]. Guan et al. [112] have demonstrated that the formation of dimers is due to the action of oxygen and phenol free radicals, while the cracking of the phenol ring is the result of oxygen attacking the ring. The latter process results in the formation of maleic acid, which is subsequently oxidized into the unstable oxalic acid in SCW and then transformed into gaseous CO_2. Figure 8 illustrates the distribution of degradation products resulting from the SCWO treatment of phenol. It has been demonstrated that aldehydes and acids, including formic acid, glyoxal, oxalic acid, propionic acid, succinic acid, and maleic acid, are produced as ring-opening intermediate products. The principal monocyclic products are p-benzoquinone, catechol, hydroquinone, resorcinol, and so forth. The major dimer products formed by polymerization are diphenyl ether and dibenzofuran, among others [113].

Over the past decade, there has been considerable interest in the potential of the phenol SCWG for hydrogen production. Indeed, phenol is an intermediate product that presents significant challenges in terms of complete gasification. In the absence of a catalyst, the conversion rate of phenol in SCWG at 750 °C can reach up to 98% [15]. Consequently, numerous researchers have conducted catalytic SCWG of phenol, with the results presented in Table 2 Ni-based catalysts have been demonstrated to be effective in preventing the polymerization of phenol and its intermediates [33]. The complete gasification of phenol is achievable even at relatively low temperatures, such as 450 °C, when a Ni–Co(Ni–Zn)/graphitized carbon black bimetallic nanoparticle catalyst is employed [114]. The $Ni_{20}Zn_{15}$ catalyst is conducive to methane production, whereas the $Ni_{20}Co_{15}$ catalyst is more conducive to hydrogen production,

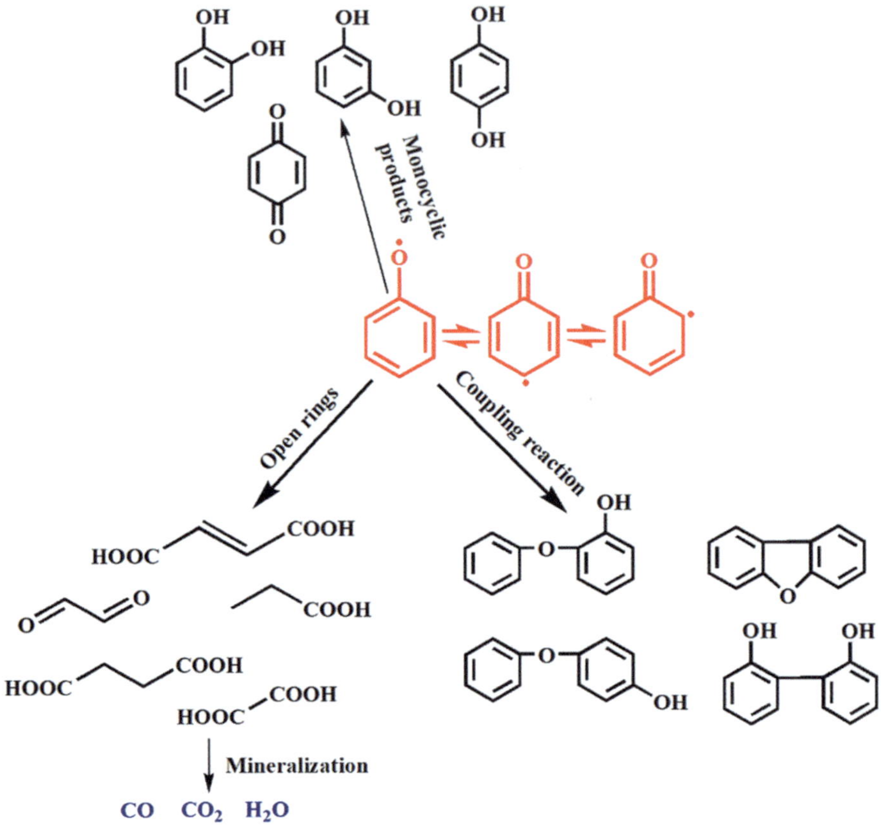

Fig. 8 Degradation products distribution of phenol SCWO

exhibiting markedly elevated phenol gasification activity. Nevertheless, the agglomeration of nanoparticles and the reduction in atom concentration on the catalyst surface can result in a significant decline in catalyst activity following the completion of the reaction [97, 114]. In the future, research into phenol degradation catalysts should focus on identifying a more stable carrier support for Ni-based catalysts. With regard to other catalysts, it has been demonstrated that Ru/graphitized carbon black and Ru/CeO_2 can stably exist in SCW and are not easily inactivated in SCWG [100, 115]. The Ru/CeO_2 catalyst has been demonstrated to significantly enhance the hydrogenation reaction and impede the formation of dimers, including dibenzofuran and polycyclic aromatic hydrocarbons (PAHs). Furthermore, Bogdan [15] put forth a non-catalytic approach, whereby the initial conversion of phenol into cyclohexanol via catalytic hydrogenation facilitates the formation of gaseous products.

The liquid products of phenol, SCWG, include benzene, toluene, o-cresol, 2-ethylphenol, naphthalene, biphenyl, phenylphenols, and dibenzofuran, among others [2]. One category of intermediate compounds comprises decomposition products

Table 2 Gasification efficiency of catalytic SCWG of phenol

No	Catalyst	Temperature (°C)	Time (min)	Carbon gasification efficiency	Reference
1	K_2CO_3	400–600	60	~ 82%	[5]
2	Ru/CeO_2	450–550	30	90%	[100]
3	Ru/graphitized carbon black	500	32	Nearly 100%	[115]
4	Ni–Co(Ni–Zn)/graphitized carbon black	450	32	Nearly 100%	[114]
5	Ni/CeO_2	525	60	80%	[116]
6	Co–Ni/AC	500	30	95%	[117]
7	Ni/Carbon nanotube, Ni/AC, $Ni/\gamma\text{-}Al_2O_3$	480	1200	~ 96%	[33]
8	$Ni\text{-}Ru/Al_2O_3$	450	30	–	[118]

with a relatively simple structure, exemplified by compounds such as benzene. Another category includes phenol dimers, such as dibenzofuran. It is noteworthy that benzene and dibenzofuran have the highest content among the aforementioned compounds, which are formed by dehydrogenation and dimerization, respectively. The formation of dibenzofuran can be explained by the combination of phenol and o-phenoxy, which then undergoes intramolecular dehydration. Benzene is the principal product of phenol pyrolysis, which is generated through the substitution of a hydroxyl group with a hydrogen atom derived from the formation of other free radicals (e.g., phenoxy) [2]. Figure 9 illustrates the potential reaction pathway of phenol SCWG. Consequently, the complete process of phenol SCWG is characterized by cracking, ring-opening, polymerization, and other associated reactions. Benzene plays a pivotal role in the combination of numerous rings, thereby facilitating the formation of intricate ring networks. As the temperature and reaction time increase, polymerization products that are difficult to decompose are formed, resulting in the formation of tar [119]. Additionally, the presence of N-containing substances in phenol SCWG should be avoided, as ammonia inhibits the formation of hydrogen and the degradation of phenol. This is primarily due to the formation of more stable nitrogen polycyclic compounds [103].

A significant number of endocrine-disrupting chemicals (EDCs) are phenols with intricate structures that are capable of mimicking hormones and disrupting the endocrine system, even at trace concentrations. Examples of such chemicals include bisphenol A (BPA), bisphenol E (BPE), nonylphenol (NP), and octylphenol (OP). SCW can provide an optimal environment for the degradation or transformation of these compounds into environmentally benign substances. The SCWO of BPA, NP, and OP at 400 °C, 24 MPa for 0.5–6 min exhibits considerably higher degradation rates than other processes. The degradation rate constants of TOC follow the order NP (0.331 min^{-1}) > OP (0.271 min^{-1}) > BPA (0.236 min^{-1}) [121]. BPA can

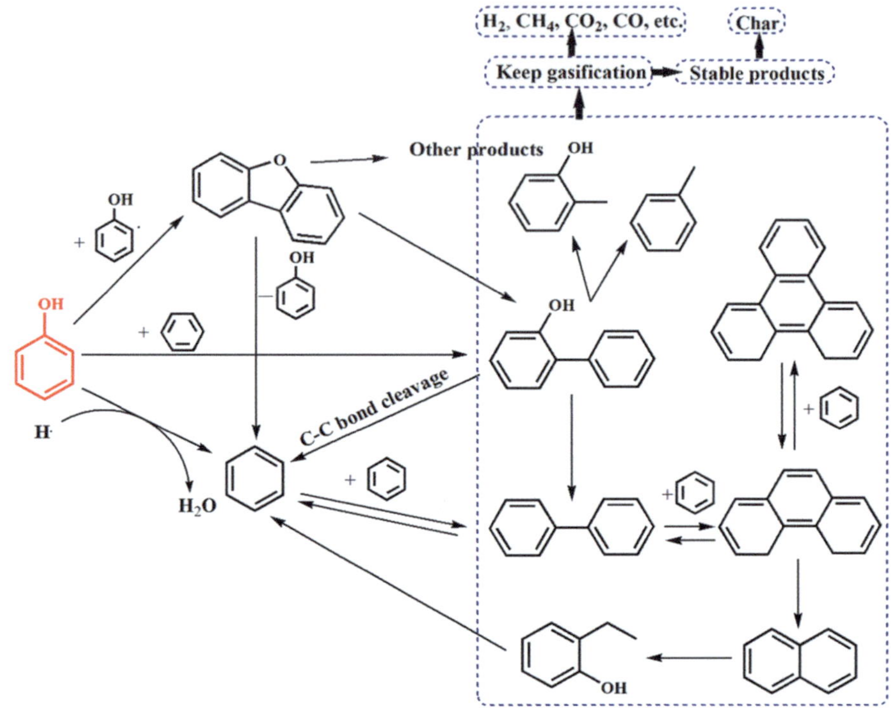

Fig. 9 Potential reaction pathway of phenol SCWG [2, 119, 120]

generate a number of other compounds, including phenol, isopropyl phenol, hydroquinone, benzoquinone, and hydroxybenzoic acid, following the attack of OH· on the hydroxyl groups on the benzene ring and the C–C bond between two aromatic rings [122]. Other researchers have also demonstrated the effectiveness and environmental compatibility of the decomposition of BPA and BPE in high-temperature water (250–350 °C) as a means of synthesizing 4-isopropenylphenol and 4-vinylphenol, respectively [123, 124].

3.3 *Alcohols*

A number of studies have reported on the oxidation mechanisms and kinetics of typical alcohols in supercritical water oxidation (SCWO), including methanol [28, 125], ethanol [21, 126], and isopropanol [127]. Table 3 presents the principal liquid and gas products, along with kinetic data, for the oxidation of various alcohols in supercritical water oxidation (SCWO). It can be concluded that alcohols are readily converted into ketones, acids, or aldehydes in SCWO. In particular, formaldehyde represents the principal intermediate product in the case of methanol SCWO, whereas acetic acid, acetaldehyde, and formaldehyde are the predominant intermediate products in ethanol SCWO. However, in the case of isopropanol SCWO, the intermediate products are acetic acid, acetaldehyde, formaldehyde, acetone, and low-carbon-number alcohols. In the study conducted by Vogel et al. on methanol SCWO, methanol can also be synthesized into ether through acid catalysis during the preheating process in subcritical water [28]. As the number of carbon atoms in alcohols increases, the composition of intermediate products of alcohols in SCWO becomes more complex, and the carbon chain length of products may either decrease or remain constant. In general, the SCWO of alcohols results in a reduction of the carbon chain length, with the final products being CO_2 and water, accompanied by a series of intermediate products. Methanol is the most challenging of the alcohols to oxidize [27], with a conversion rate of only 65% at 538 °C and 24.6 MPa [125]. Under identical conditions, the conversion rate of methanol is inferior to that of ethanol and isopropanol. The influence of the –OH hydroxyl group on the C atom connected to the alcohol renders it more active and susceptible to attack by O_2, forming $HO_2\cdot$ free radicals. In SCWO of the three alcohols, substantial quantities of active free radicals engage in interactions, which are manifested as dehydrogenation, cracking, polymerization, and other reaction forms.

Figure 10 provides a detailed illustration of the oxidation degradation pathway of typical alcohols in SCW. Methanol primarily reacts with $OH\cdot$ though it also exhibits some reactivity with $HO_2\cdot$ [128]. At a typical SCW reaction pressure in SCWO with H_2O_2, $OH\cdot$ for the degradation of alcohol is primarily derived from the decomposition of H_2O_2. However, Fujii et al. [36] have highlighted that, in addition to H_2O_2, the free radical reaction (i.e., $HO_2\cdot + H_2O = OH\cdot + H_2O_2$) is an effective means of enhancing $OH\cdot$ formation when pressure continues to rise. A reaction pathway analysis indicates that the most rapid consumption of alcohol is due to the attack of $OH\cdot$, which subsequently generates various free radicals. These radicals undergo propagation and termination reactions, resulting in the formation of ketones, aldehydes, and acids. Further oxidation of ketones by $HO_2\cdot$ results in the formation of acids, while aldehydes undergo conversion by $OH\cdot$ and $HO_2\cdot$ to yield acids or $HCO\cdot$. Acids may undergo decarboxylation, resulting in the production of CO_2; alternatively, $HCO\cdot$ may generate CO, which is subsequently oxidized into CO_2 (Fujii et al. 1998). These radicals are unstable intermediates that react at a much faster rate with alcohols than water or oxygen, thereby accelerating the propagation of reaction chains.

In recent years, research into alcohols has shifted its focus toward hydrogen production by SCWG or as an auxiliary fuel in hydrothermal flames in SCWO. In SCWG, the primary reactions are the breaking of bonds in the reactants and the initiation of chain reactions, which are predominantly facilitated by $H\cdot$ and $OH\cdot$. The electron-donating properties of the $OH\cdot$ have the potential to enhance the efficiency of gasification by stabilizing the formation of intermediate free radicals. Concurrently, the hydroxyl radical ($\cdot OH$) can attack the hydroxyl group on molecular species, resulting in the formation of water and other free radicals. Additionally, $\cdot OH$ can attack the hydrogen atom or combine with other $\cdot OH$ to produce hydrogen gas (H_2).

Table 3 Typical conversions of alcohols in SCW

Reactant	Main conditions in SCWO	Main liquid products	Main gaseous products	Activation energy	Pre-exponential factor	Reference
Methanol	500–589 °C, 24.6 MPa	Formaldehyde	CO, CO_2	78 ± 20 kcal mol^{-1}	$10^{21.3\pm5.3}$ s^{-1}	[125]
Ethanol	433–494 °C, 24.6 MPa	Acetaldehyde, formaldehyde, acetic acid	CO, CO_2, CH_4, C_2H_4	213.9 ± 18.3 kJ mol^{-1}	$10^{17.23\pm1.65}$ s^{-1}	[21]
Ethanol	260–350 °C, 25 MPa	Acetaldehyde, acetic acid	CO, CO_2	53.8 ± 4.6 kJ mol^{-1}	$10^{2.9\pm0.4}$ s^{-1}	[126]
Isopropanol	488 °C, 23 MPa	Acetaldehyde, formaldehyde, acetic acid, acetone, ethanol, methanol	CO, CO_2	89.441 ± 2.457 (kJ mol^{-1})	$9.308 \pm 3.989 \times 10^7$($m^3$ s^{-1} $kmol^{-1}$)	[127]

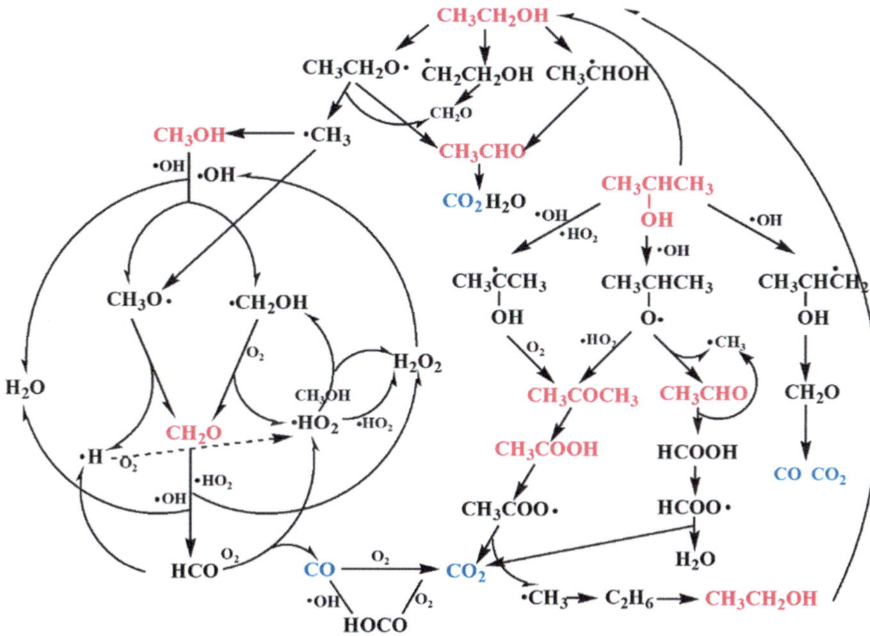

Fig. 10 Reaction network of typical alcohols SCWO [21, 125, 128, 129]

Given the challenging nature of the gasification process, carbonyl compounds, particularly aldehydes, which are anticipated to be the intermediates in alcohol gasification, are prone to instability under such extreme conditions [130]. At elevated temperatures and pressures, they readily decompose into hydrogen and carbon monoxide, which then undergo a water–gas shift, and methanation reaction. To illustrate the reaction pathway of methanol SCWG, Fig. 11 is provided. In the context of SCWG of alcohols with varying carbon chain lengths, the gasification efficiency of methanol is observed to be the lowest, while that of ethanol is the highest. The former can reach 25% at 550 °C for 140 min [131], while the latter can reach 76.3% at 550 °C and 25 MPa for 50 min [132]. Nevertheless, a slight decline in gasification efficiency is observed with an increase in carbon chain length beyond $C > 2$. This may be attributed to the formation of intermediate compounds, tar, and char via polymerization reactions [8]. Moreover, Ni/Al_2O_3 and $Ni/CeZrO_2/Al_2O_3$ have been demonstrated to exert a considerable influence on a number of key processes, including ethanol reforming decomposition, acetaldehyde decomposition, the water–gas shift reaction, and carbon monoxide methanation [133].

Glycerol contains three hydroxyl groups, which results in a more complex degradation process and the formation of a greater number of products in SCWG. A simple kinetics study was conducted by Guo et al. [134], and the apparent activation energy and apparent pre-exponential factor for glycerol SCWG at temperatures between 487 and 600 °C and pressures of 25 MPa were determined to be 104.5 kJ/

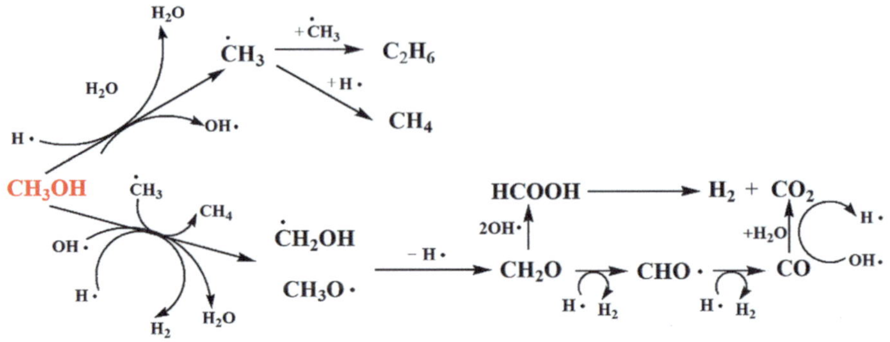

Fig. 11 Free radical reaction pathway of methanol SCWG [8, 20]

mol and 105.90 s^{-1}, respectively. In SCWG, glycerol dehydrates initially to form acrolein and hydroxyacetone. Subsequently, the C–C bond breaks during pyrolysis, resulting in the generation of C1 and C2 compounds, including formaldehyde and acetaldehyde [134]. Pairojpiriyakul et al. [135] conducted a comprehensive analysis of the products generated in glycerol SCWG at temperatures between 450 and 575 °C and pressures of 25 MPa. The predominant gaseous products were identified as H_2, CO_2, CO, and CH_4, with the presence of minor quantities of C_2H_6, C_2H_4, C_3H_8, and C_3H_6. The liquid products included acetaldehyde, acetol, methanol, acetic acid, propionaldehyde, allyl alcohol, acetone, acrolein, ethanol, ethylene glycol, and acrylic acid, among others. However, the main components were acetaldehyde and acetol. An increase in the number of hydroxyl groups can enhance the efficiency of carbon gasification, as it can mitigate the formation of hydrocarbons and promote the generation of CO, which in turn facilitates the subsequent production of CO_2 and H_2 [8]. It can therefore be concluded that this kind of alcohol with multiple hydroxyl groups is an excellent raw material for hydrogen production. Glycerin, a polyhydroxy alcohol, has emerged as a prominent research topic in the field of hydrogen production by SCWG in recent years.

The treatment process in SCW has a considerable energy requirement at the outset of the operational cycle. In the event that the organic waste treated does not possess sufficient intrinsic heat value, the utilization of auxiliary heat will result in a considerable increase in operational costs. The presence of active alcohols has been observed to result in the release of considerable quantities of $HO_2\cdot$ and $OH\cdot$ free radicals, which in turn give rise to a synergistic oxidation effect and a substantial increase in reaction heat [136]. This phenomenon has been demonstrated to facilitate the acceleration of SCWO rates for a range of refractory substances. This phenomenon is referred to as co-oxidation, and the additional alcohols are designated as co-fuels. The utilization of an organic solvent as an auxiliary fuel to facilitate the start-up phase or as a co-fuel to enhance the degradation of reactants in SCWO of certain organic wastes has emerged as an effective strategy to mitigate the high cost associated with SCWO.

It has been demonstrated that common alcohols, including methanol [137, 138], ethanol [139], and isopropanol [129, 140, 141], can accelerate the oxidation degradation of reactants, particularly in the context of enhancing the removal rate of refractory intermediates in SCWO. In a study by Shimoda et al. [137] the mixing effect of ammonia and methanol in SCWO was investigated. The initial concentrations of both substances were 2.9 mmol/L, and it was found that methanol could promote the oxidation of NH_3–N to form N_2O. Figure 12 illustrates the principal reaction pathway of methanol in SCW, which serves to accelerate the oxidative degradation of NH_3–N. The fastest reaction is that of $NH_3 + OH\cdot \rightarrow NH_2\cdot + H_2O$, as illustrated in the accompanying figure. Therefore, ammonia and methanol share generated free radicals and promote mutual oxidation reactions, thereby facilitating the decomposition of NH_3–N. In a comparative study, Zhang et al. [142] investigated the positive effect and exothermic performance of methanol and isopropanol as auxiliary fuels for ammonia. The findings indicated that methanol exhibited superior characteristics compared to isopropanol. Qian et al. [143] highlighted that, in addition to the role of free radicals, alcohols not only inhibited the formation of refractory products (e.g., non-nitrogen and nitrogen aromatic compounds), but also promoted the formation of easily degradable substances (e.g., non-nitrogen open chain compounds). Furthermore, Zhang et al. [138] demonstrated that methanol plays a role in co-oxidation during SCWO of methanol, acetic acid, and phenol, thereby enhancing the decomposition of phenol. In their report, methanol was initially oxidized to generate substantial quantities of active intermediates, including $HO_2\cdot$ and $OH\cdot$. This resulted in a series of subsequent proliferation reactions, including the oxidative decarboxylation of acid and the ring-opening reaction of phenol. Ethanol releases the greatest quantity of $HO_2\cdot$ in a given time period when compared with ethanol and isopropanol. Furthermore, $HO_2\cdot$ can influence the concentration of $OH\cdot$, according to the equation $HO_2\cdot + H_2O = OH\cdot + H_2O_2$, which leads to the conclusion that ethanol is the most effective of the three alcohols in terms of co-oxidation [143].

3.4 *Aldehydes, Ketones, and Acids*

Aldehydes, ketones, and acids are regarded as typical intermediate products in selective catalytic wood gasification (SCWG) of model compounds and real biomass (see Eq. (3.35)). They function as the precursors for the generation of CO_2 and H_2O in SCWO and H_2 and CO_2 in SCWG. Therefore, this section provides a comprehensive description of these three types of compounds.

$$\begin{aligned}\text{Hydrocarbons} + H_2O &\rightarrow \text{aldehydes} + \text{ketones} + \text{acids} + \text{other products} \\ &\rightarrow CO + H_2 + CH_4 + CO_2 + \text{other gases and intermediates} \quad (35)\end{aligned}$$

There is a paucity of studies examining the reactions of aldehydes and ketones as reactants in SCW. Furthermore, these reactions often manifest as intermediates in the

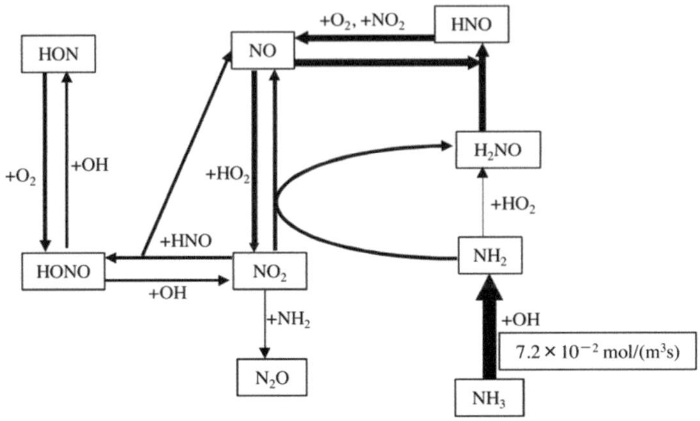

Fig. 12 Major reaction pathway of methanol in SCW to accelerate oxidative degradation of NH_3–N (the thickness of the arrows indicates the magnitude of the relative reaction rate) [137]

literature. Ketones exhibit robust reactivity and can be transformed into polymers, coke, resin, carbon monoxide, and carbon dioxide, among other products. Alcohols and ketones can be converted into one another through the processes of dehydrogenation and hydrogenation [144]. Acetic acid can be transformed into acetone via bimolecular decarboxylation. Concurrently, ketones can be transformed into carboxylic acids through C–C cleavage in the presence of an oxidant [144].

While the majority of reaction products in SCW are generated through free radical reactions, the formation of aldehydes (e.g., acetaldehyde and formaldehyde) is attributed to a high density of ion reactions [30]. Additionally, aldehydes are unique compounds that represent an intermediate redox state of organic carbon. They can often serve as precursors to carboxylic acids through disproportionation. Akgül et al. [145] have reported that formaldehyde participates in two competitive disproportionation reactions in subcritical water. One such process is self-disproportionation (Eq. (36)), whereby two types of formaldehyde engage in an oxidation–reduction reaction, resulting in the formation of methanol and formic acid. Another disproportionation reaction, designated as cross-disproportionation (Eq. (37)), results in the formation of methanol and carbonic acid through an oxidation–reduction reaction between formaldehyde and formic acid. Subsequently, the carbonic acid undergoes facile decomposition into CO_2 and H_2O. In this context, formic acid serves as the hydrogenation agent for formaldehyde reduction, and the presence of excessive formic acid has been observed to promote cross-disproportionation. In SCWG, aldehydes represent a significant source of hydrogen (Eq. (39)) [77]. It is of significant importance to increase the content of aldehydes in the intermediate products of SCWG, as this will help to improve the final production of hydrogen. Fan [146] put forth the proposition that the introduction of formic acid to formaldehyde SCWG would serve to enhance both the efficiency of gasification and that of hydrogen production. This can be explained by the fact that formic acid promotes the self-decomposition of formaldehyde (Eq. (39)) to generate H_2 and induces free radical reactions of generated methanol (Eqs. (36)–(37)) to produce more H_2 (Eqs. (40)–(41)) [146].

$$2HCHO + H_2O \rightarrow CH_3OH + HCOOH \tag{36}$$

$$HCHO + HCOOH + H_2O \rightarrow CH_3OH + CO_2 + H_2O \tag{37}$$

$$RCH_2CHO \rightarrow RCH_3 \cdot +CO \tag{38}$$

$$HCHO \rightarrow H_2 + CO \tag{39}$$

$$CH_3OH + H_2O \rightarrow CO_2 + 3H_2 \tag{40}$$

$$CH_3OH \rightarrow CO + 2H_2 \tag{41}$$

Decarboxylation and dehydration are two main types of reaction pathways of organic acids in SCW [89, 147] and are decarboxylation and dehydration, as illustrated by Eqs. (42)–(43), respectively. Nevertheless, decarboxylation is the predominant reaction in SCWG, facilitating the removal of O from acid in the form of CO_2 (Eq. (44)) [89] is a consequence of the reaction of formic acid in SCWG, whereas the reaction of acetic acid in SCWG results in the production of methane and carbon dioxide (Eq. (45)). In general, as the proportion of carbon atoms in the acids increases, the overall trend in hydrocarbon yield is characterized by oscillating and rising values following SCWG [8]. Chakinala [8] observed that as the length of the acid chain increased from C_1 to C_8, the efficiency of gasification exhibited a notable decline, followed by a period of relative stability. Acetic acid can be identified in the liquid products of long-chain acids with a carbon number above C_2, as well as in other products such as acetone, benzene, and toluene [8]. This demonstrates that formic acid can be fully gasified, whereas long-chain acids ($C \geq 2$) are challenging to gasify completely and tend to be converted into short-chain acids through C–C bond breaking in reactions. This indicates a conversion trend toward the formation of acetic acid.

$$RCOOH \rightarrow CO_2 + RH \tag{42}$$

$$RCOOH \rightarrow H_2O + R/CO \tag{43}$$

$$HCOOH \rightarrow CO_2 + H_2 \tag{44}$$

$$CH_3COOH \rightarrow CH_4 + CO_2 \tag{45}$$

Furthermore, the production of benzene and toluene provides evidence that an aromatization reaction is occurring during the degradation of long-chain acids. Acetic acid is indeed a highly stable intermediate product, rendering it quite difficult to degrade in SCWG [59, 148]. Nevertheless, numerous researchers have found that the addition of acetic acid to SCWG reactions can facilitate the degradation of recalcitrant compounds, including naphthalene and phenol. This is primarily due to the fact that the active free radicals $CH_3COO\cdot$, $CH_3\cdot$, $\cdot COOH$, and $\cdot COCH_3$, which are generated during the decomposition of acetic acid, have the capacity to stimulate the decomposition of other refractory substances [59]. The gasification of linear carboxylic acids in SCW is initiated primarily by the cleavage of the O–H bond in the carboxyl group, followed by β-scission, which is the predominant reaction governing fractionation and gas formation [8]. The impact of the catalyst is frequently investigated with a view to enhancing the degradation of other complex acids. Gong [149] employed the use of $FeCl_3$ as a catalyst to enhance the production of hydrogen from humic acid, which is inherently challenging to degrade in SCWG. The combination of PtSnx/C and activated carbon has been demonstrated to facilitate the decarboxylation of saturated and unsaturated fatty acids into fuel series hydrocarbons in subcritical or supercritical water [150, 151].

4 N-Containing Organic Compounds Conversions

N-containing organic compounds frequently manifest in the conversion products of algae/protein and other substances. Given the hydrothermally stable nature of N-containing model compounds, there is a paucity of studies examining the SCWG of such compounds. It is our understanding that the SCWG of N-containing model compounds necessitates exceedingly high temperatures (550–750 °C), reaching up to 500 °C even with the presence of catalysts [152–154]. The study of the SCWG of these compounds presents a significant challenge for future research. In SCWO, a number of N-containing groups have the capacity to stabilize free radicals. For instance, the free radical RNO· of nitrogen oxide is particularly stable. N-containing cyclic organic compounds may be described as free radical scavengers, which impede the progression of further free radical reactions [61]. Consequently, denitrogenation represents a pivotal research avenue within the domain of SCWO. There are numerous forms of nitrogen present in organic compounds, including NH_2–, –NO_2, N-heterocyclic, –N = N–, and mixtures of two or more of these forms. The oxidation of different N-containing functional groups under SCWO conditions results in the formation of various oxidation products. However, the ultimate oxidation product is expected to be N_2. Ammonia is an example of a stable intermediate product that is challenging to degrade in the majority of nitrogen compounds subjected to SCWO [155]. The degradation of NH_3-N is contingent upon temperature. At temperatures exceeding

500 °C [83], and its conversion rate increases rapidly above 550 °C [156]. Consequently, the subsequent oxidation of ammonia represents the rate-limiting step in the complete oxidation of N-containing compounds, resulting in the formation of N_2. The addition of an auxiliary fuel may prove an effective method. When the methanol concentration is in excess of twice the initial NH_3 concentration, the conversion of NH_3 to N_2O can be increased by 50–60 wt%, and the presence of methanol can also enhance the yields of NO_3^- and NO_2^- [141].

Various N-containing functional groups or existing forms make reaction pathways different. SCWO products of N-containing substances are mainly nitrogen, nitrate, ammonia, almost without nitrite and nitrogen oxide [23], and they can be shown by the following Eqs. (46)–(50):

$$-NH_2 \rightarrow NH_4^+ \tag{46}$$

$$-NO_2 \rightarrow NO_3^- \tag{47}$$

$$-NO_2 + -NH_2 \rightarrow N_2 + \text{others} \tag{48}$$

$$-N = N - +OH\cdot \rightarrow -NO\cdot + \text{others} \tag{49}$$

$$\text{N - heterocyclic compounds} \rightarrow NH_4^+ + \text{others} \tag{50}$$

The C–N bond energy is less than that of the C–C bond; thus, the C–N bond in nitrobenzene or aniline initially undergoes cleavage, resulting in the formation of a benzene ring [157]. The OH· free radical is regarded as the principal oxidant in SCWO, with a diffusion coefficient that is markedly superior to that of nitrobenzene in SCW [158]. The effective collision frequency of free radicals, such as OH· and $HO_2\cdot$, in SCW is increased, thereby facilitating the occurrence of free radical reactions and promoting the oxidation and degradation of nitrobenzene. In a study by Tan et al. [159], it was demonstrated that N-containing organic compounds exhibited superior denitrification effects compared to inorganic compounds, due to their capacity to generate a greater number of active radicals. This indicates that the nitro and amino groups are susceptible to attack by OH·, resulting in the production of amino radicals ($NH_2\cdot$) and nitro radicals ($NO_2\cdot$). Yang et al. [23, 157] put forth the hypothesis that the amino functional group is more susceptible to attack than the nitro functional group when both are present simultaneously. Accordingly, the potential degradation pathway of 2-nitroaniline SCWO can be illustrated in Fig. 13. The NH_2 group in 2-nitroaniline initially reacts with OH· to form $NH_2\cdot$, after which the $-NO_2$ group is removed in the form of $NO_2\cdot$. The two active radicals then continue to react, resulting in the generation of N_2 gas. This finding is consistent with the conclusion of Yang et al. that organic nitrogen plays a predominant role in the formation of N_2 at the early stages of SCWO, while inorganic nitrogen (in the forms of NH_4^+ and NO_3^-) contributes only minimally to the conversion of N_2 at the late stages.

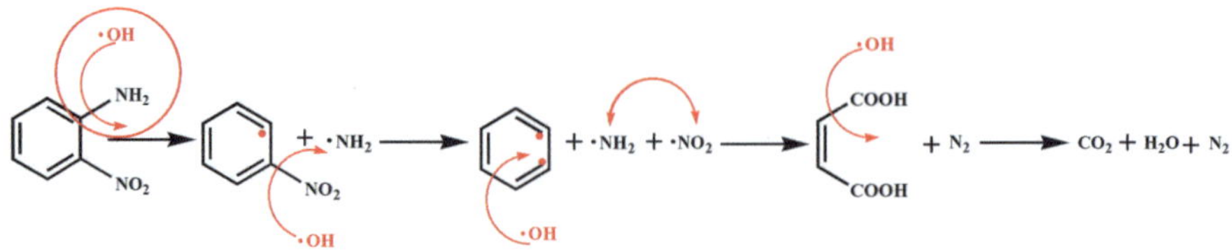

Fig. 13 Conceivable reaction pathway of 2-nitroaniline SCWO [157]

Pyridine, quinoline, quinazoline, and pyrimidine are significant N-containing heterocyclic compounds. Typically, they are initially oxidized into corresponding derivatives in SCWO, after which polycyclic compounds are decomposed into monocyclic compounds and subsequently split into open rings. Consequently, the liquid-phase composition is highly intricate, and the formation of select polycyclic intermediates may also occur through the coupling of free radicals [160]. For example, naphthalene, biphenyl and phenanthrene are formed by the polymerization of phenyl and benzyl, among other compounds [161]. The oxidative degradation of these heterocyclic compounds can be summarized in four steps. Initially, hydroxylation generates derivative alcohols. Subsequently, oxidative reforming forms carbonyls, while cyclolysis (which involves the reattack of OH·) results in deamination, decarboxylation, and rearrangement reactions. Finally, mineralization occurs, producing CO_2, H_2O, and NH_3 [161].

The F(0) value, representing the magnitude of the attack power of OH· free radical, has been calculated for C and N atoms in pyridine. The F(0) value of the N atom is considerably higher than that of the C atom, indicating that the N atom is more susceptible to attack by OH·, which can result in fracture [162]. The bond energy of the C–N bond is 305 kJ mol/L, which is lower than the bond energy of the C–C bond, which is 332 kJ mol/L. It may therefore be surmised that the fracture of the C–N bond plays a dominant role in the ring-opening process of N-heterocyclic compounds. Conversely, the N atom exhibits a strong electron attraction ability, which results in a relatively low electron cloud density within the pyridine ring [161]. This results in a nucleophilic substitution reaction occurring at the α- or γ-position of the N atom. Pyrimidine, with one additional nitrogen atom compared to pyridine, is more susceptible to oxidation and degradation due to its reduced electron cloud density. Similarly, the benzene ring in quinazoline and quinoline can be readily oxidized due to the low electron cloud density of the pyridine/pyrimidine ring. The benzene ring in quinazoline has a greater number of N atoms than that in quinoline, which results in greater ease of oxidation and opening. Figure 14 illustrates the potential reaction pathways of quinazoline and pyrimidine in SCWO, based on the identification of intermediate products. It would appear that there are two potential pathways for the ring opening of quinazoline in SCWO. The initial reaction pathway is initiated by the higher electron cloud density of the pyrimidine ring in quinazoline, which is susceptible to attack by OH· and O_2, resulting in the opening of the benzene ring and the formation of pyrimidine. Subsequently, the pyrimidine ring undergoes cleavage, resulting in the formation of N-containing chain products, such as carbamate and oxalic acid, and ultimately, gases. The alternative ring-opening pathway is that the pyrimidine ring is opened prior to the benzene ring, as evidenced by the appearance of intermediates such as benzene, naphthalene, and methylbenzene. Gong et al. [161] postulated two potential scenarios for the ring-opening process. Nevertheless, in the study of supercritical water oxidation of quinoline by Ren et al. [38], a single possibility exists whereby the benzene ring is opened first, with the opening of the pyridine ring occurring at a slightly slower rate. Accordingly, the discrepancy between the various proposed ring-opening orders necessitates further investigation in subsequent studies.

To date, there has been a paucity of reports on the decomposition of diazonium group compounds in SCWO. In comparison with the aromatic group, the azo bond (–N = N–) is more readily broken, resulting in the OH· radical being more prone to attack –N = N– initially in reactions. This subsequently leads to the fragmentation of macromolecular compounds into smaller units [23]. Concurrently, the presence of chlorine in compounds has been observed to facilitate the degradation of N-containing compounds. Furthermore, it has been demonstrated that an elevated chlorine content in amino-N compounds results in a greater production of ammonia (NH_4^+).[23]. One potential explanation is that chloride ions can react with certain

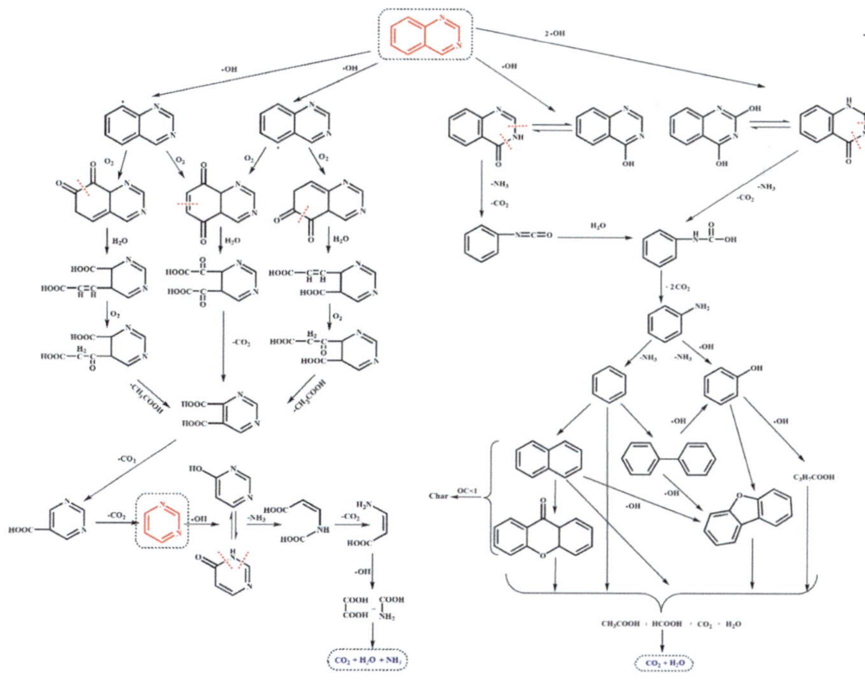

Fig. 14 Potential reaction pathways of quinazoline and pyrimidine in SCWO [161]

metals, resulting in the release of transition metal ions (e.g., Fe^{3+}, Mn^{2+} and Cu^{2+}) play a catalytic role in reactions.

The SCWO processes of N-containing substances derived from industrial or living sources, such as sludge and wastewater, are more complex than those of model compounds [156, 163–165]. The liquid-phase intermediates that emerge following SCWO can be classified into five distinct categories: non-nitrogen open chain compounds, nitrogen open chain compounds, non-nitrogen aromatic compounds, nitrogen aromatic compounds, and non-aromatic ring compounds. Among these, the non-nitrogen open chain compounds are composed of aliphatic acids and small amounts of aliphatic hydrocarbons and esters [83]. The proportion of nitrogen open chain compounds is relatively low, with the majority of these compounds being amide compounds. Non-nitrogen aromatic compounds are primarily oxygen-containing aromatic derivatives, comprising phenols (phenol and its methyl substituents), acids (phenylacetic acid), alcohols (phenylethanol), and ketones (acetophenone) [81]. N-containing aromatic compounds can be decomposed from macromolecules to yield pyridine, indole, pyrazine, and derivatives substituted with methyl and amino groups [78, 166]. However, pyridine is resistant to oxidation, resulting in a relatively high concentration of pyridine and its derivatives in N-aromatic compounds in the liquid phase. N-heterocyclic compounds, such as pyridine, can be formed by the Maillard reaction between protein degradation products and cellulose decomposition products. They can serve as a steady source of free radical scavengers, reducing the number of free radicals in the system. As the temperature and oxidant level increase, pyridine decomposes, forming ammonia [156]. Additionally, pyridine can combine with the benzene ring to form fused ring compounds, such as quinoline. The remaining liquid-phase products of wastewater SCWO are non-aromatic ring compounds, though their concentration is considerably lower. The majority of these non-aromatic ring compounds are cyclic ketonic compounds, including cyclopentanone, cyclohexanone, pyrrolidone, and piperidone. These compounds serve as intermediates in both the Maillard reaction and the sugar hydrolysis reaction. Their subsequent dehydrogenation can result in the formation of phenols [83].

5 Halogen Atom-, S- and P-Containing Organic Compounds Conversions

5.1 Cl-, Br-, and F-Containing Compounds

Organic compounds containing chlorine, such as chloroalkyls, chloroaromatic hydrocarbons, and macromolecule polymers, are significant chemical entities or intermediates within industrial processes. The high toxicity and low biodegradability of these compounds can result in significant environmental damage. Subcritical or supercritical fluids represent an effective medium for the removal of chlorine from organic matter. Catalytic dechlorination can selectively break one or more C–Cl bonds, thereby reducing their toxicity [167]. The dechlorination effect of chlorinated compounds in subcritical or supercritical water has been the subject of study by numerous researchers [168–170].

The SCW process has been demonstrated to be an effective method for the dechlorination of Cl-containing organic compounds. SCWO can facilitate the conversion of organic chlorine into HCl, which is then dissolved in water. Concurrently, organic carbon is transformed into a gaseous product, CO_2. Some researchers have employed hydrogen peroxide as an oxidant to investigate the degradation kinetics of o-dichlorobenzene (o-DCB) in SCW [24, 171]. During o-DCB SCWO, a number of intermediates are produced, including 5-chloro-2-pentanone, 4-hydroxy-4-methyl-2-pentanone, 2-hexanone, 2-pentanone, and 3,4-dichlorophenol. In SCWO experiments involving chlorobenzene, Liu [172] identified some intermediate by-products that were not fully oxidized, including benzene, 4-methyl-3-penten-2-one and 4-hydroxy-4-methyl-2-pentanone. Conversely, Shimokawa [173] employed TiO_2 nanoparticles photocatalysis to generate intermediate products with reduced molecular weight, namely aliphatic C_2–C_4 ketone, aldehyde, alcohol, and carboxylic acid, without the presence of aromatic hydrocarbons or their derivatives. The Fe/ZrO_2 system has been demonstrated to catalyze the reductive dechlorination of chlorobenzene in SCW at temperatures between 410 and 500 °C, without the addition of any oxidant. Fe/ZrO_2 has the capacity to catalyze the reductive dechlorination of chlorobenzene in SCW at temperatures between 410 and 500 °C, without the addition of any oxidant. Furthermore, it has been demonstrated that Fe [174], as shown in Eq. (3.51). Furthermore, the Lewis acid/base pair formed by Zr^{4+}–O^{2-} significantly enhances the cleavage of the C–Cl bond in the chlorobenzene aromatic ring, thereby accelerating the formation of HCl (Eq. (3.52)) [174]. This shows that catalyst can greatly improve the degradation efficiency of Cl-containing compounds in SCW.

$$2H_2O \overset{Fe/ZrO_2}{\longrightarrow} 2H\cdot + 2OH\cdot \tag{51}$$

$$2H\cdot + C_6H_5Cl \rightarrow C_6H_6 + HCl \tag{52}$$

With regard to Cl- and O-containing organic compounds, Hashimoto [148] investigated the oxidative degradation of 2,4-dichlorophenoxyacetic acid in sub- and supercritical water. The findings revealed that the dechlorination effect increased with reaction time and temperature. Moreover, the study identified 2,4-dichlorophenol (2,4-DCP) and acetic acid as the primary degradation intermediates. Nevertheless, 2,4-DCP can undergo almost complete disappearance and transformation into inorganic chlorine in supercritical water at 400 °C. The initial reaction pathway of 2,4-DCP decomposition results in the formation of dimers and a range of macromolecular compounds, including dichlorophenoxyphenols and 5-dichloro-2-hydroxyacetophene. Subsequently, in SCWO, these compounds undergo secondary oxidation to form ring-opening products and monocyclic compounds, such as phenol, 2-methylphenol, 2,4-dichloro-6-methylphenol, which can ultimately be oxidized into CO_2 [175]. Furthermore, the use of NaOH as a catalyst has been demonstrated to be an effective method for the dechlorination of 2-chlorophenol, with the potential to significantly reduce the formation of dimers in SCWO [176].

The dechlorination of complex compounds is a process that commonly occurs in subcritical water. This is due to the fact that it is a more straightforward process to achieve dechlorination in subcritical water than in supercritical water [170]. Takeshita et al. [177] treated polyvinyl chloride with subcritical water (200–350 °C) and observed a notable dechlorination effect, with no evidence of the formation of harmful chlorinated organic compounds in the liquid or gas phases. The corresponding reaction principle is believed to result in the release of HCl in polyvinyl chloride and the formation of a polyene structure of the form $–(CH = CH)_n–$. Subsequently, the formed HCl immediately dissolves in water to form hydrochloric acid. The polyene structure then undergoes further reactions, including cyclisation, hydrolysis, and other processes, which generate low-molecular-weight aromatic and aliphatic compounds. Soler et al. [167] conducted dechlorination experiments of polyvinyl chloride electric wires at temperatures between 200 and 300 °C. They observed that the emission of HCl and the formation of chlorobenzene and chlorophenol decreased when the residues after dechlorination were subjected to high-temperature pyrolysis. Conversely, the formation of polycyclic aromatic hydrocarbons increased with the dechlorination temperature. This evidence demonstrates that subcritical water treatment is an efficacious dechlorination method, which can diminish the formation of chlorinated compounds in the subsequent heat treatment process. This finding is also consistent with their experiments on the debromination of printed circuit boards at temperatures between 225 and 275 °C [178]. The debromination and dichlorination efficiencies increase with an increase in process temperature. The maximum debromination efficiency of 63.6% and the maximum dechlorination efficiency of 95.96% were achieved at temperatures of 275 °C and 300 °C, respectively [167, 178].

It is evident that the final conversion products of halogen elements during the degradation of compounds in SCW are predominantly HCl and HBr, among other compounds. Due to steric hindrance, monovalent atoms are more susceptible to exposure and attack by free radicals than high-valent atoms in the majority of seizure reactions. Consequently, the formation of HCl and HBr is possible through the combination of H· and Cl· or Br·.

Additionally, fluorine is a halogen element, which is the most electronegative element, thereby conferring upon fluorochemicals unique physiochemical properties. Perfluorochemicals (PFCs), including perfluorooctanesulfonate (PFOS), perfluorohexanesulfonate (PFHS), and perfluorooctanoic acid (PFOA), have been extensively employed in numerous industrial processes and commercial products, such as polymer additives, surfactants, fire-fighting foams and pesticides. PFCs have recently been the subject of considerable scrutiny due to their classification as refractory environmental contaminants and their prevalence in wastewater sludge [179, 180].

The utilization of SCW has been demonstrated to be an effective method for the degradation of PFCs and the realization of defluorination. Hori et al. [181] the decomposition of PFHS in sub/supercritical water with a Fe catalyst at temperatures between 150 and 380 °C and pressures between 21.2 and 26.4 MPa for a period of six hours was observed. The results demonstrated that PFHS could be decomposed to F– and SO_4^{2-} ions in the aqueous phase, while small amounts of H_2, CO_2, and CF_3H were detected in the gas phase. Furthermore, the addition of oxygen can also facilitate the decomposition of PFCs. The addition of oxygen to the decomposition of $[(C_6H_5)_3S][CF_3SO_3]$ (a substitute for PFOS) in subcritical water at 372 °C for 6 h has been shown to significantly enhance the yield of fluoride ions, reaching up to 93.6% [182]. This indicates that the majority of fluorine atoms in $CF_3SO_3^-$ are effectively transformed into fluoride ions. Furthermore, the addition of oxygen can inhibit the formation of CF_3H and benzene. Furthermore, $Ca(OH)_2$ has been demonstrated to be an effective defluorinating agent for PFCs under thermal conditions, whereby CaF_2 is formed [183]. In general, the objective of using SCW for defluorination is to release fluorine from complex fluoroorganic compounds and convert it into fluoride ions, which can then be recycled.

5.2 S-Containing Compounds

In recent years, research into the use of coal for the production of hydrogen has attracted considerable attention. Nevertheless, the utilization of S-containing compounds in coal processes has the potential to result in air pollution and corrosion of the reactor [184, 185]. Moreover, residual oil contains a plethora of S-containing compounds, including homologues of thiophene [186]. The utilization of SCW for the removal of sulfur from heavy oil, asphalt, and other raw materials represents a significant area of potential development. Demirbas [187] employed a partial oxidation method to desulfurize residual oil with a catalyst in SCW, achieving a desulfurization efficiency of up to 60%. The transformation of the original sulfide into smaller sulfides by SCW provides an important theoretical basis for the control of S-containing pollutants in subsequent utilization, as the pathways of transformation of the S element in SCW can be studied.

Figure 15 illustrates the distribution of sulfur in the gas, liquid, and solid phases following the coal SCWG and pyrolysis processes. Consequently, the S-containing gases resulting from coal pyrolysis encompass both hydrogen sulfide (H_2S) and sulfur dioxide (SO_2). Nevertheless, SCW is capable of providing a plentiful supply of H· free radicals, which can facilitate the generation of additional H_2S while simultaneously inhibiting the formation of SO_2 and organic sulfur. Additionally, SCW facilitates the decomposition of sulfate and pyrite, while OH· free radicals enhance the oxidation of S^{-2} and coke in supercritical fluid [188]. These findings are in accordance with the conclusions presented by Chen et al. [189]. The influence of pressure, temperature, and oxidation coefficient on the distribution of S in solid, liquid, and gas products after lignite SCWG was investigated, and it was found that SO_4^{2-} and $S_2O_3^{2-}$ were the main sulfur components in the liquid fraction, while the contents of S^{2-}, HS-, SO_3^{2-}, and HSO_3^- were very low, and H_2S was the only sulfur compound detected in the gaseous products. Pathardhan et al. [32] also highlighted that SCW could effectively transform sulfur in the form of sulfides (e.g., aliphatic, aromatic, and

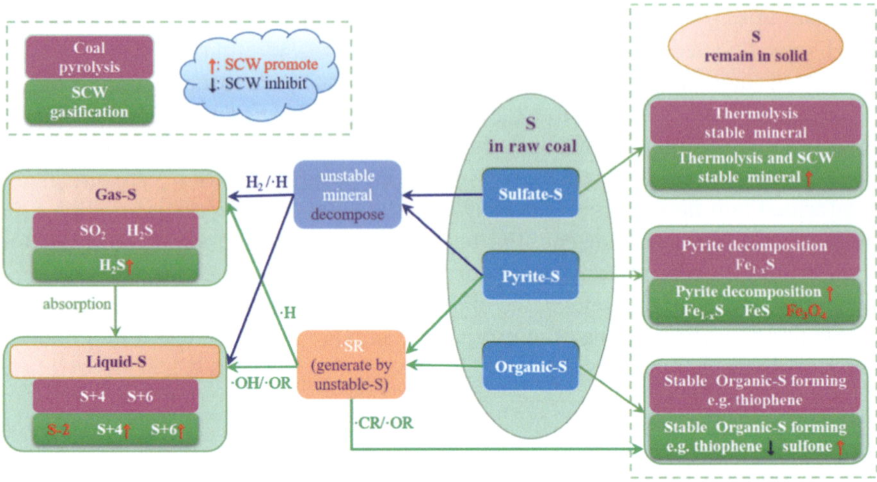

Fig. 15 Conversion of sulfur in coal SCWG and pyrolysis [188]

cyclic) into H_2S in the presence of α-hydrogen, thereby weakening the S-R bond, as illustrated by Eqs. (53)–(57).

At the initiation stage, C–S bonds are usually the weakest bonds in the system.

$$R_1 - CH_2S - R_2 \rightarrow R_1 - CH_2S \cdot + R_2 \cdot \quad (53)$$

$$(R_1 - CH_2S - R_2 \rightarrow R_1 - CH_2 \cdot + R_2S \cdot \quad (54)$$

At the propagation stage, α C–H is the easiest to abstract.

$$R_2 + RH \rightarrow R_2H + R \cdot \quad (55)$$

$$R \cdot + R_1 - CH_2S - R_2 \rightarrow RH + R_1 - C \cdot HS - R_2 \quad (56)$$

$$R - C \cdot HS - R_2 \rightarrow R_1 - CH = S + R_2 \cdot \quad (57)$$

The present study also aims to elucidate the desulfurization mechanism of alkyl sulfur compounds in SCW, including hexyl sulfide [190, 191], hexanethiol [192], and isopropylthiobenzene [193]. The role of water in the decomposition of alkyl sulfides in SCW is of significant importance. It serves not only as a reactant for the hydrolysis of significant intermediates (e.g., thioaldehyde and thioacetone) but also as a source of numerous H atoms required for desulfurization to form H_2S. Additionally, it functions as a H-transfer catalyst for desulfurization and decomposition of some intermediate thiols. It is proposed that alkanes can lead to the selective cleavage of the C–S bond, thus promoting the decomposition and desulfurization process by forming aliphatic radicals that attack α-H on isopropylthiobenzene [193]. Similarly, the isopropyl group in isopropylthiobenzene can act in a manner analogous to that of an alkane, thereby enhancing the desulfurization of isopropylthiobenzene in SCW.

To date, there has been a paucity of SCW desulfurization catalysts. Ates et al. [194] investigated the impact of ZnO, MoO_3, and MoS_2 on the desulfurization of heavy oil in SCW. Their findings revealed that ZnO exhibited the most effective removal of dibenzothiophene; however, its primary structure and composition underwent alterations, leading to a decline in activity due to the sulfurization reaction. It would therefore be beneficial for researchers to focus their attention on the catalytic desulfurization effect and the activity and stability of the catalyst in future studies. Following the SCWO process, sulfur is exclusively released into liquid products, with minimal sulfur-containing substances present in the gas products. The principal sulfur components in the liquid products are sulfide, thiosulfate, sulfite and sulfate, with sulfate being the dominant form. The content of sulfide or thiosulfate is sometimes high when reactants include methylmercaptan, iron sulfide [195], sulfur coal [196], and thiirane. If SCWO is conducted at a high temperature over an extended period of time, it is probable that complete conversion of sulfur into sulfate will occur.

5.3 P-Containing Compounds

Representative P-containing organic compounds, including tributyl phosphate (TBP) and other organophosphate flame retardants (OPFRs), are extensively utilized across a range of sectors. For example, TBP is typically employed as a raw material for herbicides, fungicides, plasticizers, and other products, whereas the majority of OPFRs can be found in furniture, textiles, electronics, and other processing chemicals. However, with the advancement of the phosphorus chemical industry, environmental contamination is becoming increasingly severe. In light of this, a number of treatment methods for P-containing organic compounds have been proposed in recent years, including ultraviolet-persulfate, advanced oxidation processes (AOPs), photocatalytic oxidation, and supercritical water oxidation (SCWO) [197–199]. This section presents a summary of studies on the SCWO of tributyl phosphate (TBP) and other organophosphate flame retardants (OPFRs).

TBP can be successfully decomposed and achieve a TOC removal rate of greater than 99% in SCWO at 480 °C and less than 50 MPa for 100 min without the use of a catalyst [198]. The decomposition efficiency of TBP in SCWO can reach a maximum value of 79.4% when Fe_2O_3 is employed at 405 °C for 70 min. The order of catalytic performance is as follows: $Fe_2O_3 > CuO > MgO > Ag_2O > ZnO$ [198]. The most abundant by-product of TBP SCWO is H_3PO_4, followed by other compounds such as butanol, monobutylphosphate (MBP), and dibutylphosphate (DBP). Besides, Golmohammadi et al. [200] reported that the order of catalytic performance was $CeO_2 > Co_3O_4 > Cr_2O_3 > MnO_2$.

To date, there has been a paucity of research conducted on the subject of SCWO of other OPFRs. Yang et al. [201] investigated the decomposition of 14 OPFRs in SCWO with temperature varying from 350 to 550 °C at 24 MPa for 1–6 min with 300% of excess oxygen. The research revealed that aryl and chloro OPFRs, including 2-ethylhexyl diphenyl phosphate and tris(1,3-dichloro)phosphate, exhibited higher TOC degradation rates than alkyl OPFRs, such as tripropyl phosphate. In conclusion, the potential degradation pathways of OPFRs can be broadly categorized as hydroxylation, ring cleavage and mineralization for aryl compounds, in comparison with hydroxylation, carbonylation, and mineralization for alkyl and chloro compounds [201].

The production of hydrogen by SCWG is a promising avenue of research for the utilization of heavy hydrocarbons and O- and N-containing organic compounds. In the case of certain compounds containing heteroatoms (such as chlorine, bromine, fluorine, sulfur, and phosphorus), the research focus is on utilizing the superior characteristics of SCW to achieve the removal of heteroatoms and to attain a more environmentally friendly outcome. SCWO represents an extremely effective method for the degradation of the majority of organic compounds when conducted under optimal conditions. The role of free radicals in SCWG and SCWO is of great significance, as they are involved in a multitude of complex chemical reactions, the formation of intermediate products, the pathways of reactions, and the underlying mechanisms of these processes. The development of effective catalysts based on the catalytic mechanisms of typical compounds is of great significance, particularly in the context of SCWG. This will provide invaluable guidance for the conversion of actual substances (e.g., biomass and refractory organic wastewater) in SCW.

References

1. N. Ding, R. Azargohar, A.K. Dalai, J.A. Kozinski, Catalytic gasification of glucose to H_2 in supercritical water. Fuel Process. Technol. **127**, 33–40 (2014)
2. C.M. Huelsman, P.E. Savage, Intermediates and kinetics for phenol gasification in supercritical water. Phys. Chem. Chem. Phys. **14**, 2900–2910 (2012)
3. Y. Karakuş, F. Aynaci, E. Kipçak, M. Akgün, Hydrogen production from 2-propanol over Pt/Al_2O_3 and Ru/Al_2O_3 catalysts in supercritical water. Int. J. Hydrogen Energy **38**, 7298–7306 (2013)
4. P.E. Savage, Organic chemical reactions in supercritical water. Chem. Rev. **99**, 603–622 (1999)
5. D. Selvi Gökkaya, M. Saglam, M. Yüksel, L. Ballice, Supercritical water gasification of phenol as a model for plant biomass. Int. J. Hydrogen Energy **40**, 11133–11139 (2015)
6. M. Akizuki, K. Tomita, Y. Oshima, Kinetics of solid acid catalyzed 1-octene reactions with TiO_2 in sub- and supercritical water. J. Supercrit. Fluids **56**, 14–20 (2011)
7. Y. Wang, F. Gao, J. Yang, Y. Guo, Y. Zhu, G. Zhao, C. Fang, Y. Guo, S. Wang, N. Guo, Partial oxidation kinetics of the mixture of acetic acid, phenol and naphthalene in supercritical water for hydrogen production. Int. J. Hydrogen Energy **42**, 19843–19850 (2017)
8. A.G. Chakinala, S. Kumar, A. Kruse, S.R.A. Kersten, W.P.M. Van Swaaij, D.W.F. Brilman, Supercritical water gasification of organic acids and alcohols: the effect of chain length. J. Supercrit. Fluids **74**, 8–21 (2013)

9. M. Akizuki, T. Fujii, R. Hayashi, Y. Oshima, Effects of water on reactions for waste treatment, organic synthesis, and bio-refinery in sub- and supercritical water. J. Biosci. Bioeng. **117**, 10–18 (2014)
10. V. Anikeev, W. Tsang, J. Manion, Density effects in the reaction of 2-propanol in supercritical water. Combust. Sci. Technol. **178**, 417–441 (2006)
11. A.G. Kalinichev, J.D. Bass, Hydrogen bonding in supercritical water. 2. Computer simulations. J. Phys. Chem. A **101**, 9720–9727 (1997)
12. A. Loppinet-Serani, C. Aymonier, F. Cansell, Current and foreseeable applications of supercritical water for energy and the environment. Chemsuschem **1**, 486–503 (2008)
13. J.V. Sengers, J.T.R. Watson, R.S. Basu, B. Kamgar parsi, R.C. Hendricks, Representative equations for the thermal conductivity of water substance. J. Phys. Chem. Ref. Data **13**, 893–933 (1984)
14. A.A. Peterson, F. Vogel, R.P. Lachance, M. Fröling, M.J. Antal Jr., J.W. Tester, Thermochemical biofuel production in hydrothermal media: a review of sub- and supercritical water technologies. Energy Environ. Sci. **1**, 32–65 (2008)
15. V.I. Bogdan, A.V. Kondratyuk, A.E. Koklin, V.V. Lunin, Interaction of phenol and cyclohexanol with supercritical water. Russ. J. Phys. Chem. B **11**, 1207–1213 (2017)
16. N. Ding, R. Azargohar, A.K. Dalai, J.A. Kozinski, Catalytic gasification of cellulose and pinewood to H_2 in supercritical water. Fuel **118**, 416–425 (2014)
17. S. Nanda, S.N. Reddy, H.N. Hunter, D.-V.N. Vo, J.A. Kozinski, I. Gökalp, Catalytic subcritical and supercritical water gasification as a resource recovery approach from waste tires for hydrogen-rich syngas production. J. Supercrit. Fluids **154**, 104627 (2019)
18. E. Weiss-Hortala, A. Kruse, C. Ceccarelli, R. Barna, Influence of phenol on glucose degradation during supercritical water gasification. J. Supercrit. Fluids **53**, 42–47 (2010)
19. H. Jin, Y. Wu, L. Guo, X. Su, Molecular dynamic investigation on hydrogen production by polycyclic aromatic hydrocarbon gasification in supercritical water. Int. J. Hydrogen Energy **41**, 3837–3843 (2016)
20. D. Castello, L. Fiori, Kinetics modeling and main reaction schemes for the supercritical water gasification of methanol. J. Supercrit. Fluids **69**, 64–74 (2012)
21. J. Schanzenbächer, J.D. Taylor, J.W. Tester, Ethanol oxidation and hydrolysis rates in supercritical water. J. Supercrit. Fluids **22**, 139–147 (2002)
22. B. Al-Duri, F. Alsoqyiani, I. Kings, Supercritical water oxidation (SCWO) for the removal of N-containing heterocyclic hydrocarbon wastes. Part I: process enhancement by addition of isopropyl alcohol. J. Supercrit. Fluids **116**, 155–163 (2016)
23. B. Yang, Z. Shen, Z. Cheng, W. Ji, Total nitrogen removal, products and molecular characteristics of 14 N-containing compounds in supercritical water oxidation. Chemosphere **188**, 642–649 (2017)
24. C.M. Lin, H. Liu, W.S. Zhen, K. Bei, Z.Y. Pan, Study on supercritical water oxidation of o-dichlorobenzene in a quartz micro-reactor with in situ microscope and Raman spectroscopy. Chem. Pap. **73**, 1257–1262 (2019)
25. B. Veriansyah, J.D. Kim, J.C. Lee, Supercritical water oxidation of thiodiglycol. Ind. Eng. Chem. Res. **44**, 9014–9019 (2005)
26. F. Wang, X. Lu, K. Shih, C. Liu, Influence of calcium hydroxide on the fate of perfluorooctanesulfonate under thermal conditions. J. Hazard. Mater. **192**, 1067–1071 (2011)
27. L.T. Boock, M.T. Klein, Lumping strategy for modeling the oxidation of C1–C3 alcohols and acetic acid in high-temperature water. Ind. Eng. Chem. Res. **32**, 2464–2473 (1993)
28. F. Vogel, J.L.D. Blanchard, P.A. Marrone, S.F. Rice, P.A. Webley, W.A. Peters, K.A. Smith, J.W. Tester, Critical review of kinetic data for the oxidation of methanol in supercritical water. J. Supercrit. Fluids **34**, 249–286 (2005)
29. A. Kruse, E. Dinjus, Hot compressed water as reaction medium and reactant. Properties and synthesis reactions. J. Supercrit. Fluids **39**, 362–380 (2007)
30. W. Bühler, E. Dinjus, H.J. Ederer, A. Kruse, C. Mas, Ionic reactions and pyrolysis of glycerol as competing reaction pathways in near- and supercritical water. J. Supercrit. Fluids **22**, 37–53 (2002)

31. B. Al-Duri, F. Alsoqyani, Supercritical water oxidation (SCWO) for the removal of nitrogen containing heterocyclic waste hydrocarbons. Part II: system kinetics. J. Supercrit. Fluids **128**, 412–418 (2017)
32. P.R. Patwardhan, M.T. Timko, C.A. Class, R.E. Bonomi, Y. Kida, H.H. Hernandez, J.W. Tester, W.H. Green, Supercritical water desulfurization of organic sulfides is consistent with free-radical kinetics. Energy Fuels **27**, 6108–6117 (2013)
33. Y. Wang, Y. Zhu, Z. Liu, L. Wang, D. Xu, C. Fang, S. Wang, Catalytic performances of Ni-based catalysts on supercritical water gasification of phenol solution and coal-gasification wastewater. Int. J. Hydrogen Energy **44**, 3470–3480 (2019)
34. D. Xu, L. Liu, N. Wei, Y. Guo, S. Wang, Z. Wu, P. Duan, Catalytic supercritical water gasification of aqueous phase directly derived from microalgae hydrothermal liquefaction. Int. J. Hydrogen Energy **44**, 26181–26192 (2019)
35. J. Zhang, C. Ma, Y. Sun, X. Ren, Hydroxyl radical reactions with 2-chlorophenol as a model for oxidation in supercritical water. Res. Chem. Intermed. **40**, 973–990 (2014)
36. T. Fujii, R. Hayashi, S.I. Kawasaki, A. Suzuki, Y. Oshima, Water density effects on methanol oxidation in supercritical water at high pressure up to 100 MPa. J. Supercrit. Fluids **58**, 142–149 (2011)
37. Y.M. Gong, S.Z. Wang, X.Y. Tang, D.H. Xu, H.H. Ma, Supercritical water oxidation of acrylic acid production wastewater. Environ. Technol. (U.K.) **35**, 907–916 (2014)
38. M. Ren, S. Wang, C. Yang, H. Xu, Y. Guo, D. Roekaerts, Supercritical water oxidation of quinoline with moderate preheat temperature and initial concentration. Fuel **236**, 1408–1414 (2019)
39. Y.M. Alshammari, K. Hellgardt, Partial oxidation of n-hexadecane through decomposition of hydrogen peroxide in supercritical water. Chem. Eng. Res. Des. **93**, 565–575 (2015)
40. A. Zaker, P. Guerra, Y. Wang, G.A. Tompsett, X. Huang, J.Q. Bond, M.T. Timko, Evidence of heterogeneous catalytic activity of ZSM-5 in supercritical water for dodecane cracking. Catal. Today **317**, 2–11 (2018)
41. Y.M. Alshammari, K. Hellgardt, Sub and supercritical water reforming of n-hexadecane in a tubular flow reactor. J. Supercrit. Fluids **107**, 723–732 (2016)
42. Q. Guan, P.E. Savage, C. Wei, Gasification of alga *Nannochloropsis* sp. in supercritical water. J. Supercrit. Fluids **61**, 139–145 (2012)
43. J. Picou, J. Wenzel, A. Niemoeller, S. Lee, H.B. Lanterman, A kinetic model based on the sequential reaction mechanism for the noncatalytic reformation of jet fuel in supercritical water. Energy Sour, Part A: Recovery, Utilization Environ. Eff. **33**, 785–794 (2011)
44. P. Arcelus-Arrillaga, K. Hellgardt, M. Millan, Effect of process conditions on the hydrothermal partial oxidation of phenanthrene as a heavy oil model structure. Fuel **209**, 434–441 (2017)
45. B. Bai, W. Wang, H. Jin, Experimental study on gasification performance of polypropylene (PP) plastics in supercritical water. Energy **191** (2020)
46. J. Yu, Q. Chen, Q. Guan, B. Li, P. Ning, J. Gu, X. Lu, Characterization of the liquid intermediates from gasification of lignite in supercritical water: insights into the gasification process for hydrogen production. Int. J. Hydrogen Energy **41**, 17309–17322 (2016)
47. Y. Han, T. Ma, F. Chen, W. Li, J. Zhang, Supercritical water gasification of naphthalene over iron oxide catalyst: a ReaxFF molecular dynamics study. Int. J. Hydrogen Energy **44**, 30486–30498 (2019)
48. H. Jin, S. Liu, W. Wei, D. Zhang, Z. Cheng, L. Guo, Experimental investigation on hydrogen production by anthracene gasification in supercritical water. Energy Fuels **29**, 6342–6346 (2015)
49. J. Zhang, X. Weng, Y. Han, W. Li, J. Cheng, Z. Gan, J. Gu, The effect of supercritical water on coal pyrolysis and hydrogen production: a combined ReaxFF and DFT study. Fuel **108**, 682–690 (2013)
50. D. Jiang, Y. Wang, M. Zhang, J. Zhang, W. Li, Y. Han, H_2 and CO production through coking wastewater in supercritical water condition: ReaxFF reactive molecular dynamics simulation. Int. J. Hydrogen Energy **42**, 9667–9678 (2017)

51. J.A. Onwudili, P.T. Williams, Reaction mechanisms for the decomposition of phenanthrene and naphthalene under hydrothermal conditions. J. Supercrit. Fluids **39**, 399–408 (2007)
52. S. Xu, I. Butler, I. Gökalp, J.A. Kozinski, Evolution of naphthalene and its intermediates during oxidation in subcritical/supercritical water. Proc. Combust. Inst. **33**, 3185–3194 (2011)
53. S. Xu, Z. Fang, J.A. Koziński, Oxidation of naphthalene in supercritical water up to 420 °C and 30 MPa. Combust. Sci. Technol. **175**, 291–318 (2003)
54. Y. Yang, F. Hildebrand, Phenanthrene degradation in subcritical water. Anal. Chim. Acta **555**, 364–369 (2006)
55. Q. Wang, H. Fan, S. Wu, Z. Zhang, P. Zhang, B. Han, Water as an additive to enhance the ring opening of naphthalene. Green Chem. **14**, 1152–1158 (2012)
56. Y. Cheng, H. Fan, S. Wu, Q. Wang, J. Guo, L. Gao, B. Zong, B. Han, Enhancing the selectivity of the hydrogenation of naphthalene to tetralin by high temperature water. Green Chem. **11**, 1061–1065 (2009)
57. P.M. Yeletsky, T.R. Reina, O.A. Bulavchenko, A.A. Saraev, E.Y. Gerasimov, O.O. Zaikina, J.M. Bermúdez, P. Arcelus-Arrillaga, V.A. Yakovlev, M. Millan, Phenanthrene catalytic cracking in supercritical water: effect of the reaction medium on $NiMo/SiO_2$ catalysts. Catal. Today **329**, 197–205 (2019)
58. Y. Wang, S. Wang, G. Zhao, Y. Guo, Y. Guo, Hydrogen production by partial oxidation gasification of a phenol, naphthalene, and acetic acid mixture in supercritical water. Int. J. Hydrogen Energy **41**, 2238–2246 (2016)
59. Y. Wang, F. Gao, J. Yang, G. Zhao, C. Fang, Y. Guo, Y. Guo, S. Wang, Interactive effects of phenol, naphthalene and acetic acid during the supercritical water gasification process. Environ. Prog. Sustain. Energy **37**, 1140–1148 (2018)
60. M. Kerolli-Mustafa, V. Lajqi-Makolli, S. Berisha-Shala, L. Latifi, I. Malollari, I. Morina, Biomass and biofuel overview. A global sustainability challenge. J. Environ. Prot. Ecol. **16**, 222–232 (2015)
61. G. Caputo, M. Dispenza, P. Rubio, F. Scargiali, G. Marotta, A. Brucato, Supercritical water gasification of microalgae and their constituents in a continuous reactor. J. Supercrit. Fluids **118**, 163–170 (2016)
62. K. Kang, R. Azargohar, A.K. Dalai, H. Wang, Systematic screening and modification of Ni based catalysts for hydrogen generation from supercritical water gasification of lignin. Chem. Eng. J. **283**, 1019–1032 (2016)
63. C. Zhu, L. Guo, H. Jin, Z. Ou, W. Wei, J. Huang, Gasification of guaiacol in supercritical water: detailed reaction pathway and mechanisms. Int. J. Hydrogen Energy **43**, 14078–14086 (2018)
64. C. Zhu, L. Guo, H. Jin, J. Huang, S. Li, X. Lian, Effects of reaction time and catalyst on gasification of glucose in supercritical water: detailed reaction pathway and mechanisms. Int. J. Hydrogen Energy 6630–6639 (2016)
65. S. Kumar, R.B. Gupta, Hydrolysis of microcrystalline cellulose in subcritical and supercritical water in a continuous flow reactor. Ind. Eng. Chem. Res. **47**, 9321–9329 (2008)
66. P.L. Dhepe, A. Fukuoka, Cellulose conversion under heterogeneous catalysis. Chemsuschem **1**, 969–975 (2008)
67. J. Jalak, P. Väljamäe, Mechanism of initial rapid rate retardation in cellobiohydrolase catalyzed cellulose hydrolysis. Biotechnol. Bioeng. **106**, 871–883 (2010)
68. M. Tyufekchiev, P. Duan, K. Schmidt-Rohr, S. Granados Focil, M.T. Timko, M.H. Emmert, Cellulase-inspired solid acids for cellulose hydrolysis: structural explanations for high catalytic activity. ACS Catal. **8**, 1464–1468 (2018)
69. P. Yang, H. Kobayashi, A. Fukuoka, Recent developments in the catalytic conversion of cellulose into valuable chemicals, Cuihua Xuebao/Chinese. J. Catal. **32**, 716–722 (2011)
70. Z. Fang, T. Minowa, C. Fang, J.R.L. Smith, H. Inomata, J.A. Kozinski, Catalytic hydrothermal gasification of cellulose and glucose. Int. J. Hydrogen Energy **33**, 981–990 (2008)
71. J. Yin, Z. Cheng, L. Guo, S. Li, H. Jin, Products distribution and influence of nickel catalyst on glucose hydrothermal decomposition. Int. J. Hydrogen Energy **42**, 4642–4650 (2017)

72. R.F. Susanti, L.W. Dianningrum, T. Yum, Y. Kim, B.G. Lee, J. Kim, High-yield hydrogen production from glucose by supercritical water gasification without added catalyst. Int. J. Hydrogen Energy **37**, 11677–11690 (2012)
73. J.A. Onwudili, P.T. Williams, Role of sodium hydroxide in the production of hydrogen gas from the hydrothermal gasification of biomass. Int. J. Hydrogen Energy **34**, 5645–5656 (2009)
74. J. Huang, C. Zhu, X. Lian, H. Feng, J. Sun, L. Wang, H. Jin, Catalytic supercritical water gasification of glucose with in-situ generated nickel nanoparticles for hydrogen production. Int. J. Hydrogen Energy **44**, 21020–21029 (2019)
75. H. Jin, Y. Lu, L. Guo, C. Cao, X. Zhang, Hydrogen production by partial oxidative gasification of biomass and its model compounds in supercritical water. Int. J. Hydrogen Energy **35**, 3001–3010 (2010)
76. Y.J. Lu, H. Jin, L.J. Guo, X.M. Zhang, C.Q. Cao, X. Guo, Hydrogen production by biomass gasification in supercritical water with a fluidized bed reactor. Int. J. Hydrogen Energy **33**, 6066–6075 (2008)
77. K. Kobiro, K. Sumoto, Y. Okimoto, P. Wang, Saccharides as new hydrogen sources for one-pot and single-step reduction of alcohols and catalytic hydrogenation of olefins in supercritical water. J. Supercrit. Fluids **77**, 63–69 (2013)
78. S. Changi, M. Zhu, P.E. Savage, Hydrothermal reaction kinetics and pathways of phenylalanine alone and in binary mixtures. Chemsuschem **5**, 1743–1757 (2012)
79. N. Sato, A.T. Quitain, K. Kang, H. Daimon, K. Fujie, Reaction kinetics of amino acid decomposition in high-temperature and high-pressure water. Ind. Eng. Chem. Res. **43**, 3217–3222 (2004)
80. J.D. Sheehan, P.E. Savage, Reaction pathways and kinetics of tryptophan in hot, compressed water. Chem. Eng. J. **390**, 124600 (2020)
81. W. Ma, G. Du, J. Li, Y. Fang, L. Hou, G. Chen, D. Ma, Supercritical water pyrolysis of sewage sludge. Waste Manage. **59**, 371–378 (2017)
82. A.A. Peterson, R.P. Lachance, J.W. Tester, Kinetic evidence of the maillard reaction in hydrothermal biomass processing: glucose-glycine interactions in high-temperature, high-pressure water. Ind. Eng. Chem. Res. **49**, 2107–2117 (2010)
83. L. Qian, S. Wang, D. Xu, Y. Guo, X. Tang, L. Wang, Treatment of sewage sludge in supercritical water and evaluation of the combined process of supercritical water gasification and oxidation. Biores. Technol. **176**, 218–224 (2015)
84. G.J. DiLeo, M.E. Neff, S. Kim, P.E. Savage, Supercritical water gasification of phenol and glycine as models for plant and protein biomass. Energy Fuels **22**, 871–877 (2008)
85. T. Samanmulya, S. Inoue, T. Inoue, Y. Kawai, H. Kubota, H. Munetsuna, T. Noguchi, Y. Matsumura, Gasification characteristics of alanine in supercritical water. J. Jpn. Petrol. Inst. **57**, 225–229 (2014)
86. Y. Su, W. Zhu, M. Gong, H. Zhou, Y. Fan, B. Amuzu-Sefordzi, Interaction between sewage sludge components lignin (phenol) and proteins (alanine) in supercritical water gasification. Int. J. Hydrogen Energy **40**, 9125–9136 (2015)
87. T. Samanmulya, O. Farobie, Y. Matsumura, Gasification characteristics of aminobutyric acid and serine as model compounds of proteins under supercritical water conditions. J. Jpn. Petrol. Inst. **60**, 34–40 (2017)
88. A.G. Chakinala, D.W.F. Brilman, W.P.M. Van Swaaij, S.R.A. Kersten, Catalytic and non-catalytic supercritical water gasification of microalgae and glycerol. Ind. Eng. Chem. Res. 1113–1122 (2010)
89. D. Xu, S. Wang, X. Hu, C. Chen, Q. Zhang, Y. Gong, Catalytic gasification of glycine and glycerol in supercritical water. Int. J. Hydrogen Energy **34**, 5357–5364 (2009)
90. P. Azadi, E. Afif, H. Foroughi, T. Dai, F. Azadi, R. Farnood, Catalytic reforming of activated sludge model compounds in supercritical water using nickel and ruthenium catalysts. Appl. Catal. B **134–135**, 265–273 (2013)
91. I.G. Lee, A. Nowacka, C.H. Yuan, S.J. Park, J.B. Yang, Hydrogen production by supercritical water gasification of valine over Ni/activated charcoal catalyst modified with Y, Pt, and Pd. Int. J. Hydrogen Energy **40**, 12078–12087 (2015)

92. T.L.K. Yong, Y. Matsumura, Reaction kinetics of the lignin conversion in supercritical water. Ind. Eng. Chem. Res. **51**, 11975–11988 (2012)
93. L. Zhang, C. Xu, P. Champagne, Energy recovery from secondary pulp/paper-mill sludge and sewage sludge with supercritical water treatment. Biores. Technol. **101**, 2713–2721 (2010)
94. D. Castello, A. Kruse, L. Fiori, Low temperature supercritical water gasification of biomass constituents: Glucose/phenol mixtures. Biomass Bioenerg. **73**, 84–94 (2015)
95. J. Zhang, S. Wang, Y. Guo, D. Xu, X. Li, X. Tang, Co-oxidation effects of methanol on acetic acid and phenol in supercritical water. Ind. Eng. Chem. Res. **52**, 10609–10618 (2013)
96. M.Z. Hossain, M.B.I. Chowdhury, A.K. Jhawar, P.A. Charpentier, Supercritical water gasification of glucose using bimetallic aerogel Ru-Ni-Al_2O_3 catalyst for H_2 production. Biomass Bioenerg. **107**, 39–51 (2017)
97. M.Z. Hossain, M.B.I. Chowdhury, P.A. Charpentier, Effect of surface acidity of Al_2O_3 supported metal catalysts on catalytic activity and carbon deposition during SCWG of glucose. Biomass Bioenerg. **124**, 142–150 (2019)
98. A.G. Chakinala, J.K. Chinthaginjala, K. Seshan, W.P.M. Van Swaaij, S.R.A. Kersten, D.W.F. Brilman, Catalyst screening for the hydrothermal gasification of aqueous phase of bio-oil. Catal. Today **195**, 83–92 (2012)
99. G. Akgül, A. Kruse, Influence of salts on the subcritical water-gas shift reaction. J. Supercrit. Fluids **66**, 207–214 (2012)
100. Q. Guan, X. Huang, J. Liu, J. Gu, R. Miao, Q. Chen, P. Ning, Supercritical water gasification of phenol using a Ru/CeO_2 catalyst. Chem. Eng. J. **283**, 358–365 (2016)
101. F. Xia, S. Tian, P. Ning, J. Gu, Q. Guan, R. Miao, Y. Wang, Catalytic gasification of lignite with KOH in supercritical water. Can. J. Chem. Eng. **92**, 421–425 (2014)
102. N. Martin-Sanchez, M. Jesus Sanchez-Montero, C. Izquierdo, F. Salvador, Conversion of a resistant pollutant, phenol, into green fuels by gasification using supercritical water compressed up to 1000 bar. React. Chem. Eng. **2**, 799–810 (2017)
103. Y. Wang, Y. Zhu, Z. Liu, C. Fang, J. Yang, Y. Guo, S. Wang, Effect of ammonia on gasification performances of phenol in supercritical water. Chem. Eng. Res. Des. **136**, 242–250 (2018)
104. M. Yang, J. Zhang, Y. Guo, Supercritical water gasification of phenol over Ni-Ru bimetallic catalyst: Intermediates and kinetics. J. Supercrit. Fluids **160**, 104810 (2020)
105. Y. Matsumura, T. Urase, K. Yamamoto, T. Nunoura, Carbon catalyzed supercritical water oxidation of phenol. J. Supercrit. Fluids **22**, 149–156 (2002)
106. J. Yu, P.E. Savage, Phenol oxidation over CuO/Al_2O_3 in supercritical water. Appl. Catal. B **28**, 275–288 (2000)
107. C.L. De Silva, R.K. Garlapalli, J.P. Trembly, Removal of phenol from oil/gas produced water using supercritical water treatment with TiO_2 supported MnO_2 ctalyst. J. Environ. Chem. Eng. **5**, 488–493 (2017)
108. X. Zhang, P.E. Savage, Fast catalytic oxidation of phenol in supercritical water. Catal. Today **40**, 333–342 (1998)
109. J. Yu, P.E. Savage, Catalyst activity, stability, and transformations during oxidation in supercritical water. Appl. Catal. B **31**, 123–132 (2001)
110. S. Gopalan, P.E. Savage, Reaction mechanism for phenol oxidation in supercritical water. J. Phys. Chem. **98**, 12646–12652 (1994)
111. Y. Matsumura, T. Nunoura, T. Urase, K. Yamamoto, Supercritical water oxidation of high concentrations of phenol. J. Hazard. Mater. **73**, 245–254 (2000)
112. Q. Guan, C. Wei, X.S. Chai, Pathways and kinetics of partial oxidation of phenol in supercritical water. Chem. Eng. J. **175**, 201–206 (2011)
113. S.M. Ghoreishi, S.M.S. Mortazavi, A. Hedayati, Modeling of non-catalytic supercritical water oxidation of phenol. Chem. Prod. Process. Model. **10**, 243–251 (2015)
114. L. Jia, J. Yu, Y. Chen, P. Ning, Q. Guan, J. Gu, R. Miao, Q. Chen, Noble-metal-free bimetallic alloy nanoparticle-catalytic gasification of phenol in supercritical water. J. Supercrit. Fluids **126**, 79–88 (2017)
115. J. Yu, Q. Guan, Q. Chen, P. Ning, J. Gu, X. Huang, Y. Shi, R. Miao, Catalytic gasification of phenol in supercritical water with Ru/graphitized carbon black. RSC Adv. **6**, 75512–75521 (2016)

116. B. Li, B. Zhang, Q. Guan, S. Chen, P. Ning, Activity of Ni/CeO_2 catalyst for gasification of phenol in supercritical water. Int. J. Hydrogen Energy **43**, 19010–19018 (2018)
117. Q. Guan, Z. Kong, Z. Xie, Y. Chen, S. Chen, S. Tian, P. Ning, Catalytic gasification of phenol in supercritical water over bimetallic Co–Ni/AC catalyst. Environ. Technol. (U.K.) **40**, 2182–2190 (2019)
118. J.P. Zhang, C. Pendenque, K.L. Guen, R. Delaunay, I. Vickridge, D. Schmaus, Q.G. Fu, P. Jonnard, Kossel interferences of proton-induced X-ray emission lines to study thin film waveguides. Nucl. Instrum. Methods Phys. Res., Sect. B **452**, 12–20 (2019)
119. T.L.K. Yong, Y. Matsumura, Kinetics analysis of phenol and benzene decomposition in supercritical water. J. Supercrit. Fluids **87**, 73–82 (2014)
120. C.M. Huelsman, P.E. Savage, Reaction pathways and kinetic modeling for phenol gasification in supercritical water. J. Supercrit. Fluids **81**, 200–209 (2013)
121. B. Yang, Z. Cheng, X. Gao, T. Yuan, Z. Shen, Decomposition of 15 aromatic compounds in supercritical water oxidation. Chemosphere **218**, 384–390 (2019)
122. B. Yang, Z. Cheng, T. Yuan, Z. Shen, Catalytic oxidation of various aromatic compounds in supercritical water: experimental and DFT study. J. Taiwan Inst. Chem. Eng. **100**, 47–55 (2019)
123. S.E. Hunter, C.A. Felczak, P.E. Savage, Synthesis of p-isopropenylphenol in high-temperature water. Green Chem. **6**, 222–226 (2004)
124. P.E. Savage, S.E. Hunter, K.L. Hoffee, T.J. Schuelke, M.J. Smith, Bisphenol E decomposition in high-temperature water. Ind. Eng. Chem. Res. **45**, 7775–7780 (2006)
125. E.E. Brock, Y. Oshima, P.E. Savage, J.R. Barker, Kinetics and mechanism of methanol oxidation in supercritical water. J. Phys. Chem. **100**, 15834–15842 (1996)
126. K. Koido, Y. Ishida, K. Kumabe, K. Matsumoto, T. Hasegawa, Kinetics of ethanol oxidation in subcritical water. J. Supercrit. Fluids **55**, 246–251 (2010)
127. J.P.S. Queiroz, M.D. Bermejo, M.J. Cocero, Kinetic model for isopropanol oxidation in supercritical water in hydrothermal flame regime and analysis. J. Supercrit. Fluids **76**, 41–47 (2013)
128. P. Dagaut, M. Cathonnet, J.C. Boettner, Chemical kinetic modeling of the supercritical-water oxidation of methanol. J. Supercrit. Fluids **9**, 33–42 (1996)
129. J. Abelleira, J. Sánchez-Oneto, J.R. Portela, E.J. Martínez De La Ossa, Kinetics of supercritical water oxidation of isopropanol as an auxiliary fuel and co-fuel. Fuel **111**, 574–583 (2013)
130. P. Wang, X. Shi, K. Kataoka, Y. Maeda, K. Kobiro, Non-catalytic Oppenauer oxidation of alcohols in supercritical water. J. Supercrit. Fluids **52**, 222–227 (2010)
131. G.J. DiLeo, P.E. Savage, Catalysis during methanol gasification in supercritical water. J. Supercrit. Fluids **39**, 228–232 (2006)
132. S. Therdthianwong, N. Srisiriwat, A. Therdthianwong, E. Croiset, Hydrogen production from bioethanol reforming in supercritical water. J. Supercrit. Fluids **57**, 58–65 (2011)
133. S. Therdthianwong, N. Srisiriwat, A. Therdthianwong, E. Croiset, Reforming of bioethanol over Ni/Al_2O_3 and Ni/$CeZrO_2$/Al_2O_3 catalysts in supercritical water for hydrogen production. Int. J. Hydrogen Energy **36**, 2877–2886 (2011)
134. S. Guo, L. Guo, C. Cao, J. Yin, Y. Lu, X. Zhang, Hydrogen production from glycerol by supercritical water gasification in a continuous flow tubular reactor. Int. J. Hydrogen Energy **37**, 5559–5568 (2012)
135. T. Pairojpiriyakul, W. Kiatkittipong, S. Assabumrungrat, E. Croiset, Hydrogen production from supercritical water reforming of glycerol in an empty Inconel 625 reactor. Int. J. Hydrogen Energy **39**, 159–170 (2014)
136. J. Zhang, S. Wang, D. Xu, Y. Guo, M. Ren, J. Lu, Kinetics study on hydrothermal combustion of methanol in supercritical water. Chem. Eng. Res. Des. **98**, 220–230 (2015)
137. E. Shimoda, T. Fujii, R. Hayashi, Y. Oshima, Kinetic analysis of the mixture effect in supercritical water oxidation of ammonia/methanol. J. Supercrit. Fluids **116**, 232–238 (2016)
138. J. Zhang, J. Lu, S. Chen, D. He, R. Huang, X. Luo, Experimental and kinetics study on oxidation of three-component in supercritical water. Can. J. Chem. Eng. **97**, 1871–1880 (2019)

139. J.M. Ploeger, W.H. Green, J.W. Tester, Co-oxidation of methylphosphonic acid and ethanol in supercritical water. II: elementary reaction rate model. J. Supercrit. Fluids **39**, 239–245 (2006)
140. M.D. Bermejo, F. Cantero, M.J. Cocero, Supercritical water oxidation of feeds with high ammonia concentrations. Pilot plant experimental results and modeling. Chem. Eng. J. **137**, 542–549 (2008)
141. P. Cabeza, B. Al-Duri, M.D. Bermejo, M.J. Cocero, Co-oxidation of ammonia and isopropanol in supercritical water in a tubular reactor. Chem. Eng. Res. Des. **92**, 2568–2574 (2014)
142. J. Zhang, P. Li, J. Lu, F. Xin, X. Zheng, S. Chen, Supercritical water oxidation of ammonia with methanol as the auxiliary fuel: comparing with isopropanol. Chem. Eng. Res. Des. **147**, 160–170 (2019)
143. L. Qian, S. Wang, M. Ren, S. Wang, Co-oxidation effects and mechanisms between sludge and alcohols (methanol, ethanol and isopropanol) in supercritical water. Chem. Eng. J. **366**, 223–234 (2019)
144. Z. Yang, I.R. Gould, L.B. Williams, H.E. Hartnett, E.L. Shock, The central role of ketones in reversible and irreversible hydrothermal organic functional group transformations. Geochim. Cosmochim. Acta **98**, 48–65 (2012)
145. G. Akgül, A. Kruse, Hydrothermal disproportionation of formaldehyde at subcritical conditions. J. Supercrit. Fluids **73**, 43–50 (2013)
146. Y.J. Fan, W. Zhu, M. Gong, Y. Su, C.Y. Wang, Investigation of the interaction between intermediates from gasification of biomass in supercritical water: formaldehyde/formic acid mixtures. Int. J. Hydrogen Energy **43**, 13090–13097 (2018)
147. M. Watanabe, H. Inomata, R.L. Smith Jr., K. Arai, Catalytic decarboxylation of acetic acid with zirconia catalyst in supercritical water. Appl. Catal. A **219**, 149–156 (2001)
148. M. Hashimoto, S. Taniguchi, R. Takanami, R.R. Giri, H. Ozaki, Oxidative degradation of 2,4-dichlorophenoxyacetic acid (2,4-D) in subcritical and supercritical waters. Water Sci. Technol. **62**, 484–490 (2010)
149. M. Gong, S. Nanda, M.J. Romero, W. Zhu, J.A. Kozinski, Subcritical and supercritical water gasification of humic acid as a model compound of humic substances in sewage sludge. J. Supercrit. Fluids **119**, 130–138 (2017)
150. J. Fu, F. Shi, L.T. Thompson Jr., X. Lu, P.E. Savage, Activated carbons for hydrothermal decarboxylation of fatty acids. ACS Catal. **1**, 227–231 (2011)
151. T.M. Yeh, R.L. Hockstad, S. Linic, P.E. Savage, Hydrothermal decarboxylation of unsaturated fatty acids over PtSnx/C catalysts. Fuel **156**, 219–224 (2015)
152. Y. Guo, S. Wang, C.M. Huelsman, P.E. Savage, Products, pathways, and kinetics for reactions of indole under supercritical water gasification conditions. J. Supercrit. Fluids **73**, 161–170 (2013)
153. Y. Guo, S. Wang, T. Yeh, P.E. Savage, Catalytic gasification of indole in supercritical water. Appl. Catal. B **166–167**, 202–210 (2015)
154. S. Liu, H. Jin, W. Wei, L. Guo, Gasification of indole in supercritical water: nitrogen transformation mechanisms and kinetics. Int. J. Hydrogen Energy **41**, 15985–15997 (2016)
155. K.M. Benjamin, P.E. Savage, Supercritical water oxidation of methylamine. Ind. Eng. Chem. Res. **44**, 5318–5324 (2005)
156. X. Du, R. Zhang, Z. Gan, J. Bi, Treatment of high strength coking wastewater by supercritical water oxidation. Fuel **104**, 77–82 (2013)
157. B. Yang, Z. Cheng, M. Fan, J. Jia, T. Yuan, Z. Shen, Supercritical water oxidation of 2-, 3- and 4-nitroaniline: a study on nitrogen transformation mechanism. Chemosphere **205**, 426–432 (2018)
158. Y. Huang, X. Dong, Y. Dong, Y. Yu, A molecular dynamics simulation study on nitrobenzene and ·OH radical in supercritical water. J. Mol. Liq. **206**, 278–284 (2015)
159. K.S. Tan, Power, play and pedagogy through the power point performance-lecture. Int. J. Manage. Appl. Res. **7**, 382–394 (2020)
160. X. Dong, Z. Gan, X. Lu, W. Jin, Y. Yu, M. Zhang, Study on catalytic and non-catalytic supercritical water oxidation of p-nitrophenol wastewater. Chem. Eng. J. **277**, 30–39 (2015)

161. Y. Gong, Y. Guo, S. Wang, W. Song, Supercritical water oxidation of Quinazoline: effects of conversion parameters and reaction mechanism. Water Res. **100**, 116–125 (2016)
162. B. Yang, Z. Shen, Supercritical water oxidation of pyridine and 3-cyanopyridine: TOC removal, kinetics, and degradation pathway. J. Environ. Eng. **145**, 04019012 (2019)
163. P. Cabeza, J.P.S. Queiroz, S. Arca, C. Jiménez, A. Gutiérrez, M.D. Bermejo, M.J. Cocero, Sludge destruction by means of a hydrothermal flame. Optim. Ammonia Destruction Conditions, Chem. Eng. J. **232**, 1–9 (2013)
164. J. Li, J. Liu, J. Zhang, J. Ma, Y. Hu, G. Xie, J. Zhang, G. Tu, Efficient non-fullerene organic solar cells based on thickness-insensitive conjugated small molecule cathode interface. Sol. Energy **191**, 219–226 (2019)
165. G. Lin, D. Xu, Z. Ma, S. Wang, Y. Guo, Effect of oxidation coefficient on products of sewage sludge treatment in supercritical water. Energy Procedia 357–362 (2017)
166. L. Qian, S. Wang, D. Xu, Y. Guo, X. Tang, L. Wang, Treatment of municipal sewage sludge in supercritical water: a review. Water Res. **89**, 118–131 (2016)
167. A. Soler, J.A. Conesa, N. Ortuño, Application of subcritical water to dechlorinate polyvinyl chloride electric wires. Energies **11**, 2612 (2018)
168. M. Akimoto, K. Ninomiya, S. Takami, M. Ishikawa, M. Sato, K. Washio, Hydrothermal dechlorination and denitrogenation of municipal-waste-plastics-derived fuel oil under sub- and supercritical conditions. Ind. Eng. Chem. Res. **41**, 5393–5400 (2002)
169. A. Kubátová, J. Herman, T.S. Steckler, M. De Veij, D.J. Miller, E.B. Klunder, C.M. Wai, S.B. Hawthorne, Subcritical (hot/liquid) water dechlorination of PCBs (Aroclor 1254) with metal additives and in waste paint. Environ. Sci. Technol. **37**, 5757–5762 (2003)
170. A. Kubátová, A.J.M. Lagadec, S.B. Hawthorne, Dechlorination of lindane, dieldrin, tetrachloroethane, trichloroethene, and PVC in subcritical water. Environ. Sci. Technol. **36**, 1337–1343 (2002)
171. J. Wang, Y. Zhang, W. Zheng, I.M. Chou, C. Lin, Q. Wang, Z. Pan, Using Raman spectroscopy and a fused quartz tube reactor to study the oxidation of o-dichlorobenzene in hot compressed water. J. Supercrit. Fluids **140**, 380–386 (2018)
172. H. Liu, Z. Pan, Visual observations and raman spectroscopic studies of supercritical water oxidation of chlorobenzene in an anticorrosive fused-silica capillary reactor. Environ. Sci. Technol. **46**, 3384–3389 (2012)
173. A. Shimokawa, N. Kometani, Y. Yonezawa, Degradation of chlorobenzene by the hybrid process of supercritical water oxidation and TiO_2 photocatalysis. Sep. Sci. Technol. **45**, 1538–1545 (2010)
174. G.-T. Wei, C.-H. Wei, C.-F. Wu, F.-M. He, Reductive dechlorination of chlorobenzene in supercritical water catalyzed by Fe/ZrO_2. Environ. Chem. Lett. **8**, 165–169 (2010)
175. H.-S. Lee, J.-H. Yoon, J.-S. Park, Y.-M. Yi, A study on failure characteristic of spherical pressure vessel. J. Mater. Process. Technol. **164–165**, 882–888 (2005)
176. G. Lee, T. Nunoura, Y. Matsumura, K. Yamamoto, Comparison of the effects of the addition of NaOH on the decomposition of 2-chlorophenol and phenol in supercritical water and under supercritical water oxidation conditions. J. Supercrit. Fluids **24**, 239–250 (2002)
177. Y. Takeshita, K. Kato, K. Takahashi, Y. Sato, S. Nishi, Basic study on treatment of waste polyvinyl chloride plastics by hydrothermal decomposition in subcritical and supercritical regions. J. Supercrit. Fluids **31**, 185–193 (2004)
178. A. Soler, J.A. Conesa, N. Ortuño, Emissions of brominated compounds and polycyclic aromatic hydrocarbons during pyrolysis of E-waste debrominated in subcritical water. Chemosphere **186**, 167–176 (2017)
179. M.M. Schultz, D.F. Barofsky, J.A. Field, Fluorinated alkyl surfactants. Environ. Eng. Sci. **20**, 487–501 (2003)
180. J.X. Zhu, Y.J. Xie, F.S. Zhang, Life cycle analysis for PFOS decomposition in supercritical water. Procedia Environ. Sci. **16**, 304–310 (2012)
181. H. Hori, Y. Nagaoka, T. Sano, S. Kutsuna, Iron-induced decomposition of perfluorohexanesulfonate in sub- and supercritical water. Chemosphere **70**, 800–806 (2008)

182. H. Hori, H. Yokota, Efficient oxygen-induced decomposition of triphenylsulfonium trifluoromethanesulfonate to fluoride ions in subcritical water. J. Fluorine Chem. **178**, 1–5 (2015)
183. S. Wang, Y. Guo, L. Wang, Y. Wang, D. Xu, H. Ma, Supercritical water oxidation of coal: investigation of operating parameters effects, reaction kinetics and mechanism. Fuel Process. Technol. **92**, 291–297 (2011)
184. Y. Li, S. Wang, X. Tang, D. Xu, Y. Guo, J. Zhang, L. Qian, Effects of sulfides on the corrosion behavior of Inconel 600 and Incoloy 825 in supercritical water. Oxid. Met. **84**, 509–526 (2015)
185. J. Yang, C. Wang, Q. Zhang, B. Chang, F. Wang, X. Wang, M. Wu, Modeling of laneway environment and locating method of roadheader based on self-coupling and hector SLAM, 2020 5th International Conference on Electromechanical Control Technology and Transportation (ICECTT), pp. 263–268 (2020)
186. P.Q. Yuan, Z.M. Cheng, W.L. Jiang, R. Zhang, W.K. Yuan, Catalytic desulfurization of residual oil through partial oxidation in supercritical water. J. Supercrit. Fluids **35**, 70–75 (2005)
187. A. Demirbas, Sulfur removal from crude oil using supercritical water. Pet. Sci. Technol. **34**, 622–626 (2016)
188. N. Meng, D. Jiang, Y. Liu, Z. Gao, Y. Cao, J. Zhang, J. Gu, Y. Han, Sulfur transformation in coal during supercritical water gasification. Fuel **186**, 394–404 (2016)
189. G. Chen, S. Chen, Z. Song, X. Zhao, C. Ma, Lignite sulfur transformation during the supercritical water gasification process. J. Anal. Appl. Pyrol. **116**, 161–167 (2015)
190. C.A. Class, A.K. Vasiliou, Y. Kida, M.T. Timko, W.H. Green, Detailed kinetic model for hexyl sulfide pyrolysis and its desulfurization by supercritical water. Phys. Chem. Chem. Phys. **21**, 10311–10324 (2019)
191. Y. Kida, C.A. Class, A.J. Concepcion, M.T. Timko, W.H. Green, Combining experiment and theory to elucidate the role of supercritical water in sulfide decomposition. Phys. Chem. Chem. Phys. **16**, 9220–9228 (2014)
192. J. Kang, N.L. Doh, Automatic targetless camera–LIDAR calibration by aligning edge with Gaussian mixture model. J. Field Robot. **37**, 158–179 (2020)
193. J. Wang, F. He, Y. Li, H. Sun, New insights into isopropylthiobenzene decomposition in supercritical water. RSC Adv. **6**, 93260–93266 (2016)
194. A. Ates, G. Azimi, K.H. Choi, W.H. Green, M.T. Timko, The role of catalyst in supercritical water desulfurization. Appl. Catal. B **147**, 144–155 (2014)
195. H. Ma, S. Wang, L. Zhou, Kinetics behavior and sulfur transformations of iron sulfide during supercritical water oxidation. Adv. Mater. Res. 1939–1942 (2012)
196. T. Wang, X. Zhu, Sulfur transformations during supercritical water oxidation of a Chinese coal. Fuel **82**, 2267–2272 (2003)
197. M. Antonopoulou, A. Giannakas, F. Bairamis, M. Papadaki, I. Konstantinou, Degradation of organophosphorus flame retardant tris (1-chloro-2-propyl) phosphate (TCPP) by visible light N, S-codoped TiO_2 photocatalysts. Chem. Eng. J. **318**, 231–239 (2017)
198. M. Kosari, M. Golmohammadi, J. Towfighi, S.J. Ahmadi, Decomposition of tributhyl phosphate at supercritical water oxidation conditions: non-catalytic, catalytic, and kinetic reaction studies. J. Supercrit. Fluids **133**, 103–113 (2018)
199. H.S. Ou, J. Liu, J.S. Ye, L.L. Wang, N.Y. Gao, J. Ke, Degradation of tris(2-chloroethyl) phosphate by ultraviolet-persulfate: kinetics, pathway and intermediate impact on proteome of Escherichia coli. Chem. Eng. J. **308**, 386–395 (2017)
200. M. Golmohammadi, S.J. Ahmadi, J. Towfighi, Catalytic supercritical water destructive oxidation of tributyl phosphate: study on the effect of operational parameters. J. Supercrit. FluidsSupercrit. Fluids **140**, 32–40 (2018)
201. B. Yang, Z. Cheng, Z. Shen, Decomposition of 14 organophosphate flame retardants during supercritical water oxidation. J. Taiwan Inst. Chem. Eng. **95**, 40–47 (2019)

Catalytic Supercritical Water Gasification for Hydrogen Production from Biomass

1 Introduction

Energy shortages and environmental pollution are two significant challenges, and we will face in the future. As fossil fuel consumption continues to rise, the finite nature of these resources means they cannot be replenished through sustainable methods. Among the various alternatives, hydrogen stands out as a renewable and eco-friendly energy source, garnering considerable global interest. Hydrogen can be utilized in machinery with zero emissions and can power highly efficient hydrogen fuel cells. Its alignment with growing environmental protection regulations further enhances its appeal, leading to increased research and development efforts in many countries.

In general, conventional hydrogen production methods and their shortcomings can be classified as follows:

(i) The cost of producing hydrogen through water electrolysis is currently too high.
(ii) Water photolysis is still primarily in the experimental stage in laboratories.
(iii) In the thermal pyrolysis and reforming of organic compounds, it is essential to control moisture content, necessitating a dehydration process for the pretreatment of wet feedstocks.
(iv) The methanation of microorganisms for hydrogen production suffers from a slow reaction rate and low conversion efficiency.

In contrast to the conventional methods mentioned above, SCWG process offers high reaction efficiency and selectivity for hydrogen production. It uses water as the reaction medium and can accommodate feedstocks with high moisture content. Many researchers consider SCWG to be one of the most promising hydrogen production methods among various bioconversion techniques [1, 2].

However, SCWG process has drawbacks, such as limited material compatibility and the requirement for high temperatures and pressures. Additionally, the distribution of products in the SCWG process is heavily influenced by the operating

Y. Guo et al., *Hydrothermal Processing of Biomass for Hydrogen and Bio-oil Production*, https://doi.org/10.1007/978-981-96-6336-1_3

conditions. Early studies have indicated that a high yield of hydrogen can only be achieved when the temperature exceeds 600 °C [3, 4]. At low temperatures (<450 °C), methane is the predominant product. As the density of water increases, the yield of methane rises while the yield of hydrogen decreases. Unlike many exothermic reactions, the SCWG process cannot sustain the minimum required temperature through self-heating. Consequently, economic efficiency has emerged as the primary barrier to the further advancement of SCWG technology [5].

To mitigate extreme reaction conditions and decrease both equipment investment and operating costs, a viable approach is to lower the activation energy by introducing appropriate hydrothermal catalysts. Therefore, researching catalysts suitable for hydrothermal environments is crucial for this technique. Catalytic supercritical water gasification (CSCWG) is poised to become a significant area of research in this field.

2 Supercritical Water

2.1 Thermophysical Property of SCW

When the temperature and pressure of water reach its critical point (Tc $\geq$ 374 °C and Pc $\geq$ 22.1 MPa), it enters a supercritical state. In this state, the physicochemical properties of water, including the ion product, density, dielectric constant, and viscosity, differ significantly from those in either the gaseous or liquid phases.

Hydrogen bonding plays a crucial role in illustrating the differences between the macroscopic and microscopic properties of water. An increase in temperature can alter or break the chemical structure of water at room temperature. Under standard temperature and pressure, water has a relatively high dielectric constant (about 80) due to the strong influence of hydrogen bonds. However, as temperature and pressure rise, the dielectric constant of water decreases significantly, dropping to around 5 at the critical point.

The change in dielectric constant affects the dissolving capacity of water. In the supercritical state, the dielectric constant of water is similar to that of many organic compounds, allowing it to behave like a nonpolar solvent. This gives supercritical water excellent solubility for nonpolar organic compounds. However, its ability to dissolve highly polar inorganic compounds decreases significantly, which leads to the separation of dissolved organic compounds from water.

An increase in temperature also affects the viscosity of water. At the critical point, the viscosity of water (approximately 2.98×10^{-5} Pa s) approaches that of vapor (2×10^{-5} Pa s). This lower viscosity enhances the diffusion coefficient of water, facilitating optimal reaction conditions for achieving high reaction rates.

As shown in Fig. 1, near the critical point, both the density and ion product of water are lower than under normal conditions. Density changes from liquid (1 g/cm^3) to low-pressure vapor (0.0011 g/cm^3) with variations in temperature and pressure.

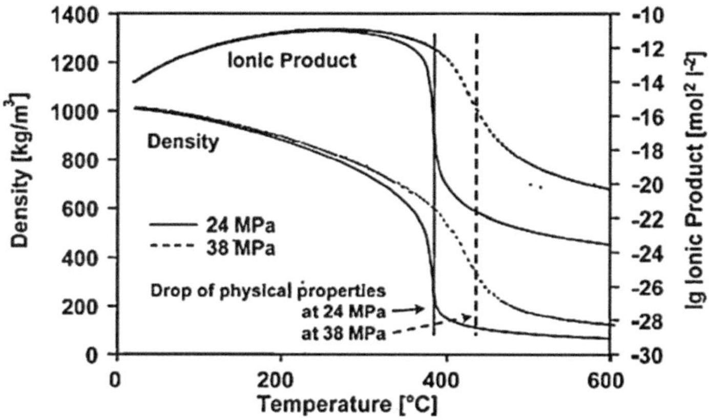

Fig. 1 Physical properties of HTW at different pressures [6]

The ion product depends on both temperature and density, with density exerting a greater influence. Below the critical point, higher density correlates with a higher ion product, making ionic reaction mechanisms more favorable. Conversely, as temperature increases beyond the critical point and density decreases, the ion product drops, indicating that free radical mechanisms become more predominant in this scenario [7].

Reactions in supercritical water provide several benefits compared to traditional gaseous and liquid reaction methods:

(1) A high reaction rate enables the production of gaseous products at elevated concentrations [1].
(2) Since there are no interphase mass transfer limitations present in conventional reaction systems, reactions occur rapidly and proceed to completion [8].
(3) Convenient separation of water and products can be accomplished by simply adjusting operating parameters like temperature and pressure, making it a more favorable method than distillation or extraction [9, 10].
(4) The reaction process offers higher dispersivity and improved heat transfer compared to gas and liquid phases [11].
(5) The excellent flowability and carrying capacity of SCW can reduce coke formation, thereby extending the catalyst's service life [2, 12–15].

Supercritical water is widely considered an attractive and promising reaction medium due to its unique properties and strong environmental compatibility.

2.2 *Role of Water in the Reaction*

High-temperature water (HTW), which exists at temperatures above the normal boiling point, exhibits a distinct structure compared to water at ambient temperatures, particularly in its supercritical state. As mentioned, the ion product of supercritical water is two orders of magnitude greater than that at ambient temperature. This results in a high concentration of H^+ and OH^- ions in supercritical water, creating an ideal environment for acid- or base-catalyzed reactions [6]. SCWG involves both hydrolysis and pyrolysis reactions [16]. Water plays a dual role as both a reactant and a catalyst, significantly influencing the SCWG process. Therefore, understanding the contributions of water in supercritical water reactions is essential for researching both homogeneous and heterogeneous catalytic mechanisms.

2.2.1 Water as a Participant in Reaction

(1) **Contributions in Hydrolysis Reaction**

HTW or SCW acts not only as a reaction medium but also as a reactant. In hydrolysis reactions, water and salts can generate acids or bases, which influence the bond scission of organic compounds. Numerous experiments using model compounds have been conducted to evaluate the impact of water in SCW or HTW reactions. Ogunsola et al. [17] proposed that the protons supplied by SCW can facilitate the saturation of heterocycles and the formation of alkyl and C–N radicals. The interaction with SCW or HTW accelerates the elimination of sulfur and the ring opening of heterocycles compared to standard hydrolysis processes.

Several researchers have reported that the conversion of amines and amides involves a combination of pyrolysis and hydrolysis processes [18, 19]. The pyrolysis mechanism is favored at high temperatures, while the hydrolysis mechanism becomes more advantageous at high water densities. Additionally, the reaction of organic compounds in subcritical water can be enhanced by the autocatalytic effects of hydrolysates [20, 21]. The acids and bases produced during hydrolysis can catalyze the reaction pathway, further influencing the overall reaction dynamics.

(2) **Resource of Hydrogen**

At high temperatures, the structure of water is altered, resulting in the weakening of intramolecular and intermolecular hydrogen bonding. This change enables water to more effectively provide hydrogen. To demonstrate the ability of HTW to deliver hydrogen, a study was conducted using D_2O instead of H_2O as the reaction medium in the hydrothermal treatment of hydrocarbons. The results indicated that hydrogen–deuterium exchange occurred during the process, with some deuterium atoms incorporating into the product [22]. Park et al. [23] used D_2O instead of H_2O as a reaction medium for the SCWG of some organic compounds with RuO_2 catalyst. They found that the methane and hydrogen in gas production are not CHD_3, H_2 or HD, but CD_4 and D_2, which indicated that all the H in gas product is from water. The reaction mechanism is demonstrated in Fig. 2.

The ability of water to provide hydrogen significantly impacts the distribution of pyrolytic products. In free radical chemistry, hydrogenation serves as a chain-terminating step, reducing the molecular weight of organic compounds, such as

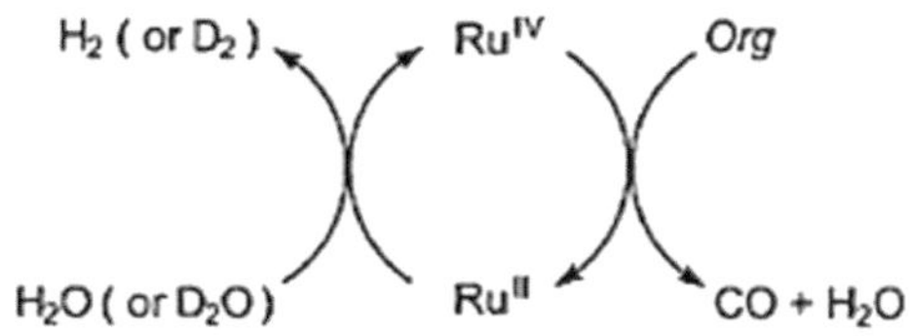

Fig. 2 SCW-induced redox cycle between RuIV and RuII catalyzing the gasification of organic compounds [23]

polymers, and leading to the formation of numerous small molecules. Watanabe et al. [24] demonstrated that HTW can minimize coking and inhibit the formation of condensation products during the hydrolysis of polyethylene in supercritical water. Similarly, Kruse et al. [25] studied the SCWG of catechol and found that the increase in hydrogen yield correlates with the amount of H_2O present in the reaction. In contrast, a higher concentration of catechol (indicating reduced water concentration) inversely affected hydrogen production. This suggests that water molecules release hydrogen atoms, facilitating the rupture of intramolecular bonds in the reactants.

Additionally, water can generate H_2 through the water–gas shift reaction. Numerous studies have confirmed the occurrence of this phenomenon:

$$CO + H_2O \leftrightarrow CO_2 + H_2 \quad \Delta H = -41 \ \text{kJ/mol} \tag{1}$$

This reaction is weakly exothermic and thermodynamically constrained at the process's operating temperatures; however, it enables significant CO reduction and additional hydrogen production [26]. Lee et al. [27] examined the SCWG of glucose without a catalyst at temperatures between 480–750 °C and pressures of 28 MPa for 10–50 s. They noted that CO yield was high during the initial conversion stages. However, when the temperature exceeded 650 °C, as illustrated in Fig. 3, CO concentration decreased while H_2 production increased, indicating the onset of the water–gas shift reaction. Byrd et al. [28] investigated the SCWG of glycerol and found that the CO yield was lower than the expected equilibrium value, while H_2 yield exceeded the anticipated equilibrium value. This suggests that the water–gas shift reaction takes place in the later stages of the process when using a Ru/Al_2O_3 catalyst.

(3) **Resource of Free Radicals**

The conversion of organic compounds at high temperatures is primarily governed by a free radical mechanism, which consists of two phases: an induction period and a rapid free radical reaction phase. Both phases are crucial for understanding reaction kinetics. The duration of the induction period and the concentration of free radicals are influenced by factors such as temperature, catalysts, and reactants [29]. Generally, temperature significantly affects the induction period. For instance, Al-Duri et al. [30] found that the induction times were 3.6 s at 400 °C and 1.5 s at 450 °C during their study on the high-temperature degradation of nitrogen compounds. This finding suggests that the formation of free radicals can be controlled by temperature.

Low water density is advantageous for free radical reactions. Henrikson et al. [7] observed that varying water density can either promote or hinder reactions in

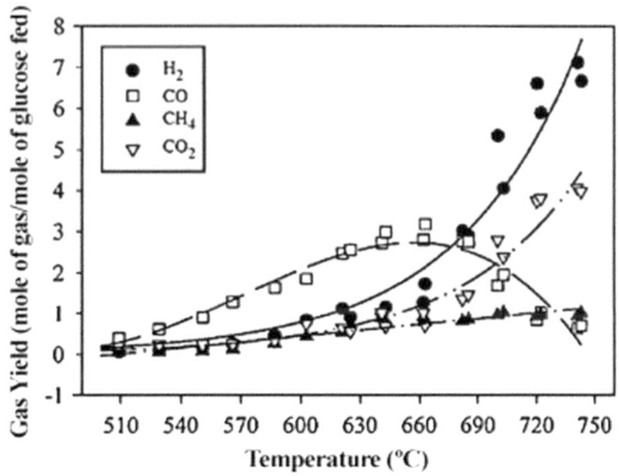

Fig. 3 Variation of product distribution with reactor temperature in supercritical water gasification (glucose 0.6 M pressure 28 MPa and reactor residence time 30 s) [27]

supercritical water at a constant temperature. One potential explanation is that at high water density, the reaction may proceed via an ionic mechanism, while at low water density, it relies more on the free radical mechanism.

The H and OH species produced from SCW ionization can facilitate the heterolytic reactions of aromatic compounds. Concurrently, the process suppresses coupling reactions. As hydroxyl groups are replaced, π-electrons become delocalized, leading to the instability of the benzene ring. This instability allows for further substitution and attack, resulting in ring cleavage reactions and the formation of relatively low-molecular weight compounds [29].

The hydroxyl radical is recognized as the most crucial and effective oxidant in SCW reactions. In many instances, its formation is a rate-controlling step in the overall reaction kinetics. Dagaut et al. [31] developed a comprehensive kinetic model for free radical reactions of methanol using a tubular flow plug reactor under SCW oxidation conditions. They indicated that the kinetics of methanol reactions are governed by the formation and decomposition of H_2O_2, as H_2O_2 can decompose into hydroxyl radicals, which then control methanol consumption. Similarly, Savage et al. [32] reported that the generation of hydroxyl radicals is a key control step in the reaction kinetics of methane during supercritical water oxidation.

2.2.2 Water as a Catalyst

Water molecules also act as catalysts in SCW reactions. Under SCW conditions, water can dissociate into high concentrations of H^+ and OH^- ions, creating an optimal environment that favors acid–base catalytic reactions [33–35].

(1) Contribution in Acid Catalytic Reactions

Acid-catalyzed reactions of organic compounds can occur in both pure subcritical and supercritical water without the need for additional catalysts. Kuhlmann et al.

[36] observed that cyclohexene dehydrates under subcritical conditions without any added catalyst, suggesting that the reaction is catalyzed by H_3O^+ ions generated from water at elevated temperatures. This aligns with findings from Xu et al. [37], who reported that tert-butyl alcohol reacts at 250 °C in water without any external catalyst. A scheme illustrating the catalytic reaction mechanism is provided in Fig. 4.

Chandler et al. [38] discovered that Friedel–Crafts alkylation reactions can occur in subcritical water without the need for a catalyst. They found that temperature significantly affects the ionization of water, allowing it to act as an effective acid catalyst under subcritical conditions. Additionally, Ikushima et al. [34] reported that SCW can effectively serve as an acid catalyst, enhancing reactions such as pinacol and Beckmann rearrangements.

(2) **Contribution in Base-Catalyzed Reactions**

In SCW, numerous chemical reactions that typically require the addition of base-catalyzed materials under normal conditions can occur spontaneously. To investigate the base catalytic effect of SCW, Ikushima et al. [35] studied the disproportionation of benzaldehyde without an external catalyst. The proposed mechanism is illustrated

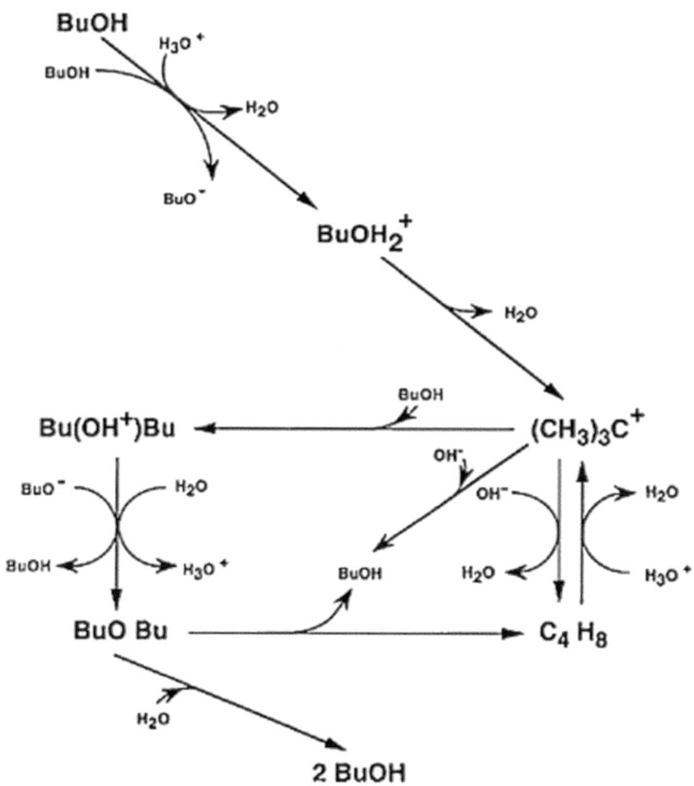

Fig. 4 Acid-catalyzed reactions of tert-BuOH in HTW [37]

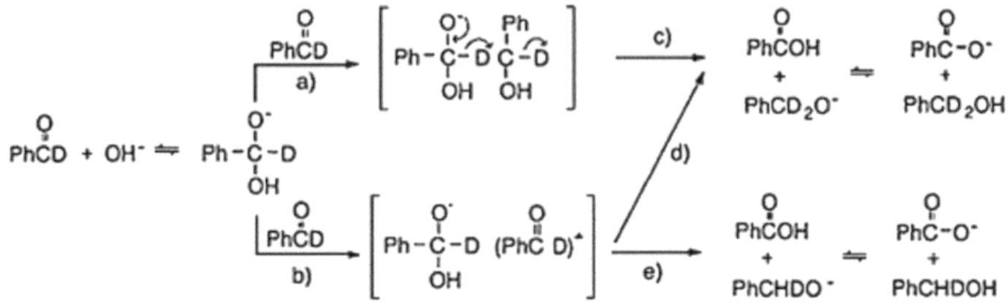

Fig. 5 A postulated mechanism for the disproportionation of benzaldehyde in SCW; (d) D˙ transfer; (e) H˙ transfer [35]

in Fig. 5. In this mechanism, the presence of the OH^- ion is essential for producing the alcohol product. The involvement of the OH^- ion in the disproportionation reaction using SCW is clearly evidenced in their findings.

Ethanol and formic acid can be produced from Cannizzaro-type reactions of formaldehyde in subcritical water without any catalyst, which is typically known to require a substantial amount of base catalyst under ambient conditions [39]. Ramayya et al. [40] demonstrated that the aldol condensation of acetaldehyde can be catalyzed by OH^- ions generated from high-temperature water. Furthermore, while intramolecular aldol condensation can also be facilitated by diluted NaOH, the reaction is inhibited when H_2SO_4 is added [41].

3 Catalytic Supercritical Water Gasification

Over the past two decades, research institutions such as the U.S. Pacific Northwest Laboratory, Hawaii Natural Energy Institute, Forschungszentrum Karlsruhe in Germany, and the National Institute for Resources and Environment in Japan have conducted extensive studies on hydrogen production via supercritical water gasification (SCWG) of various organic compounds without catalysts. Their research focused on several typical model compounds, including glucose [26, 42–46], methanol [47, 48], cellulose [49–52], lignin [51, 53], and other real biomass materials [44, 54–58], as well as organic waste/water mixtures [59–61]. As successful demonstrations accumulated, a solid foundation for subsequent research was established, encompassing detailed studies on reaction mechanisms, kinetics, and thermodynamics.

To enhance the selectivity of hydrogen production in SCWG, high activation energy is required for reactions without catalysts. Consequently, the significant costs associated with equipment and operation have emerged as the main barriers to the advancement of this technology. As a result, considerable attention has been directed towards the study of catalytic supercritical water gasification. This work will not cover the preparation of catalysts; instead, it will focus exclusively on the application of catalysts to facilitate chemical transformations within the SCWG process. A review of the literature regarding catalytic supercritical water gasification research over the past five years is presented in Table 1.

3.1 Homogenous Catalyst

The catalytic effect of alkali metal catalysts, such as Na_2CO_3, $KHCO_3$, K_2CO_3, and NaOH, on the SCWG of biomass has been well-documented in numerous studies. A key feature of these catalysts is their ability to enhance the water–gas shift reaction, thereby improving overall hydrogen production efficiency.

Kruse et al. [25] investigated the catalytic effect of KOH on the SCWG of pyrocatechol. Their findings indicated that increasing the KOH concentration from 0 to 5% led to a reduction in CO yield while simultaneously enhancing the production of H_2 and CO_2. This change is attributed to KOH's catalytic influence on the water–gas shift reaction. In comparison, LiOH exhibited similar activity regarding the gas-phase composition, albeit to a lesser extent than KOH. Additionally, Garcia Jarana et al. [74] assessed the catalytic role of KOH in the SCWG of industrial organic waste, finding that the addition of KOH accelerated the water–gas shift reaction. They proposed that formic acid serves as an intermediate product in this reaction process, with the generation of H_2 and CO_2 resulting from the decomposition of formic acid.

$$CO + H_2O \leftrightarrow \text{“HCOOH”} \leftrightarrow CO_2 + H_2 \tag{2}$$

Watanabe et al. [33] demonstrated the catalytic effect of NaOH in their studies, revealing that the yield of CO decreased as the concentration of OH^- increased. This finding suggests that the reactions involving formaldehyde (HCHO) are influenced by the OH^- concentration. In further experiments on n-hexadecane and lignin conducted under conditions of 400 °C and 30 MPa, they found that adding NaOH increased H_2 production by four times compared to reactions without it. Additionally, the presence of NaOH effectively reduced coke formation [75]. The observed decrease in gas yield during algae gasification in the presence of NaOH was attributed to the reaction between carbon monoxide and NaOH, resulting in the formation of carbonate and bicarbonate [76]. In comparison, KOH provided a higher hydrogen yield than NaOH when gasifying fructose [76] and sugarcane bagasse [77]. Moreover, K_2CO_3 outperformed Na_2CO_3 in hydrogen yield during the gasification of sugarcane bagasse [77] and notably *Nannochloropsis gaditana* [78].

In their investigation of Ru/TiO_2 catalytic properties in SCWG process, researchers found that the addition of Na or K cations as assistants can enhance the water–gas shift reaction without impacting the carbon conversion rate [72]. Yanik et al. [4] utilized K_2CO_3 and Trona ($NaHCO_3 \cdot Na_2CO_3 \cdot 2H_2O$) catalysts in their CSCWG studies involving various feedstocks, including lignocellulosic materials like cotton stalk and corncob, as well as tannery waste. Their findings indicated a significant increase in H_2 yield when these catalysts were employed. The enhanced

Table 1 Summary of the investigations on some hydrogen production via catalytic supercritical water gasification

Target reactant	Catalyst	Reactor type	Experiment condition	Experimental results and comments	References
Glucose	Raney nickel and homogeneous organometallic catalysts	Union tee reactor constructed of 316 stainless steel	310–350 °C and 10–21 MPa	Raney nickel was a more effective catalyst compared to homogeneous $Ni(acac)_2$,$Co(acac)_2$, and $Fe(acac)_3$ catalysts	[62]
Glucose	Ni/activated charcoal	Hastelloy C-276 packed bed reactor	575–725 °C; 28 MPa; feed concentration 0.3–0.9 M; LHSV 6–24 h^{-1}	Ni/AC showed high hydrogen selectivity and stability; deactivated due to coke deposition at low temperature and also due to sintering of nickel particles	[63]
Glycerol	Ru/Al_2O_3	Tubular fixed-bed flow reactor	700–800 °C; reaction time < 5 s; feed concentration 40 wt%	Ruthenium catalyst showed a high activity for C–C bond scission; glycerol was completely gasified; H_2 yield decreases with an increase in the feed concentration	[28]
Lignin	Ruthenium trivalent-salts supported on titanium oxide and charcoal	Sus 316 tube reactor	400 °C; 37.1 MPa; water density 0.5 g/cm^3	The order of gasification activity was Ru/C ≈ $Ru(NO)(NO_3)_3/C$ ≈ $Ru(NO)(NO_3)_3/TiO_2$ > $RuCl_3/C$ ≈ $RuCl_3/TiO_2$; small ruthenium metal particles were formed during the lignin gasification in supercritical water	[64]
Phenol	Ni/C	Fixed-bed reactor heated by fluidized sand bed	350 °C; 20 MPa; 50 h^{-1} LHSV; feed concentration of 2% TOC	The metal load after carbonization was about 47% and surface area was 178 m^2/g; no effect of carbonization time on particle size was observed; conversion or activity decreased as particle size increased	[65]

(continued)

Table 1 (continued)

Target reactant	Catalyst	Reactor type	Experiment condition	Experimental results and comments	References
Lignin	Ru/TiO_2	316 stainless-steel bomb reactor	400 °C; 0.1 g of lignin; 5.0 × 10^{-4} g of elemental sulfur; 7.4 × 10^{-5} mol of metal in catalyst; 0.50 g/cm^{-3} water density	The gas yield decreased with the amount of sulfur added; the sulfur species which poisoned the ruthenium sites were found to be ruthenium sulfide, ruthenium sulfite, and ruthenium sulfate	[66, 67]
Lignin and 4-propylphenol	Ru/TiO_2, Ru/C, Ru/Al_2O_3, Rh/C, Pt/C, Pt/Al_2O_3, Pd/C, Pd/Al_2O_3, Ni/Al_2O_3	316 stainless-steel tube bomb reactor	400 °C; water density 0–0.5 g/cm^3	Decomposition to low-molecular weight compounds in the lignin gasification was enhanced by increasing the water density (4-propylphenol is not); gasification of the low-molecular weight compounds was accelerated over metal catalysts	[68]
Methanol	In situ generated copper nanoparticles	Stainless-steel 316 tubular reactor	700 °C and 27.6 MPa; reaction time < 1 min	The concept is successfully demonstrated for the production of hydrogen from methanol via copper catalysis in supercritical water; the generated nanoparticle surface is highly catalytic and reforms methanol in about 15 s; such scheme is also suitable for sulfur-containing feeds	[69]
Lignin	Ni/MgO	Stainless-steel tube bomb reactors	250–400 °C; water density 0.3 g/cm^3	The amount of gases produced increased with an increase in nickel loading on magnesium oxide; MgO decomposed lignin to reactive intermediates and nickel promoted reaction between intermediates and water to form gases; water density affected the reaction kinetics	[70]

(continued)

Table 1 (continued)

Target reactant	Catalyst	Reactor type	Experiment condition	Experimental results and comments	References
Methanol	Ni catalyst	Quartz tube	500 and 550 °C; methanol 10 wt.%; fluid density 0.079 g/mL	The conversion rates are greatly increased by added Ni catalyst; Ni metal is subject to deactivation with repeated use	[71]
Glycerol, glucose and wood	Ru/TiO_2	Quartz capillary reactor	450–700 °C; 5–50 MPa; feedstock 1–20 wt%	Complete conversion of solutions with 1–17 wt% glucose is achieved by adding a 3 wt% Ru/TiO_2 catalyst	[72]
Methanol	Wall of the reactor made of Inconel 600 (73% Ni, 18% Cr, and 9% Fe), K_2CO_3, KOH	Tubular reactor	700 °C; 27.6 MPa	The reactor made of Ni–Cu tubing minimizes the formation of CH_4; methanation can also be reduced with the addition of K_2CO_3 or KOH in the aqueous methanol feed	[47]
Cellulose	Ru/C, Pd/C, CeO_2 particles, nano-CeO_2 and nano-$(CeZr)_xO_2$	140 ml, high-pressure autoclave	650 °C; 35 MPa; 10 wt% cellulose or sawdust	Catalytic activities were Ru/C > Pd/C > nano-$(CeZr)xO_2$ > nano-CeO_2 > CeO_2 particle in turn	[73]

catalytic activity of K_2CO_3 is attributed to the formation of formate (HCOOK) salt within the reaction pathway [79, 80]:

$$K_2CO_3 + H_2 \rightarrow KHCO_3 + KOH \tag{3}$$

$$KOH + CO \rightarrow HCOOK \tag{4}$$

Hydrogen is obtained by reacting of formate with water.

$$HCOOK + H_2O \rightarrow KHCO_3 + H_2 \tag{5}$$

The catalytic cycle is completed by the formation of CO_2 and K_2CO_3.

$$2KHCO_3 \rightarrow H_2O + K_2CO_3 + CO_2 \tag{6}$$

The overall reaction (2) is obtained by the integration of the above formulas.

K_2CO_3 clearly alters the reaction pathway to enhance the water–gas shift reaction, leading to a reduction in the activation energy of the process. Additionally, it is generally observed that the hydrogen yield from these catalysts follows the order: hydroxides > carbonates > bicarbonates [77, 81].

$FeCl_3$ has been studied for its role in SCWG of humic acid, which serves as a model for humic substances found in sewage sludge [82]. The results indicated that $FeCl_3$ is more effective than alkali or transition metal catalysts, such as K_2CO_3, nickel, and $ZnCl_2$, in enhancing both hydrogen and total gas yields. Furthermore, a machine learning-assisted literature review conducted by Li et al. [83] on the SCWG of wet waste feedstocks found that $FeCl_3$ outperformed other transition metals. While alkali catalysts are crucial for achieving high hydrogen production, they present challenges such as difficult recovery and potential issues with corrosion, clogging, or fouling [80, 84].

3.2 Heterogeneous Catalyst

Heterogeneous catalysts offer several advantages over homogeneous catalysts, including high selectivity, recyclability, and environmental friendliness. As a result, they have become a focal point of research in this field.

3.2.1 Transition Metal Catalyst

(1) Nickel Catalyst

Nickel catalysts are relatively low cost and have been widely used in various petrochemical industries. Building on previous theories regarding nickel catalysts in other chemical processes, many researchers have incorporated them into supercritical water gasification systems to better understand their hydrothermal activity and stability. Most studies have shown that the conversion of biomass is enhanced by nickel catalysts; however, both the nickel catalyst and its support tend to be unstable. Issues such as sintering and deactivation are common challenges encountered during the reaction process.

Elliott [85] conducted exploratory research on catalyst evaluation for hydrothermal gasification, highlighting that various reactive metal catalysts, including nickel, can enhance the gasification of biomass under supercritical and even subcritical water conditions. Their findings indicate that the lifespan of nickel catalysts is inherently limited, often lasting less than 100 h. Additionally, the physical and chemical structure of the catalyst support undergoes significant changes under hydrothermal conditions, leading to sintering of nickel crystallites in both batch and continuous-flow experiments [86].

Waldner et al. [87] provided insights into the catalytic effects of Raney nickel for the SCWG of woody biomass at operating conditions of 300–410 °C and 12–34 MPa. They found that if the reaction time is sufficient (90 min), complete gasification can be achieved. However, this process results in the accumulation of carbon deposits on the catalyst surface. The presumed catalytic reaction pathway is illustrated in Fig. 6.

Nickel catalysts exhibit high activity in the CSCWG processes for various real compounds, such as lignin and cellulose. However, inactivation is inevitable due to the adsorption of intermediate products on the catalyst surface during the process [88].

Furusawa et al. [89, 90] utilized a Ni/MgO catalyst for the CSCWG of lignin. Their research indicated that the carbon yield of gas products increases with the metal surface area of nickel. The 10 wt% Ni/MgO catalyst at 873 K exhibited the

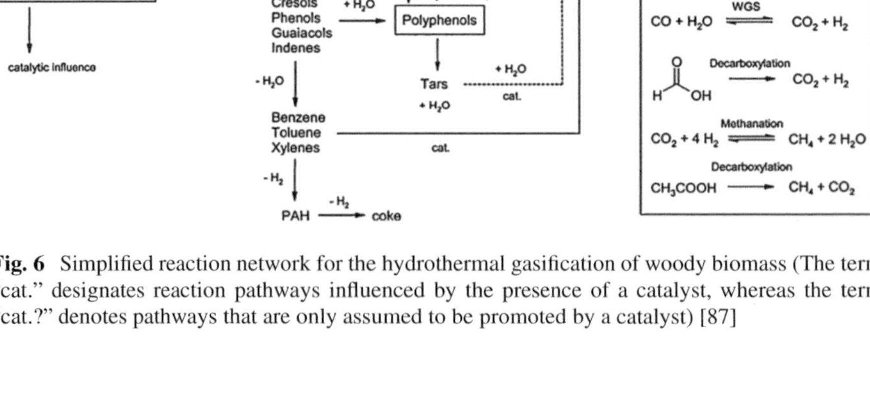

Fig. 6 Simplified reaction network for the hydrothermal gasification of woody biomass (The term "cat." designates reaction pathways influenced by the presence of a catalyst, whereas the term "cat.?" denotes pathways that are only assumed to be promoted by a catalyst) [87]

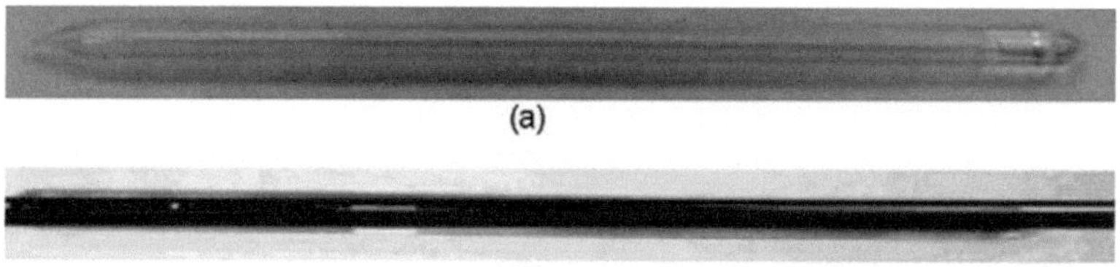

Fig. 7 Quartz reactors **a** before and **b** after SCWG of guaiacol [93]

best catalytic performance, achieving a carbon yield of 30% under the tested reaction conditions. Furthermore, Minowa et al. [91] highlighted the significance of nickel catalysts in steam reforming and methanation reactions. In a subsequent article, they noted that different support materials could influence the activity of the nickel catalyst differently [92].

Dileo et al. [93] employed a nickel wire integrated into a quartz tube reactor (Fig. 7) to catalyze the hydrothermal gasification of guaiacol and phenol at temperatures ranging from 400 to 700 °C. They found that nickel did not influence the conversion of guaiacol to phenol and o-cresol, but it significantly altered the composition of the gas products. After 1 h of reaction time, the conversion rate of phenol reached 68%, while a 100% conversion rate could be achieved with the catalyst in just 10 min.

The carbon-based nickel (Ni/C) catalyst demonstrates high catalytic activity under hydrothermal conditions [94]. This catalyst was synthesized by ion exchanging nickel ions with the exchangeable sites in an ion exchange resin, followed by carbonization of the resin. The resulting catalyst has a high surface area of 170 m^2/g, allowing it to incorporate 47% effective nickel content. A SEM image of the prepared catalyst is presented in Fig. 8. Following this work, the group conducted further evaluations on how carbonization temperature affects the nickel crystallite size of the Ni/C catalyst during the catalytic hydrothermal gasification of organic compounds.

The catalyst derived from loading nickel onto $MgO–Al_2O_3$ notably improved hydrogen yield during the supercritical water gasification of glucose (as shown in Fig. 9a). The carbon gasification efficiency increased from 23.4% without a catalyst to 53.8% and 61.8% with the catalyst (Fig. 9b). Additionally, the catalyst demonstrated robust catalytic performance, even after multiple uses [95].

Khandelwal and Dalai [96] conducted a thorough evaluation of a novel nickel-based catalyst by systematically testing six supports—activated carbon (AC), carbon nanotubes (CNT), hydrothermal carbon (HTC-HC), hydrothermal liquefaction carbon (HTL-HC), alumina (Al_2O_3), and zirconia (ZrO_2)—along with three promoters (potassium, zinc, and cerium) to determine the optimal combinations for catalytic performance. The catalytic activity was assessed for the gasification of rapeseed straw in supercritical water under conditions of 500 °C, 23–25 MPa, a mass fraction of 10 wt%, and a reaction time of 60 min. Among all the supported catalysts tested, Ni/ZrO_2 demonstrated the highest catalytic activity, achieving a rate of 10.5 mmol/g for hydrogen production. In contrast, although Ni/CNT had the highest specific surface area of 590 m^2/g, it resulted in the lowest hydrogen yield of only 6.8

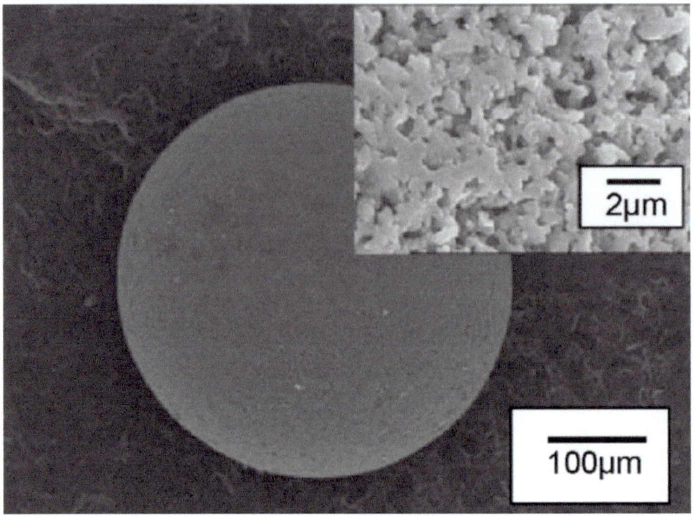

Fig. 8 SEM image of the catalyst prepared [94]

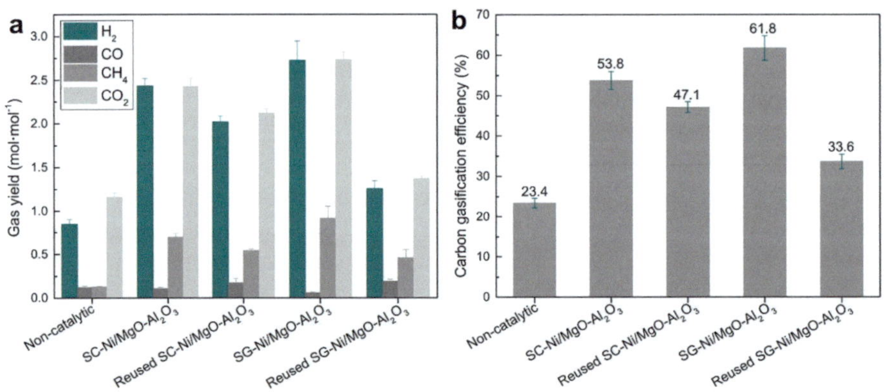

Fig. 9 Yield distribution (**a**) and carbon gasification efficiency (**b**) for non-catalytic and catalytic SCWG of 5 wt% glucose at 500 °C and 23 MPa (glucose:catalyst = 10:1) [95]

mmol/g [96]. This indicates that, in supercritical water gasification, factors such as the dispersion of active metal, the crystal phase of the active metal, and the interactions between the metal and support significantly influence catalytic activity, even when high specific surface area is present.

To enhance the performance of nickel catalysts, Elliott et al. [97] investigated the addition of trace elements such as Cu, Ag, Sn, and Ru to the nickel catalyst. Their findings indicated that Ru effectively improved the activity of the nickel catalyst. Additionally, the catalyst demonstrated prolonged durability, as evidenced by continuous laboratory tests conducted over a period of six months.

(2) **Ruthenium Catalyst**

Ruthenium is a highly effective catalyst for low-temperature catalytic gasification, with even low metal loadings resulting in strong catalytic activity [97]. Byrd et al. [98] studied the Ru/Al_2O_3 catalyst in a glucose CSCWG experiment and found that its addition improved both conversion rates and hydrogen yield, while also reducing methane production. In further studies on glycerol CSCWG [98], as shown in Table 1, the high activity of Ru/Al_2O_3 for C–C bond cleavage was confirmed. The proposed catalytic mechanism involves oxygenated compounds with hydroxyl groups adsorbing to the Ru surface via oxygen atoms. The reactant undergoes dehydrogenation, followed by C–C or C–O bond cleavage. C–C bond cleavage produces synthesis gas, which can undergo the water–gas shift or methanation reactions, while C–O bond cleavage yields organic acids and alcohols. BET analysis revealed that the catalyst undergoes morphological changes during the reaction. A similar observation was made in Osada et al.'s study [99], where after 180 min of reaction, the Al_2O_3 support transformed from γ to α phase, and some Ru metal dissolved into the water.

Ruthenium supported on graphitic carbon black (GCB) was synthesized and utilized as a catalyst for the gasification of phenol in supercritical water. The catalyst demonstrated outstanding performance, achieving complete gasification of 5.2 wt% phenol at a moderate temperature of 500 °C. A kinetic analysis using a pseudo-first-order reaction model indicated that the activation energy (E_a) was 154.0 ± 27.2 kJ mol^{-1} [100]. Characterization of both fresh and used catalysts revealed that the GCB support remains highly stable in supercritical water (Fig. 10).

Carbon is commonly utilized as a catalyst support due to the functional groups present on its surface, allowing it to function effectively as both a catalyst and support. Among the various carbonaceous materials, carbon nanotubes (CNTs) stand out because of their unique nanostructure and high specific surface area [101]. In our previous study [102], it was demonstrated that CNTs exhibit significant stability in SCW, with the gasification rate of activated carbon being 8.5 times higher than that of CNTs. Given the excellent catalytic activity of ruthenium (Ru), Ru nanoparticles supported on high-surface-area CNTs are anticipated to exhibit exceptional catalytic performance. Figure 11 presents a scanning electron microscope (SEM) image of the synthesized Ru/CNT, indicating no damage to the CNTs post-Ru deposition. In experiments, 5 wt% glucose, a typical biomass model compound, was completely gasified using just 0.15 g of a 0.5 wt% Ru catalyst supported on CNTs [103]. The catalyst demonstrated high stability in SCW at 600 °C and 25 MPa in a packed bed reactor. Notably, the amount of Ru utilized in this study was one to two orders of magnitude lower than that required for other catalyst supports, underscoring the potential of CNTs as an effective support for supercritical water gasification.

The Ru catalyst demonstrated high H_2 selectivity in CSCWG experiments involving lignin and glucose [104]. At 400 °C, formaldehyde, as an intermediate compound, rapidly decomposes into CH_4, CO_2, and H_2 when catalyzed. In contrast, without a catalyst, formaldehyde primarily converts to methanol and CO_2. Sato et al. [105] compared variety of heterogeneous catalysts for CSCWG and found that the activity order is $Ru/\gamma\text{-}Al_2O_3$ > Ru/C > Rh/C > $Pt/\gamma\text{-}Al_2O_3$, Pd/C, and $Pd/\gamma\text{-}Al_2O_3$.

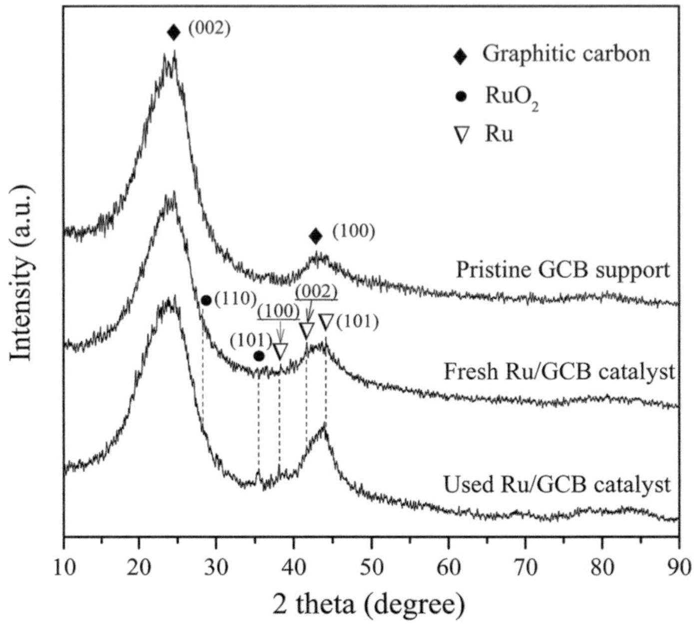

Fig. 10 XRD patterns of fresh and used Ru/GCB catalysts [100]

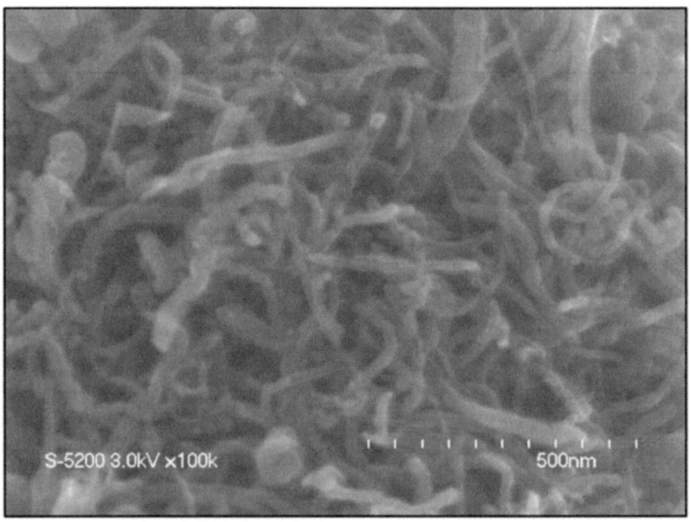

Fig. 11 A SEM image of the Ru/CNT [103]

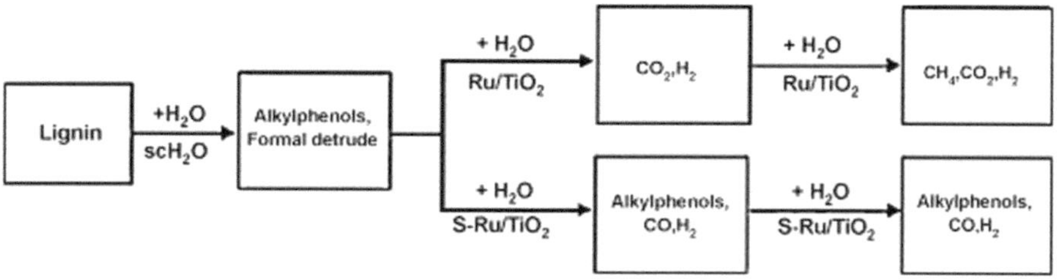

Fig. 12 Reaction pathway of lignin gasification over a supported ruthenium catalyst containing sulfur in supercritical water [67]

Ruthenium (Ru) catalyst poisoning typically occurs due to trace amounts of sulfur, which can be present in the Ru/C catalyst as $S^{2-}$and SO_4^{2-} [66, 67]. Figure 12 illustrates the reaction pathway associated with the poisoning of the ruthenium catalyst. Additionally, it is important to highlight that the introduction of even small amounts of Na_2SO_4 can contribute to catalyst poisoning [106].

(3) **Other Metal Catalyst**

Pt-based catalysts demonstrate excellent activity and selectivity for hydrogen production from sugars and alcohols through low-temperature aqueous phase reforming reactions, although further enhancements are needed for practical application [107]. A comparison of different active metal loadings on SiO_2 revealed the following order of performance: Pt > Ni > Ru > Rh > Pd > Ir [108]. Figure 13 illustrates the reaction pathway and selectivity challenges in producing H_2 from the reactions of ethylene glycol with water. During the SCWG process, Pt/Al_2O_3 showed superior selectivity for hydrogen production compared to Ru/Al_2O_3 [109]. As indicated in Fig. 14, the hydrogen selectivity for Pt/Al_2O_3 remained around 10%, while experiments without a catalyst yielded selectivity between 1 and 2%. In contrast, Pt/Al_2O_3 displayed very high selectivity for hydrogen production, benefiting from low temperatures and rapid reaction times.

CeO_2 and ZrO_2 have been widely investigated as catalysts for steam co-gasification (SCWG) of biomass and complex organic waste materials. Hao et al. [73] highlighted the catalytic effectiveness of CeO_2 and nano-CeO_2 in the SCWG of cellulose and glucose. Dejhosseini et al. [110] used nano-CeO_2 as a catalyst for the SCWG of heavy oil, leveraging its high oxygen storage capacity to improve conversion rates. Additionally, Boucard et al. [111] explored the use of CeO_2 in converting black liquor under supercritical water conditions, which led to enhanced hydrogen production and reduced coke formation. These findings enhance the understanding of the potential of CeO_2 and ZrO_2 catalysts in facilitating the conversion of biomass and organic wastes in SCWG processes.

ZrO_2 is a typical acid–base catalyst with a structure comprising both monoclinic and tetragonal crystals, where its acidity is determined by this structural composition [112]. Following SCW reactions, ZrO_2 transitions to a pure monoclinic crystal

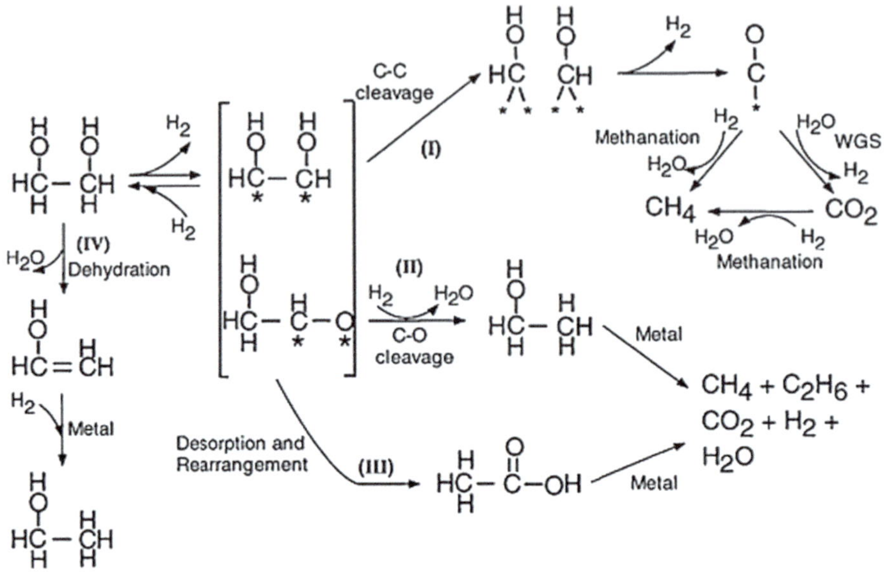

Fig. 13 Reaction pathways and selectivity challenges for production of H_2 from reactions of carbohydrates with water (* represents a surface metal site) [108]

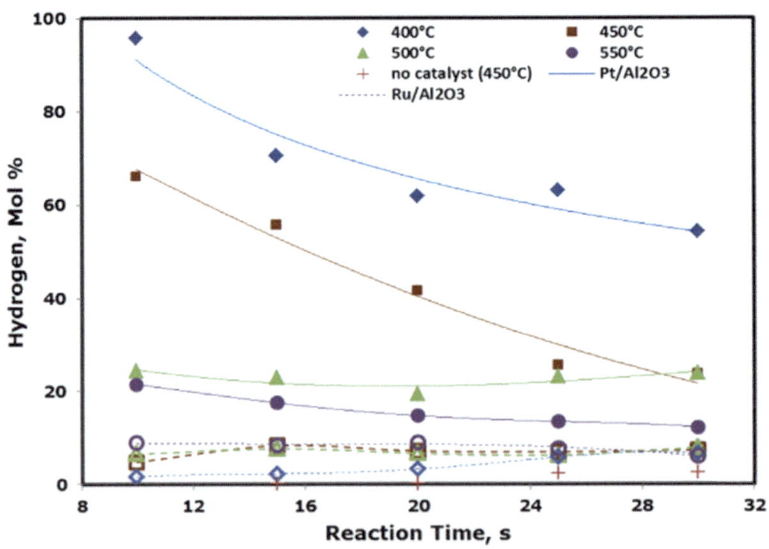

Fig. 14 Change of hydrogen content in the gas effluent with respect to the reaction temperature and time at 25 MPa [109]

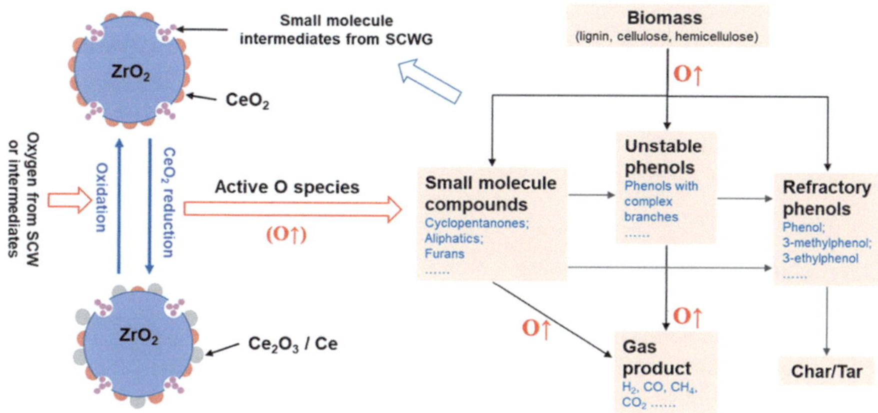

Fig. 15 Catalytic mechanism of CeO_2–ZrO_2 in SCWG of biomass [114]

system, which is stable under SCW conditions. Watanabe et al. [75, 113] investigated the catalytic activity of ZrO_2 in the SCWG of glucose and lignin, finding that it effectively inhibits CH_4 production while enhancing H_2 yield. However, its catalytic performance is significantly lower than that of NaOH.

Cao et al. [114] investigated the CeO_2–ZrO_2 catalyst prepared via the co-precipitation method. Their findings demonstrated that this catalyst significantly enhanced gasification efficiency and hydrogen (H_2) production from lignin and cellulose. In the SCWG of cellulose at 500 °C, the CeO_2–ZrO_2 catalyst improved H_2 yield by more than 2.5 times, reaching 8.50 mol/kg. The catalyst also aided in the decomposition of small molecular components, such as furan and cyclopentanone, found in the liquid residues from the SCWG of cellulose and lignin (Fig. 15). Additionally, the CeO_2–ZrO_2 catalyst was effective in reducing certain phenolic compounds with complex branched radicals, although some refractory phenolics were found to increase in its presence. These results underscore the potential of the CeO_2–ZrO_2 catalyst to enhance biomass conversion efficiency and facilitate the production of valuable products like hydrogen.

3.2.2 Activated Carbon Catalyst

Activated carbon is a promising catalyst for hydrothermal reactions, mainly due to its high catalytic activity. Additionally, it avoids the secondary pollution issues associated with metal catalysts, as it does not release dissolved metals.

Activated carbon materials, including spruce wood charcoal, macadamia shell charcoal, coal-based activated carbon, and coconut shell charcoal, were among the first evaluated for the CSCWG process of organic feedstocks by Matsumura et al. [115]. Their findings indicated that activated carbon enhances carbon gasification efficiency and also improves the water–gas shift and methanation reactions. However, deactivation of carbon gasification was observed after 4 h, while the water–gas shift reaction showed deactivation after 2 h. The presence of a vortex at the reactor entrance may aid in increasing the conversion rate. The research team also investigated the performance of the activated carbon catalyst, suggesting that the reaction pathway likely involves the adsorption of a reactant or an intermediate on the carbon surface, which significantly influences the catalyst's performance [116].

3.2.3 Catalysis of Reactor Wall

Most supercritical water reactor materials are made from alloys, meaning the metal of the reactor's inner wall significantly influences the reaction process. Inconel and "new" Hastelloy have been considered effective for the water–gas shift reaction, while "corroded Hastelloy" only facilitated the decomposition of acetic acid [45]. Lee et al. [27] found that Hastelloy C-276 enhances hydrogen yield efficiency due to its wall effect. Boukis et al. [48] investigated the reforming of methanol with supercritical water in a tubular flow reactor constructed from Inconel 625. They noted that heavy metals on the reactor's inner surface significantly affect the conversion rates and product composition. This catalytically active surface can promote both methanol decomposition and the water–gas shift reaction, maintaining catalytic activity for over 1000 h. Gadhe and Gupta [47] found that the wall of the tubular reactor made from Inconel 600, an alloy of nickel, chromium, and iron, also catalyzes the supercritical water gasification process. The main conclusions are listed in Table 1.

4 Conclusion

Supercritical water gasification is an effective method for producing hydrogen from biomass, particularly due to its suitability as a reaction medium for biomass with high water content. The physicochemical properties of supercritical water differ significantly from those of gases and liquids at normal conditions, allowing it to act both as a reactant and as a catalyst that influences the reaction pathway. However, challenges remain in reducing activation energy and managing the extreme reaction conditions. One promising approach to address these issues is the addition of suitable catalysts for hydrothermal reactions.

Previous research has classified catalysts into alkali metals, transition metals (including the "wall" effect), and activated carbon. Alkali metal catalysts can enhance the water–gas shift reaction but may cause equipment issues like corrosion and fouling. Transition metal catalysts, such as Ni, Pt, Ru, and Rh, are believed to improve reactions by accelerating steam reforming, methanation, and C–O and C–C bond cleavage. Activated carbon is also effective in water–gas shift and methanation reactions. These catalysts can increase conversion rates, hydrogen selectivity, and reduce activation energy in SCWG. However, many studies have shown that most catalysts are unstable, and their support materials may degrade.

Ongoing research is essential to gain a better understanding of the chemistry involved, which can be advanced through the development of efficient and stable catalysts. With the potential for catalysis in SCWG to drive progress in green chemistry and sustainable energy, further research and development in this expanding field are expected. If cost-effective solutions are achieved, this technology could have significant potential for large-scale commercialization in the future.

References

1. A. Kruse, A. Krupka, V. Schwarzkopf, C. Gamard, T. Henningsen, Influence of proteins on the hydrothermal gasification and liquefaction of biomass. 1. Comparison of different feedstocks. Indus. Eng. Chem. Res. **44**, 3013–3020 (2005)
2. Y. Matsumura, T. Minowa, B. Potic, S.R.A. Kersten, W. Prins, W.P.M. van Swaaij, B. van de Beld, D.C. Elliott, G.G. Neuenschwander, A. Kruse, M.J. Antal, Biomass gasification in near- and super-critical water: status and prospects. Biomass Bioenerg. **29**, 269–292 (2005)
3. H.Q. Tang, K. Kitagawa, Supercritical water gasification of biomass: thermodynamic analysis with direct Gibbs free energy minimization. Chem. Eng. J. **106**, 261–267 (2005)
4. J. Yanik, S. Ebale, A. Kruse, M. Saglam, M. Yüksel, Biomass gasification in supercritical water: II. Effect of catalyst. Int. J. Hydrogen Energy **33**, 4520–4526 (2008)
5. Y. Calzavara, C. Joussot-Dubien, G. Boissonnet, S. Sarrade, Evaluation of biomass gasification in supercritical water process for hydrogen production. Energy Convers. Manage. **46**, 615–631 (2005)
6. P. Kritzer, Corrosion in high-temperature and supercritical water and aqueous solutions: a review. J. Supercrit. Fluids **29**, 1–29 (2004)
7. J.T. Henrikson, Z. Chen, P.E. Savage, Inhibition and acceleration of phenol oxidation by supercritical water. Ind. Eng. Chem. Res. **42**, 6303–6309 (2003)
8. P. Kritzer, E. Dinjus, An assessment of supercritical water oxidation (SCWO)—existing problems, possible solutions and new reactor concepts. Chem. Eng. J. **83**, 207–214 (2001)
9. P.E. Savage, Heterogeneous catalysis in supercritical water. Catal. Today **62**, 167–173 (2000)
10. P.E. Savage, A perspective on catalysis in sub- and supercritical water. J. Supercrit. Fluids **47**, 407–414 (2009)
11. A. Loppinet-Serani, C. Aymonier, F. Cansell, Current and foreseeable applications of supercritical water for energy and the environment. Chemsuschem **1**, 486–503 (2008)
12. A. Kruse, A. Gawlik, Biomass conversion in water at 330–410 °C and 30–50 MPa: Identification of key compounds for indicating different chemical reaction pathways. Indus. Eng. Chem. Res. **42**, 267–279 (2003)
13. A. Kruse, T. Henningsen, A. Sinag, J. Pfeiffer, Biomass gasification in supercritical water: influence of the dry matter content and the formation of phenols. Ind. Eng. Chem. Res. **42**, 3711–3717 (2003)
14. T. Adschiri, T. Suzuki, K. Arai, Catalytic reforming of coal-tar pitch in supercritical fluid. Fuel **70**, 1483–1484 (1991)
15. S. Saim, Isomerization of 1-hexene over platinum/gamma-alumina catalyst at subcritical and supercritical conditions: temperature, pressure and solvent effects on catalyst coking. J. Supercrit. Fluids **3**(4), 214–221 (1990)
16. S. Saim, B. Subramaniam, Isomerization of 1-hexene over Pt/γ-Al_2O_3 catalyst: reaction mixture density and temperature effects on catalyst effectiveness factor, coke laydown, and catalyst micromeritics. J. Catal. **131**, 445–456 (1991)
17. O.M. Ogunsola, N. Berkowitz, Removal of heterocyclic-S and heterocyclic-N from oil precursors by supercritical water. Fuel **74**, 1485–1490 (1995)

18. K.M. Benjamin, P.E. Savage, Hydrothermal reactions of methylamine. J. Supercrit. Fluids **31**, 301–311 (2004)
19. K.M. Benjamin, P.E. Savage, Detailed chemical kinetic modeling of methylamine in supercritical water. Ind. Eng. Chem. Res. **44**, 9785–9793 (2005)
20. J.M.L. Penninger, R.J.A. Kersten, H.C.L. Baur, Reactions of diphenylether in supercritical water—mechanism and kinetics. J. Supercrit. Fluids **16**, 119–132 (1999)
21. J.M.L. Penninger, R.J.A. Kersten, H.C.L. Baur, Hydrolysis of diphenylether in supercritical water—effects of dissolved NaCl. J. Supercrit. Fluids **17**, 215–226 (2000)
22. B. Kuhlmann, E.M. Arnett, M. Siskin, H-D exchange in pinacolone by deuterium-oxide at high-temperature and pressure. J. Org. Chem. **59**, 5377–5380 (1994)
23. K.C. Park, H. Tomiyasu, Gasification reaction of organic compounds catalyzed by RuO_2 in supercritical water. Chem. Commun. **6**, 694–695 (2003)
24. M. Watanabe, H. Hirakoso, S. Sawamoto, T. Adschiri, K. Arai, Polyethylene conversion in supercritical water. J. Supercrit. Fluids **13**, 247–252 (1998)
25. A. Kruse, D. Meier, P. Rimbrecht, M. Schacht, Gasification of pyrocatechol in supercritical water in the presence of potassium hydroxide. Ind. Eng. Chem. Res. **39**, 4842–4848 (2000)
26. P. Corbo, F. Migliardini, Natural gas and biofuel as feedstock for hydrogen production on Ni catalysts. J. Nat. Gas Chem. **18**, 9–14 (2009)
27. I.G. Lee, M.S. Kim, S.K. Ihm, Gasification of glucose in supercritical water. Ind. Eng. Chem. Res. **41**, 1182–1188 (2002)
28. A.J. Byrd, K.K. Pant, R.B. Gupta, Hydrogen production from glycerol by reforming in supercritical water over Ru/Al_2O_3 catalyst. Fuel **87**, 2956–2960 (2008)
29. Z.Y. Ding, M.A. Frisch, L.X. Li, E.F. Gloyna, Catalytic oxidation in supercritical water. Ind. Eng. Chem. Res. **35**, 3257–3279 (1996)
30. B. Al-Duri, L. Pinto, N.H. Ashraf-Ball, R.C.D. Santos, Thermal abatement of nitrogen-containing hydrocarbons by non-catalytic supercritical water oxidation (SCWO). J. Mater. Sci. **43**, 1421–1428 (2008)
31. P. Dagaut, M. Cathonnet, J.C. Boettner, Chemical kinetic modeling of the supercritical-water oxidation of methanol. J. Supercrit. Fluids **9**, 33–42 (1996)
32. P.E. Savage, J. Yu, N. Stylski, E.E. Brock, Kinetics and mechanism of methane oxidation in supercritical water. J. Supercrit. Fluids **12**, 141–153 (1998)
33. M. Watanabe, M. Osada, H. Inomata, K. Arai, A. Kruse, Acidity and basicity of metal oxide catalysts for formaldehyde reaction in supercritical water at 673 K. Appl. Catal. A General **245**, 333–341 (2003)
34. Y. Ikushima, K. Hatakeda, O. Sato, T. Yokoyama, M. Arai, Acceleration of synthetic organic reactions using supercritical water: noncatalytic Beckmann and pinacol rearrangements. J. Am. Chem. Soc. **122**, 1908–1918 (2000)
35. Y. Ikushima, K. Hatakeda, O. Sato, T. Yokoyama, M. Arai, Structure and base catalysis of supercritical water in the noncatalytic benzaldehyde disproportionation using water at high temperatures and pressures. Angew. Chem. Int. Ed. **40**, 210–213 (2001)
36. B. Kuhlmann, E.M. Arnett, M. Siskin, Classical organic-reactions in pure superheated water. J. Org. Chem. **59**, 3098–3101 (1994)
37. X.D. Xu, M.J. Antal, D.G.M. Anderson, Mechanism and temperature-dependent kinetics of the dehydration of tert-butyl alcohol in hot compressed liquid water. Ind. Eng. Chem. Res. **36**, 23–41 (1997)
38. K. Chandler, F.H. Deng, A.K. Dillow, C.L. Liotta, C.A. Eckert, Alkylation reactions in near-critical water in the absence of acid catalysts. Ind. Eng. Chem. Res. **36**, 5175–5179 (1997)
39. Y. Tsujino, C. Wakai, N. Matubayashi, M. Nakahara, Noncatalytic Cannizzaro-type reaction of formaldehyde in hot water. Chem. Lett. **28**(4), 287–288 (1999)
40. S. Ramayya, A. Brittain, C. Dealmeida, W. Mok, M.J. Antal, Acid-catalyzed dehydration of alcohols in supercritical water. Fuel **66**, 1364–1371 (1987)
41. L. Bagnell, M. Bliese, T. Cablewski, C.R. Strauss, J. Tsanaktsidis, Environmentally benign procedures for the preparation and isolation of 3-methylcyclopent-2-en-1-one. Aust. J. Chem. **50**, 921–925 (1997)

42. P.T. Williams, J. Onwudili, Composition of products from the supercritical water gasification of glucose: a model biomass compound. Ind. Eng. Chem. Res. **44**, 8739–8749 (2005)
43. B.M. Kabyemela, T. Adschiri, R.M. Malaluan, K. Arai, Glucose and fructose decomposition in subcritical and supercritical water: detailed reaction pathway, mechanisms, and kinetics. Ind. Eng. Chem. Res. **38**, 2888–2895 (1999)
44. P.T. Williams, J. Onwudili, Subcritical and supercritical water gasification of cellulose, starch, glucose, and biomass waste. Energy Fuels **20**, 1259–1265 (2006)
45. D.H. Yu, M. Aihara, M.J. Antal, Hydrogen production by steam reforming glucose in supercritical water. Energy Fuels **7**, 574–577 (1993)
46. A.K. Goodwin, G.L. Rorrer, Conversion of glucose to hydrogen-rich gas by supercritical water in a microchannel reactor. Ind. Eng. Chem. Res. **47**, 4106–4114 (2008)
47. J.B. Gadhe, R.B. Gupta, Hydrogen production by methanol reforming in supercritical water: suppression of methane formation. Ind. Eng. Chem. Res. **44**, 4577–4585 (2005)
48. N. Boukis, V. Diem, W. Habicht, E. Dinjus, Methanol reforming in supercritical water. Ind. Eng. Chem. Res. **42**, 728–735 (2003)
49. Y. Matsumura, H. Nonaka, H. Yokura, A. Tsutsumi, K. Yoshida, Co-liquefaction of coal and cellulose in supercritical water. Fuel **78**, 1049–1056 (1999)
50. T. Yoshida, Y. Matsumura, Gasification of cellulose, xylan, and lignin mixtures in supercritical water. Ind. Eng. Chem. Res. **40**, 5469–5474 (2001)
51. M. Sasaki, T. Adschiri, K. Arai, Kinetics of cellulose conversion at 25 MPa in sub- and supercritical water. AIChE J. **50**, 192–202 (2004)
52. B.M. Kabyemela, M. Takigawa, T. Adschiri, R.M. Malaluan, K. Arai, Mechanism and kinetics of cellobiose decomposition in sub- and supercritical water. Ind. Eng. Chem. Res. **37**, 357–361 (1998)
53. F.L.P. Resende, S.A. Fraley, M.J. Berger, P.E. Savage, Noncatalytic gasification of lignin in supercritical water. Energy Fuels **22**, 1328–1334 (2008)
54. M.J. Antal, S.G. Allen, X.F. Dai, B. Shimizu, M.S. Tam, M. Gronli, Attainment of the theoretical yield of carbon from biomass. Ind. Eng. Chem. Res. **39**, 4024–4031 (2000)
55. P. D'Jesús, N. Boukis, B. Kraushaar-Czarnetzki, E. Dinjus, Influence of process variables on gasification of corn silage in supercritical water. Ind. Eng. Chem. Res. **45**, 1622–1630 (2006)
56. P. D'Jesús, N. Boukis, B. Kraushaar-Czarnetzki, E. Dinjus, Gasification of corn and clover grass in supercritical water. Fuel **85**, 1032–1038 (2006)
57. P. D'Jesús, C. Artiel, N. Boukis, B. Kraushaar-Czarnetzki, E. Dinjus, Influence of educt preparation on gasification of corn silage in supercritical water. Ind. Eng. Chem. Res. **44**, 9071–9077 (2005)
58. D. Yamaguchi, P.J. Sanderson, S. Lim, L. Aye, Supercritical water gasification of Victorian brown coal: experimental characterisation. Int. J. Hydrogen Energy **34**, 3342–3350 (2009)
59. E. Gasafi, L. Meyer, L. Schebek, Exergetic efficiency and options for improving sewage sludge gasification in supercritical water. Int. J. Energy Res. **31**, 346–363 (2007)
60. E. Gasafi, M.-Y. Reinecke, A. Kruse, L. Schebek, Economic analysis of sewage sludge gasification in supercritical water for hydrogen production. Biomass Bioenerg. **32**, 1085–1096 (2008)
61. V. Sricharoenchaikul, Assessment of black liquor gasification in supercritical water. Biores. Technol. **100**, 638–643 (2009)
62. P. Azadi, A.A. Khodadadi, Y. Mortazavi, R. Farnood, Hydrothermal gasification of glucose using Raney nickel and homogeneous organometallic catalysts. Fuel Process. Technol. **90**, 145–151 (2009)
63. I.-G. Lee, S.-K. Ihm, Catalytic gasification of glucose over Ni/activated charcoal in supercritical water. Ind. Eng. Chem. Res. **48**, 1435–1442 (2009)
64. A. Yamaguchi, N. Hiyoshi, O. Sato, M. Osada, M. Shirai, Lignin gasification over supported ruthenium trivalent salts in supercritical water. Energy Fuels **22**, 1485–1492 (2008)
65. A. Sharma, I. Saito, H. Nakagawa, K. Miura, Effect of carbonization temperature on the nickel crystallite size of a Ni/C catalyst for catalytic hydrothermal gasification of organic compounds. Fuel **86**, 915–920 (2007)

66. M. Osada, N. Hiyoshi, O. Sato, K. Arai, M. Shirai, Effect of sulfur on catalytic gasification of lignin in supercritical water. Energy Fuels **21**, 1400–1405 (2007)
67. M. Osada, N. Hiyoshi, O. Sato, K. Arai, M. Shirai, Reaction pathway for catalytic gasification of lignin in presence of sulfur in supercritical water. Energy Fuels **21**, 1854–1858 (2007)
68. M. Osada, O. Sato, M. Watanabe, K. Arai, M. Shirai, Water density effect on lignin gasification over supported noble metal catalysts in supercritical water. Energy Fuels **20**, 930–935 (2006)
69. J.B. Gadhe, R.B. Gupta, Hydrogen production by methanol reforming in supercritical water: catalysis by in-situ-generated copper nanoparticles. Int. J. Hydrogen Energy **32**, 2374–2381 (2007)
70. T. Sato, T. Furusawa, Y. Ishiyama, H. Sugito, Y. Miura, M. Sato, N. Suzuki, N. Itoh, Effect of water density on the gasification of lignin with magnesium oxide supported nickel catalysts in supercritical water. Ind. Eng. Chem. Res. **45**, 615–622 (2006)
71. G.J. DiLeo, P.E. Savage, Catalysis during methanol gasification in supercritical water. J. Supercrit. Fluids **39**, 228–232 (2006)
72. S.R.A. Kersten, B. Potic, W. Prins, W.P.M. Van Swaaij, Gasification of model compounds and wood in hot compressed water. Ind. Eng. Chem. Res. **45**, 4169–4177 (2006)
73. X.H. Hao, L.J. Guo, X.M. Zhang, Y. Guan, Hydrogen production from catalytic gasification of cellulose in supercritical water. Chem. Eng. J. **110**, 57–65 (2005)
74. M.B.G. Jarana, J. Sáanchez-Oneto, J.R. Portela, E.N. Sanz, E.J.M. de la Ossa, Supercritical water gasification of industrial organic wastes. J. Supercrit. Fluids **46**, 329–334 (2008)
75. M. Watanabe, H. Inomata, M. Osada, T. Sato, T. Adschiri, K. Arai, Catalytic effects of NaOH and ZrO_2 for partial oxidative gasification of n-hexadecane and lignin in supercritical water. Fuel **82**, 545–552 (2003)
76. J.A. Onwudili, A.R. Lea-Langton, A.B. Ross, P.T. Williams, Catalytic hydrothermal gasification of algae for hydrogen production: composition of reaction products and potential for nutrient recycling. Biores. Technol. **127**, 72–80 (2013)
77. M.J. Sheikhdavoodi, M. Almassi, M. Ebrahimi-Nik, A. Kruse, H. Bahrami, Gasification of sugarcane bagasse in supercritical water; evaluation of alkali catalysts for maximum hydrogen production. J. Energy Inst. **88**, 450–458 (2015)
78. G. Caputo, M. Dispenza, P. Rubio, F. Scargiali, G. Marotta, A. Brucato, Supercritical water gasification of microalgae and their constituents in a continuous reactor. J. Supercrit. Fluids **118**, 163–170 (2016)
79. A. Sinag, A. Kruse, V. Schwarzkopf, Key compounds of the hydropyrolysis of glucose in supercritical water in the presence of K_2CO_3. Ind. Eng. Chem. Res. **42**, 3516–3521 (2003)
80. A. Sinag, A. Kruse, J. Rathert, Influence of the heating rate and the type of catalyst on the formation of key intermediates and on the generation of gases during hydropyrolysis of glucose in supercritical water in a batch reactor. Ind. Eng. Chem. Res. **43**, 502–508 (2004)
81. D.S. Gokkaya, M. Saglam, M. Yilksel, L. Ballice, Supercritical water gasification of phenol as a model for plant biomass. Int. J. Hydrogen Energy **40**, 11133–11139 (2015)
82. M. Gong, S. Nanda, M.J. Romero, W. Zhu, J.A. Kozinski, Subcritical and supercritical water gasification of humic acid as a model compound of humic substances in sewage sludge. J. Supercrit. Fluids **119**, 130–138 (2017)
83. J. Li, L. Pan, M. Suvarna, X. Wang, Machine learning aided supercritical water gasification for H_2-rich syngas production with process optimization and catalyst screening. Chem. Eng. J. **426**, 131285 (2021)
84. C. Cao, L. Guo, J. Yin, H. Jin, W. Cao, Y. Jia, X. Yao, Supercritical water gasification of coal with waste black liquor as inexpensive additives. Energy Fuels **29**, 384–391 (2015)
85. D.C. Elliott, Catalytic hydrothermal gasification of biomass. Biofuels Bioprod Bioref Biofpr **2**, 254–265 (2008)
86. D.C. Elliott, L.J. Sealock, E.G. Baker, Chemical processing in high-pressure aqueous environments. 2. Development of catalysts for gasification. Indus. Eng. Chem. Res. **32**, 1542–1548 (1993)
87. M.H. Waldner, F. Vogel, Renewable production of methane from woody biomass by catalytic hydrothermal gasification. Ind. Eng. Chem. Res. **44**, 4543–4551 (2005)

88. T. Yoshida, Y. Oshima, Y. Matsumura, Gasification of biomass model compounds and real biomass in supercritical water. Biomass Bioenerg. **26**, 71–78 (2004)
89. T. Furusawa, T. Sato, M. Saito, Y. Ishiyama, M. Sato, N. Itoh, N. Suzuki, The evaluation of the stability of Ni/MgO catalysts for the gasification of lignin in supercritical water. Appl. Catal. A General **327**, 300–310 (2007)
90. T. Furusawa, T. Sato, H. Sugito, Y. Miura, Y. Ishiyama, M. Sato, N. Itoh, N. Suzuki, Hydrogen production from the gasification of lignin with nickel catalysts in supercritical water. Int. J. Hydrogen Energy **32**, 699–704 (2007)
91. T. Minowa, F. Zhen, T. Ogi, Cellulose decomposition in hot-compressed water with alkali or nickel catalyst. J. Supercrit. Fluids **13**, 253–259 (1998)
92. T. Minowa, T. Ogi, Hydrogen production from cellulose using a reduced nickel catalyst. Catal. Today **45**, 411–416 (1998)
93. G.J. Dileo, M.E. Neff, P.E. Savage, Gasification of guaiacol and phenol in supercritical water. Energy Fuels **21**, 2340–2345 (2007)
94. H. Nakagawa, A. Namba, M. Böhlmann, K. Miura, Hydrothermal dewatering of brown coal and catalytic hydrothermal gasification of the organic compounds dissolving in the water using a novel Ni/carbon catalyst. Fuel **83**, 719–725 (2004)
95. S. Li, B. Zhu, W. Wang, H. Zhang, Q. Li, Efficient and stable supercritical-water-synthesized Ni-based catalysts for supercritical water gasification. J. Supercrit. Fluids **160**, 104790 (2020)
96. K. Khandelwal, A.K. Dalai, Catalytic supercritical water gasification of canola straw with promoted and supported nickel-based catalysts. Molecules **29**, 911 (2024)
97. D.C. Elliott, T.R. Hart, G.G. Neuenschwander, Chemical processing in high-pressure aqueous environments. 8. Improved catalysts for hydrothermal gasification. Indus. Eng. Chem. Res. **45**, 3776–3781 (2006)
98. A.J. Byrd, K.K. Pant, R.B. Gupta, Hydrogen production from glucose using Ru/Al_2O_3 catalyst in supercritical water. Ind. Eng. Chem. Res. **46**, 3574–3579 (2007)
99. M. Osada, O. Sato, K. Arai, M. Shirai, Stability of supported ruthenium catalysts for lignin gasification in supercritical water. Energy Fuels **20**, 2337–2343 (2006)
100. J. Yu, Q. Guan, Q. Chen, P. Ning, J. Gu, X. Huang, Y. Shi, R. Miao, Catalytic gasification of phenol in supercritical water with Ru/graphitized carbon black. RSC Adv. **6**, 75512–75521 (2016)
101. Y. Matsumura, T. Kokabu, S. Inoue, T. Charinpanitkul, S. Kuboon, Effect of single-walled carbon nanotube catalysts on hydrothermal pretreatment of cellulose. J. Jpn. Petrol. Inst. **61**, 199–204 (2018)
102. M.A.M. Ali, S. Inoue, Y. Matsumura, Gasification characteristics of carbon nanotube in supercritical water. J. Supercrit. Fluids **182**, 105532 (2022)
103. M.A.M. Ali, S. Inoue, Y. Matsumura, Carbon nanotube as catalyst support in supercritical water. J. Supercrit. Fluids **190**, 105755 (2022)
104. M. Osada, T. Sato, M. Watanabe, T. Adschiri, K. Arai, Low-temperature catalytic gasification of lignin and cellulose with a ruthenium catalyst in supercritical water. Energy Fuels **18**, 327–333 (2004)
105. T. Sato, M. Osada, M. Watanabe, M. Shirai, K. Arai, Gasification of alkylphenols with supported noble metal catalysts in supercritical water. Ind. Eng. Chem. Res. **42**, 4277–4282 (2003)
106. M.H. Waldner, F. Krumeich, F. Vogel, Synthetic natural gas by hydrothermal gasification of biomass selection procedure towards a stable catalyst and its sodium sulfate tolerance. J. Supercrit. Fluids **43**, 91–105 (2007)
107. R.D. Cortright, R.R. Davda, J.A. Dumesic, Hydrogen from catalytic reforming of biomass-derived hydrocarbons in liquid water. Nature **418**, 964–967 (2002)
108. R.R. Davda, J.W. Shabaker, G.W. Huber, R.D. Cortright, J.A. Dumesic, Aqueous-phase reforming of ethylene glycol on silica-supported metal catalysts. Appl. Catal. B Environ. **43**, 13–26 (2003)
109. Y. Karakus, F. Aynaci, E. Kipçak, M. Akgün, Hydrogen production from 2-propanol over Pt/Al_2O_3 and Ru/Al_2O_3 catalysts in supercritical water. Int. J. Hydrogen Energy **38**, 7298–7306 (2013)

110. M. Dejhosseini, T. Aida, M. Watanabe, S. Takami, D. Hojo, N. Aoki, T. Arita, A. Kishita, T. Adschiri, Catalytic cracking reaction of heavy oil in the presence of cerium oxide nanoparticles in supercritical water. Energy Fuels **27**, 4624–4631 (2013)
111. H. Boucard, M. Watanabe, S. Takami, E. Weiss-Hortala, R. Barna, T. Adschiri, Beneficial use of CeO_2 nanocatalyst for black liquor conversion under sub and supercritical conditions. J. Supercrit. Fluids **105**, 66–76 (2015)
112. K. Tomishige, T. Sakaihori, Y. Ikeda, K. Fujimoto, A novel method of direct synthesis of dimethyl carbonate from methanol and carbon dioxide catalyzed by zirconia. Catal. Lett. **58**, 225–229 (1999)
113. M. Watanabe, H. Inomata, K. Arai, Catalytic hydrogen generation from biomass (glucose and cellulose) with ZrO_2 in supercritical water. Biomass Bioenerg. **22**, 405–410 (2002)
114. C. Cao, Y. Xie, L. Li, W. Wei, H. Jin, S. Wang, W. Li, Supercritical water gasification of lignin and cellulose catalyzed with co-precipitated CeO_2–ZrO_2. Energy Fuels **35**, 6030–6039 (2021)
115. X.D. Xu, Y. Matsumura, J. Stenberg, M.J. Antal, Carbon-catalyzed gasification of organic feedstocks in supercritical water. Ind. Eng. Chem. Res. **35**, 2522–2530 (1996)
116. Y. Matsumura, X. Xu, M.J. Antal, Gasification characteristics of an activated carbon in supercritical water. Carbon **35**, 819–824 (1997)

Bio-oil Production in Hydrothermal Liquefaction of Biomass and Upgrading of Biocrude

The development and application of clean and renewable energy sources is important because the overconsumption of fossil fuels has led to a number of global problems, especially environmental pollution. Therefore, renewable energy sources have attracted much attention, among which biomass energy has been widely recognized by researchers. Organic components in biomass can be extracted and converted into fuel by physical and chemical methods [1]. As an alternative to fossil fuels, liquid biofuels are emerging as the only option, as it can mitigate global warming by achieving zero carbon dioxide emissions through direct and indirect land-use changes in the life cycle of biofuels, as well as reducing energy import dependency and increasing energy security [2–5].

Feedstocks for the production of liquid biofuels have evolved from food crops to lignocellulose (straw and trees) to algae. The use of food crops and lignocellulose is limited because they consume food, arable land and water resources, while algae have the greatest potential for biofuel production [6–8]. In contrast to terrestrial plants, firstly, the growth of algae is not competitive for water resources and arable land [9–11]. Secondly, algae have a higher photosynthetic efficiency and therefore grow faster and have a higher potential to fix carbon dioxide [12, 13]. Thirdly, algae have a higher lipid content, which means that their biofuel yield is higher. With these advantages, the production of liquid biofuels from algae has proven to be economically viable and has become an energy source that has been heavily exploited [7]. Therefore, this chapter will focus on studies related to the production of oil and grease from algal biomass via subcritical hydrothermal reactions.

Hydrothermal conversion is a process in which biomass undergoes various thermochemical pathways in hot compressed water under different conditions, leading to the formation of solid, liquid, and gaseous fuels [14]. Depending on the range of reaction temperatures and pressures, hydrothermal conversion processes can be classified into HTC [15], HTL [16], and hydrothermal gasification [17].

Y. Guo et al., *Hydrothermal Processing of Biomass for Hydrogen and Bio-oil Production*, https://doi.org/10.1007/978-981-96-6336-1_4

HTL is the process of converting biomass, with or without a catalyst, into gaseous, solid, biocrude, and aqueous phases in hot compressed water at 300–350 °C and 5–20 MPa; organic solvents are then added to extract and collect the oil. This method has received much attention due to its high energy recovery and product potential for replacing fossil fuels. Based on the different organic components in biomass (i.e., proteins, non-cellulosic carbohydrates, lipids, and lignocellulose), the major chemical reactions during high-temperature pyrolysis of biomass can be classified into four categories. Conversion of biomass to biocrude oil involves hydrolysis, degradation, decarboxylation, deamination, deamination, recombination, and other chemical reactions [18]. HTL is a promising technology for the conversion of algae into primary liquid biofuels with many advantages. First, it can directly convert wet algae into biocrude oil, thus avoiding additional energy consumption during the drying process [19, 20]. Secondly, the yield of biocrude oil from HTL is higher than that of other methods (e.g., pyrolysis) because not only lipids, but also proteins and carbohydrates are mostly converted to biocrude oil in algae [21, 22]. Thirdly, nutrients (e.g., N, P, and K) in algae are converted to the corresponding water-soluble salts or acids in the HTL and can be reused for algal growth by means of aqueous phase recycling and utilization [23]. Many factors such as catalyst, residence time, reaction temperature, and solvent can change the chemical reaction process, the extent of the chemical reaction and the composition of the biocrude oil, thus affecting the yield and quality of the biocrude oil.

Thermal liquefaction of biomass produces biocrude oil with high viscosity, high acidity, high heteroatom content, and small HHV [11]. These characteristics make biocrude oil inferior to fossil fuels in terms of quality and environmental friendliness, and therefore, biocrude oil cannot be directly used as a fossil fuel [24]. Therefore, further upgrading is needed to improve the characteristics and composition of biocrude oils. There are many methods for upgrading biocrude oils, which can be categorized as thermochemical, biochemical, and physicochemical [25]. Catalytic hydrogenation upgrading—a thermochemical treatment process—is currently the preferred method for upgrading biocrude oil; it treats biocrude oil mainly through hydrogenation, desulfurization, denitrification and deoxygenation.

Currently, in order to understand the reaction process of thermal liquefaction of biomass and catalytic hydrogenation and upgrading of biocrude oil, many researchers have designed different experimental components. They explore the conversion process and results by changing the feedstock, reaction conditions, and detection methods, analyze the possible chemical reactions, and reveal the effects of parameters on chemical reactions [26–29]. In this paper, we comprehensively summarize the chemical reactions and their influencing factors during the conversion of proteins, non-cellulosic carbohydrates, lipids, and lignocellulose during high-temperature pyrolysis of biomass, mainly based on the research results in the last decade. We have also comprehensively analyzed the chemical reactions of hydrogenation, desulfurization, denitrification, and deoxygenation and their corresponding influencing factors during catalytic hydrogenation upgrading of biocrude oil. The information documented in this chapter is important for the improvement of reaction conditions and processing steps in thermal liquefaction of biomass and catalytic hydrogenation upgrading of biocrude oil.

1 Chemical Reactions in Biomass HTL

The components of biomass are mainly proteins, lipids, non-cellulosic carbohydrates and lignocellulose [2]. Under high-temperature pyrolysis conditions, these substances undergo complex chemical reactions and form different phases through complex transformation pathways [30]; therefore, experimental studies usually focus on these four types of organic substances. Researchers design experiments on typical model compounds and actual biomass and then analyze and speculate on the possible reaction pathways during high-temperature pyrolysis of biomass [31–35]. First, macromolecules are hydrolyzed, i.e., proteins are hydrolyzed to amino acids, lipids are hydrolyzed to glycerol and long-chain fatty acids, non-cellulosic carbohydrates are hydrolyzed to reducing and non-reducing sugars, and lignocellulose is hydrolyzed to glucose and alcohol monomers. Secondly, the generated small molecules are chemically unstable and are prone to decomposition or decarboxylation, deamination, dehydration, reforming, polymerization, etc., to form carboxylic acids, alkanes, olefins, ketones, nitrogen-containing compounds, and gases. In this process, the generation of OH- and HCOO- can promote the decomposition of hydrocarbon macromolecules [36]. Thirdly, some intermediates are further reformed, decomposed, dehydrated, etc., to produce stable molecules which are no longer transformed and are present in the individual phases [37, 38]. A more detailed analysis of biomass HTL and the corresponding chemical reactions are provided below.

At present, there are three main methods of producing biocrude oil, namely transesterification, pyrolysis, and pyrolysis. Transesterification refers to the process whereby triglycerides (used as feedstock) are reacted with methanol at 60 °C and atmospheric pressure, turning successively into diglycerides and monoglycerides under the action of a base catalyst, and finally forming glycerol and fatty acid methyl esters. Pyrolysis is the process of thermal decomposition of biomass in an inert atmosphere at 300–900 °C to produce a combustible gas or bio-oil [39]. Compared to these methods, HTL has the following advantages [40–43]. First, HTL is an energy-efficient process because it avoids drying of biomass and reduces energy consumption, which accounts for more than 75% of total energy consumption [44]. Second, it produces biocrude oil with low water content and high HHV value. Third, heat loss is reduced because the state of the reaction medium does not change. Fourth, the biomass utilization is high because HTL not only extracts lipids from biomass but also converts other components into biocrude oil. Therefore, HTL has great potential for producing biocrude oil from biomass [45].

The critical temperature and critical pressure of water are 374.15 °C and 22.13 MPa, respectively [46]; thus, high-temperature pyrolysis of biomass occurs in the subcritical region of water. At ambient temperatures, water is a polar substance and therefore nonpolar substances have much lower solubility. As the temperature increases, the properties of water change considerably [47], such as a decrease in dielectric constant, an increase in dissociation constant and an increase in solubility, thus providing a suitable environment for the degradation of organic molecules [48, 49]. Water plays three roles in the conversion process, i.e., a solvent that provides a reaction environment for the reactants, a reactant that participates in the reaction to increase the yield of biocrude oil, and a catalyst that facilitates the associated chemical reactions. Under such conditions, more organic substances can be decomposed into small molecules and repolymerized into biocrude oil, thus increasing the yield of biocrude oil [50].

Table 1 shows the effect of operating conditions on the biocrude oil yield of biomass high-temperature thermal liquefaction. It is clear that a wide range of biomasses can be used as feedstock for high-temperature thermal liquefaction, such as soybean, corn stover, algae, municipal sludge, and straw [51]. Different compositions and properties of biomass result in various yields. The yield of biocrude can reach up to 51.6 wt% when *S. platensis* is used as the feedstock in the presence of Na_2CO_3 as the catalyst, while the minimal value is only 17.8% with *D. tertiolecta* as the feedstock and SO_3H-Co/SBA-15 catalyst. Overall, their average value is approximately 32.4%, with an approximate average value of 32.4 wt%. The typical operating parameters are within 270–350 °C, 20–60 min, and 5–10 wt% catalyst. There is no doubt that catalysts have a significant impact on the yield and quality of biocrude oil in high-temperature liquefaction reactions by inhibiting side reactions, reducing solids formation, facilitating heteroatom removal, and improving conversion. Catalysts commonly used in high-temperature liquefaction can be categorized into homogeneous catalysts (e.g., $NaCO_3$, KOH, and CH_3COOH) and heterogeneous catalysts (e.g., Pd/C, Pt/C, Ru/C, and NiO). Alkaline catalysts promote the decomposition of large molecules into small molecules and reduce coke formation, thereby greatly increasing the yield of biocrude oil. In particular, such catalysts are more effective in increasing the conversion of carbohydrates, which has been demonstrated in many articles. However, lower conversion rates have been reported for proteins and lipids. [52].

Ross et al. [54] catalyzed HTL of microalgae using alkali and organic acids and found that these catalysts increased the yield of biocrude oil in the order of Na_2CO_3 > CH_3COOH > KOH > HCOOH. Nazari et al. [67] performed catalytic HTL of birch sawdust and confirmed that the effect of catalysts on the yield of biocrude oil showed a KOH > Na_2CO_3 > colemanite > HT/KOH > HT > $FeSO_4$ > MgO trend. An alkaline catalyst may induce saponification reactions and increase the content of solid matter [71]. Acidic catalysts may reduce the viscosity and composition of high-boiling compounds, thus improving the quality of biocrude oil; however, they are not effective in removing heteroatoms [72]. There are many types of heterogeneous catalysts and each catalyst has a different effect on the biomass high-temperature liquefaction process. Pt/C and Pt/Al_2O_3 have a better effect on the removal of O and N and the increase of H and C, while Ni/SiO_2–Al_2O_3 has a better effect on the desulphurization. Pd/C has a significant effect on the increase of biocrude oil production [73]. Currently, heterogeneous catalysts are of great interest to researchers due to their relatively easy recovery after HTL.

Table 1 Effects of operating conditions, catalysts, and biomass types on the biocrude yields of biomass HTL

Biomass	Catalyst	Major conditions	Yield (wt%)	References
S. platensis	Na_2CO_3	350 °C, 60 min, 5 wt% catalyst	51.6	[36]
	$Ca_3(PO_4)_2$	350 °C, 60 min, 5 wt% catalyst	34.5	[36]
	NiO	350 °C, 60 min, 5 wt% catalyst	30.2	[36]
	CeO_2/HZSM-5	300 °C, 5 wt% catalyst	49.7	[53]
Spirulina	Na_2CO_3	350 °C, 60 min, 95 wt% catalyst	25.6	[54]
	KOH	350 °C, 60 min, 50 wt% catalyst	21.7	[54]
	CH_3COOH	350 °C, 60 min, 54 wt% catalyst	24.3	[54]
	CeO_2	250 °C, 30 min, 0.2 g catalyst	26.0	[55]
D. tertiolecta	Co/CNTs	320 °C, 30 min, 5 wt% catalyst	40.2	[56]
	Ni/CNTs	320 °C, 30 min, 5 wt% catalyst	40.0	[56]
	Co/SBA-15	264 °C, 30 min, 5 wt% catalyst	24.1	[57]
	SO_3H-Co/SBA-15	260 °C, 30 min, 10 wt% catalyst	17.8	[58]
	Co/CNTs	320 °C, 30 min, 10 wt% catalyst	40.3	[56]
Ulva fasciata	Ni/TiO_2	270 °C, 30 min, 10 wt% catalyst	42.4	[46]
	Fe/TiO_2	270 °C, 30 min, 10 wt% catalyst	29.1	[46]
	Co/TiO_2	270 °C, 30 min, 10 wt% catalyst	32.7	[46]
Barley straw	K_2CO_3	300 °C, 15 min, 10 wt% catalyst	34.9	[59]
Swine manure	–[a]	340 °C, 30 min	25.6	[60]
Cattle manure	NaOH	310 °C, 15 min, 0.5 mol catalyst	38.5	[61]
Sludge and waste newspaper	–[a]	350 °C, 20 min	26.2	[62]
Peat	Iron ore	400 °C, 120 min, 20 wt% catalyst	40.0	[63]

(continued)

Table 1 (continued)

Biomass	Catalyst	Major conditions	Yield (wt%)	References
Sewage sludge	K_2CO_3	350 °C, 15 min, 2 wt% catalyst	26.8	[64]
	–[a]	340 °C, 10 min	22.9	[65]
	–[a]	500 °C, 1 min	28.0	[66]
Birch wood Sawdust	KOH	300 °C, 30 min, 5 wt% catalyst	39.5	[67]
	K_2CO_3	300 °C, 30 min, 5 wt% catalyst	38.5	[67]
Paulownia	–[a]	300 °C, 10 min	27.0	[68]
Pinewood sawdust	K_2CO_3	300 °C, 120 min, 10 wt% catalyst	30.8	[69]
Oil palm fruit	$ZnCl_2$	290 °C, 45 min, 10 wt% catalyst	19.6	[70]
	Na_2CO_3	290 °C, 45 min, 10 wt% catalyst	43.3	[70]

Figure 1 shows the effect of typical operating conditions on the product distribution of high-temperature pyrolysis of biomass. As shown in Fig. 1a–d, the reaction temperature has a significant effect on biocrude yield and product distribution. In general, as the temperature increases, the solid product content decreases and more material is decomposed and converted to other phases. As the temperature increases, the biocrude yield tends to increase and then decrease, with the maximum values occurring in the range of 300–350 °C. However, at temperatures of 400 °C, the biocrude yield decreases. However, the gas yield continues to increase at temperatures up to 400 °C, indicating that some of the substances in the biocrude are broken down into smaller molecules to form gas. In addition, the optimal thermal liquefaction temperature for biomass is also related to the type of biomass. For example, Tian et al. [74] conducted pyrolization experiments on *Spirulina*: the highest yields of biocrude oil occurred at about 250 °C.

As shown in Fig. 1e–h, the residence time also affects the yield and product distribution of the biofluid, but the yield of each phase changes less with increasing residence time compared to the reaction temperature. The optimum residence time for biomass high-temperature catalytic liquids is 20–60 min. Longer residence time is favorable for biomass conversion because the content of solid products is reduced. However, the biocrude yield showed a decreasing trend, but the decreasing trend was not significant with increasing residence time.

Figure 1i–l shows the relationship between biocrude yield and the ratio of biomass to water. Obviously, the smaller ratio of biomass to water is more suitable for high-temperature liquefaction of biomass. As the ratio increases, the biocrude oil yield decreases rapidly while the solid yield increases significantly. This may be due to the lack of water resulting in an inadequate reaction environment, and therefore, most

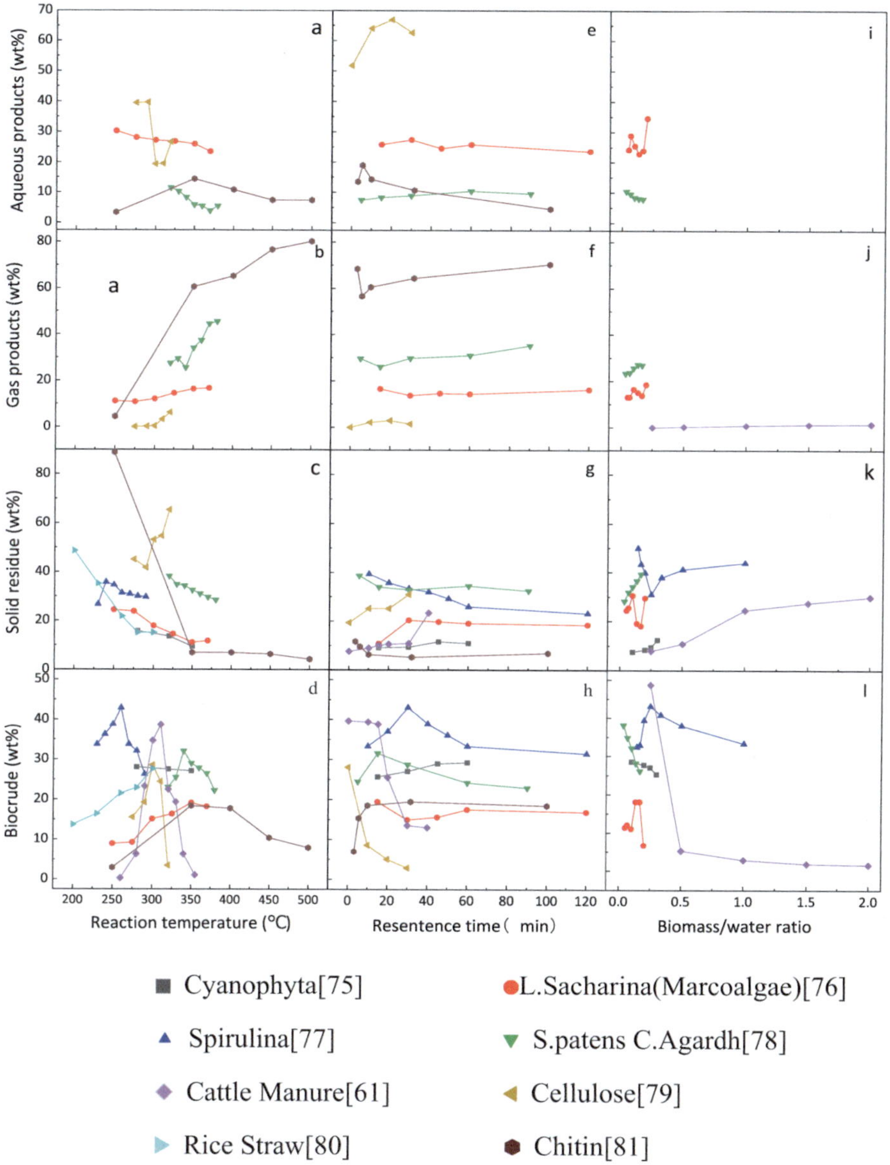

Fig. 1 Effects of typical operating conditions on product distribution of biomass HTL

of the biomass did not react or produce coke. Currently, researchers have attempted to find suitable reaction conditions (e.g., residence time, reaction temperature, and microalgae loading) and catalysts in biomass HTLs [71, 82] to improve the yield and quality of biocrude oil. This helps to make HTL more economical for industrial applications.

Typically, high-temperature pyrolysis of biomass takes place in liquid water. This means that the reaction pressure is higher than the saturation pressure of water at the corresponding reaction temperature. Therefore, most experiments explore the effect of temperature due to the correlation between reaction temperature and pressure. It is worth noting that the type and pressure of the added gas affects the HTL conversion process. Nitrogen and hydrogen are the main gases used in the experiments. Nitrogen pressure does not have a significant effect on the hydrothermal conversion of biomass because the product distribution remains essentially the same under different pressure conditions [83]. However, the biocrude yield increases with the initial addition of hydrogen pressure. As the hydrogen pressure continues to increase, the biocrude yield decreases. This may be due to hydrodeoxygenation or cleavage of some of the substances in the biocrude oil into smaller molecules.

The main role of solvents is to provide active hydrogen, facilitate feedstock decomposition, and even participate in reactions to influence the HTL process. This can greatly increase the hydrogen content of biocrude and improve the yield of biocrude [84]. Hydrogen bonds in methanol or ethanol can break down to form hydrogen radicals, thus preventing unstable intermediates from forming compounds that are more difficult to break down. This effect is enhanced with increasing solvent content [85]. In addition, algae swell in the presence of solvents; this increases the internal surface area and weakens the interconnections between the polymers, thus facilitating cleavage of macromolecules or recombination of fragments [86]. The solvents involved in the reaction are mainly alcohols, which can be used as esterification reactants to increase the yield of biocrude oil. Mixtures of multiple liquids are also used as co-solvents in experiments. The synergistic effect between different solvents allows more substances to be dissolved into the system to participate in the reaction and increase the hydrogen content, 81 thus improving the quality and yield of biocrude oil.

A reducing atmosphere is the addition of hydrogen or other reducing substances that affect conversion and extraction. For example, hydrogenation enhances the decarboxylation of fatty acids, inhibits amide formation, or hydrogenates them to produce hydrocarbons. In this process, fatty acids and fatty acid amides are reduced, while linear and branched chain hydrocarbons in lipids are increased. Hydrogenation during HTL also facilitates the formation of cyclic N-containing compounds and polycyclic aromatic compounds, and improves decarboxylation and dihydroxylation [87]. In addition, increasing the ratio of water to feedstock promotes the formation of more water-soluble substances in sugars [88], which may be related to the addition of water to provide more hydrogen. Feedstock structure also affects the HTL process. Under the same conditions, the biocrude yield of sunflower oil was lower than that of oleic acid, and the effect of reaction time on the biocrude yield varied due to changes in feedstock structure [89].

1.1 Lipids

Lipids are the main components of algae that are converted to biocrude oil after HTL. Lipids in algae contain saturated and unsaturated fatty acids known as triacylglycerols. Lipid content directly affects the yield of HTL bio-oil. Higher lipid algae produce higher bio-oil yields compared to lower lipid algae [90]. In order of contribution and trend to HTL oil production, the order is lipid > protein > carbohydrate [19]. Triacylglycerols can be rapidly hydrolyzed non-catalytically to free fatty acids in subcritical water because of their very low dielectric constants and complete miscibility with high-temperature water. Myristic, palmitoleic, cetanoic, oleic, and stearic acids are mainly hydrolyzed from triacylglycerols in *Nannochloropsis* sp. [91]. Algae also contains cholesterol, thus HTL bio-oil contains cholesterol and its derivatives [91, 92].

Figure 2 shows the possible chemical reaction pathways of lipids during HTL. Due to their low dielectric constants, lipids can undergo rapid non-catalytic hydrolysis under HTL conditions to produce glycerol and higher fatty acids. On the one hand, some of the higher fatty acids are converted to long-chain hydrocarbons by decarboxylation and then to alkanes by hydrogenation. On the other hand, some of them are converted to alcohols by deoxygenation, and then alcohols are esterified with fatty acids to form fatty acid esters. Some fatty acids react with ammonia produced during amino acid conversion to form fatty acid amides [93]. Overall, the above components constitute the biocrude oil produced from high-temperature pyrolysis of biomass. Meanwhile, glycerol can be converted to water-soluble substances (e.g., methanol, ethanol, and propionaldehyde) and gaseous products (e.g., CO, CO_2, and H_2) via complex reactions such as degradation and condensation.

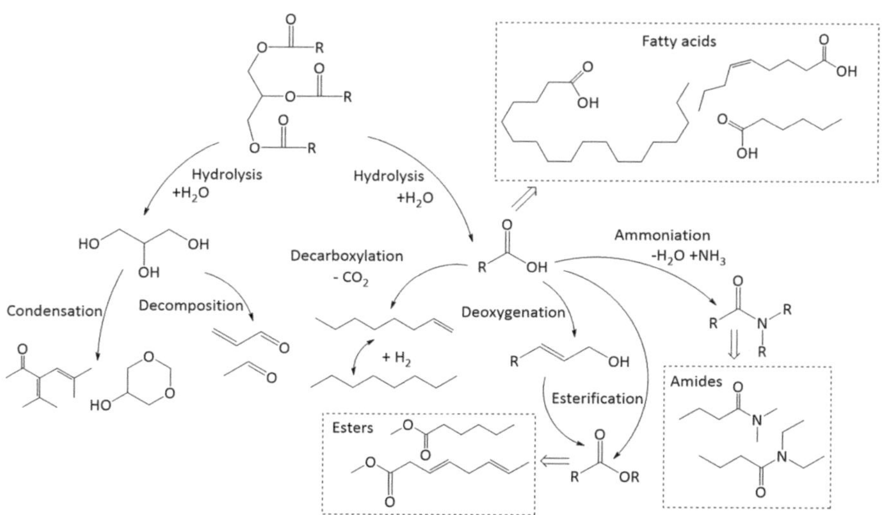

Fig. 2 Possible chemical reaction pathways of lipids in HTL [1, 94, 95]

Lipids are the main source of light components in biofuels and are not converted to solid phase at all during pyrolysis at high temperatures [96]. The contribution of lipids to the recombination fraction depends mainly on the temperature. At lower temperatures, the contribution of lipids is greater, but at higher temperatures, the contribution of lipids is similar to that of proteins. Ky Vo et al. [51] found that among the biochemical constituents of algae, lipids are the major component in the formation of biocrude oil at low temperatures. However, Valdez et al. [31]. reported that lipids play a major role in the formation of biocrude oil at temperatures in the range of 350–400 °C. Therefore, the effect of temperature on lipid conversion rate still needs to be further fully explored. In general, increasing the temperature promotes the removal of hydroxyl groups and the conversion of fatty acids to fatty acid amides or other substances [97]. For example, more amides can be formed by the condensation of weakly basic amines and fatty acids at higher temperatures [98].

Reducing atmosphere refers to the addition of hydrogen or other reducing substances, which affects the conversion and extraction of lipids. Li et al. [18] reported that hydrogenation enhances the decarboxylation of fatty acids and inhibits the formation of amides, or hydrogenates them to form hydrocarbons. In this process, fatty acids and fatty acid amides decrease while linear and branched hydrocarbons in lipids increase. Co-solvents can also act as hydrogen donors to influence the HTL process. It can greatly increase the hydrogen content in biocrude oil and improve the yield of biocrude oil [84]. In addition, the material structure also affects the HTL process. Under the same conditions, the biocrude yield of sunflower oil is lower than that of oleic acid, and the effect of reaction time on the biocrude yield is different due to the different material structures [99].

Lipids are mostly triacylglycerols, consisting of free fatty acids bound to the glycerol backbone. Therefore, glycerol is a major intermediate in the transformation of algal HTL [100]. During the Colgate-Emery process, some unstable triacylglycerols are hydrolyzed to free fatty acids and glycerol [101]. Similarly, in microalgae HTL, lipids are hydrolyzed to produce glycerol, which is the second most water-soluble reaction product after acetic acid [23]. Glycerol decomposes to light alkanes, aldehydes, and alcohols at temperatures above 310 °C, and gas products are more pronounced at temperatures above 375 °C [102]. KOH and NaOH were found to promote the conversion of glycerol, in which lactic acid is the main product [103]. The coexistence of carboxylic acids and glycerol may promote the production of esters in HTL-generated algal bio-oil. Glycerol increases ester production by promoting esterification of organic acids in bio-oils [104].

1.2 Proteins and Amino Acids

Proteins are the main components of biomass and play an important role in life activities. Amino acids are monomers of proteins and consist of a central carbon atom and amino, carboxyl, hydrogen, and side chain R groups. Different amino acids have different side chain groups, which leads to differences in reactions. However, there are some similarities in terms of overall structure. Yang et al. [105] investigated the reaction process of crude protein or crude polysaccharide and their interactions. In general, proteins are hydrolyzed under HTL conditions and peptide bonds are broken into amino acids. Subsequently, they are decarboxylated, deaminated, repolymerized, and dehydrated to form organic acids, carbonic acids, nitrogen compounds, NH_3, CO_2, and other gases. The yield of biocrude is not linearly related to the content of crude protein or crude polysaccharide. At reaction temperatures of 220 or 260 °C, polysaccharides have a negative effect on the conversion of proteins to biocrude. However, at 300 °C, polysaccharides promote the conversion of proteins to biocrude oil [105].

Proteins are the main source of nitrogen in biocrude [106]. Figure 3 shows the potential chemical reaction pathways of proteins in HTL. First, the amide bond in the peptide chain is broken and combined with a water molecule to produce an amino acid. Then the following reactions occur [86, 107]. Some amino acids are decarboxylated (–COOH) into carbon dioxide and amine compounds. Some amine compounds then tend to form water-soluble N-containing compounds and partition in the aqueous phase. Therefore, the above process can effectively improve the quality of biocrude oil by removing O and N content. Of course, some amines will be further converted to water-insoluble nitrogen-containing compounds (e.g., pyrazines and pyrroles) in biocrude oil. This nitrogen fraction will be the focus of subsequent biocrude upgrading. Some amino acids are deaminated (–NH_2) to small carboxylic acids and NH_3, which are further decarboxylated or remove a molecule of water to form short-chain alkanes and olefins. In this process, nitrogen and oxygen are also efficiently removed, thus improving the quality of the biocrude. In addition, some of the NH_3 generated can react with fatty acids from lipid hydrolysis to form amides. Aromatic hydrocarbons, indoles, phenols, and their derivatives are the main degradation products of amino acids [105]. Experiments have confirmed that the deamination and decarboxylation reactions occur almost simultaneously. However, as the amino acid concentration decreases and the solution pH increases, the decarboxylation reaction becomes the dominant reaction throughout the process. Besides pH, temperature also affects the competition between deamination and decarboxylation reactions [32]. Amino acid molecules also polymerize with each other, dehydrating and isomerizing to form cyclic nitrogen-containing compounds. For example, two valine molecules polymerize to form 3,6-diisopropylpiperazine-2,5-dione, while a valine molecule and an alanine molecule polymerize to form 2,5-piperazine dione and 3-methyl-6-(1-methylethyl)piperazine-2,5-dione [94].

Figure 3 shows that proteins themselves undergo reactions such as hydrolysis, decarboxylation, and deamination; at the same time, they interact with other substances. Amino acids in proteins undergo martensitic reactions with reducing sugars in non-cellulosic carbohydrates to produce nitrogen-containing heterocyclic compounds (e.g., pyridine, pyrazine, pyrrole, and indole) [84]. These substances can be decomposed in subsequent reactions, thus reducing the negative impact on biocrude formation [105]. The addition of acid catalysts can also reduce this negative effect by inhibiting the Maillard reaction. In addition, basic products from proteins can react with acidic products from glucose to form esters [108]. Fatty acids produced

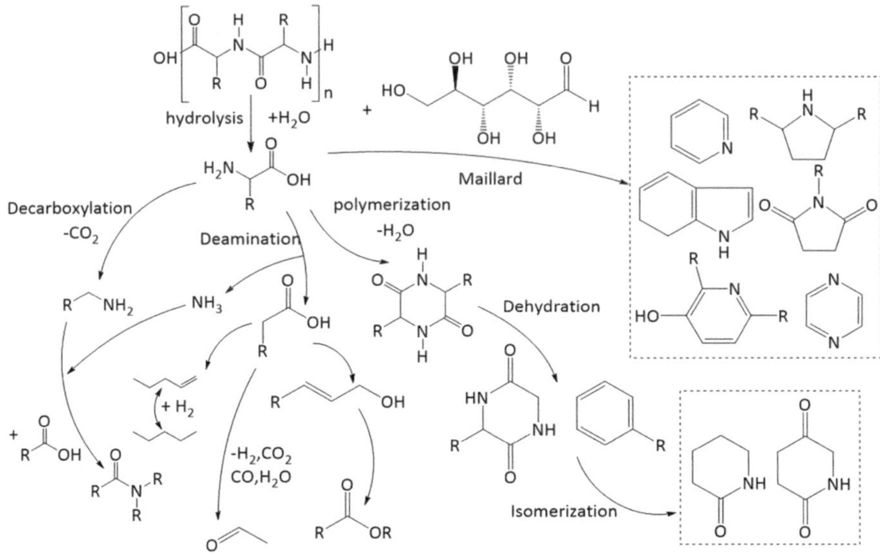

Fig. 3 Potential chemical reaction pathways of proteins in HTL [1, 22, 94, 95]

by deamination in the hydrolysis of proteins can undergo esterification with alcohols derived from the conversion of fatty acids [109]. Chemical reactions in HTL are not only dependent on the reacting substances but are also affected by reaction temperature, residence time and catalyst. Their effects are discussed in detail below.

In terms of feedstock, the type and structure of amino acids affect the degradation reaction of proteins. In order to better understand the hydrothermal liquefaction of proteins, some representative amino acids were chosen as model compounds [110]. In the process of hydrothermal liquefaction, there is a competition between decarboxylation and deamination, and the reaction rate is affected by the structure and number of amino acids [94]. Isomerized aminobutyric acids (α, β, and γ) have various deamination rates due to structural differences [111]. Branched-chain amino acids are more thermally sensitive than short-chain amino acids, and hydrophobic and hydrophilic side-chain amino acids have different degradation mechanisms. During hydrolysis of amides, O-pyridinecarboxamide is more active than meta-pyridinecarboxamide and para-pyridinecarboxamide [96]. The relative amount of protein also affects the concentration of the product. The absolute nitrogen production of heterocyclic compounds per unit of protein initially tends to increase and decreases as the ratio of protein to fructose increases [112].

Reaction temperature is an important parameter in the biomass high-temperature pyrolysis process. As the temperature increases, changes in water properties catalyze the chemical reactions during high-temperature pyrolysis, accelerating the degradation rate, increasing the degree of degradation, and converting more proteins and amino acids into intermediates. It was found that peptide bonds in proteins are stable and the degradation rate of amino acids increases with increasing temperature [37].

Hydrolysis of proteins is relatively slow when the temperature is lower than 230 °C, and the optimal cleavage temperature of peptide bonds is about 250 °C. Klingler et al. [26] investigated amino acid HTL using glycine and alanine and found that more than 70% of amino acids were degraded by decarboxylation or deamination within 30 s at 350 °C. Rogalinski et al. [113] used bovine serum protein to study the hydrolysis of amino acids in subcritical water and confirmed that amino acids were almost completely degraded after 200 s at 330 °C. Temperature not only promotes the degradation of proteins and amino acids but also affects the composition of the product. This is due to the fact that different degrees of reactivity at different temperatures result in different proportions of substances in the final product. Chen et al. [51] conducted HTL experiments using phenylalanine as a model compound and confirmed that the decomposition of phenylalanine starts at 220–250 °C. Phenylethylamine is the major product at 220–280 °C, while styrene is the major product at 310–340 °C. Excessively high reaction temperatures result in more substances being converted and more intermediates being converted to gas and coke instead of biocrude, thus reducing the yield of biocrude [75].

1.3 Carbohydrates

Carbohydrates are mostly broken down into polar water-soluble substances rather than nonpolar hydrocarbons, which are then dissolved in the aqueous phase during biomass conversion. As a result, the contribution of carbohydrates to biocrude production is negligible [95]. Due to the catalytic effect of ash in microalgae, more carbohydrates are broken down into small molecule gases, which results in the content of carbohydrate derivatives in biocrude oil showing an opposite trend to that of carbohydrates. Figure 4 shows the potential chemical reaction pathways of carbohydrates in HTL. Carbohydrates undergo hydrolysis to form monosaccharides. Glucose and fructose undergo isomerization. The rate of isomerization is lower than the rate of degradation, so glucose and fructose degradation produce different major substances. For example, glucose and fructose can be converted by rearrangement reactions. The major degradation product of fructose is 5-hydroxymethylfurfural, whereas other monosaccharide degradation products contain more acetaldehyde [108]. As the temperature increases, the C–C bonds in some monosaccharides break to form small molecules (e.g., aldehydes and acids), and some monosaccharides dehydrate and cyclize to form cyclic compounds (e.g., furans and phenols). They then react with each other to form biocrude [114]. Reducing sugars also react with amino acids to form nitrogen-containing heterocyclic compounds.

The conversion of non-cellulosic carbohydrates is also affected by the temperature of the reaction, and these effects are reflected in the products and efficiencies. At different reaction temperatures, carbohydrates can be converted into a wide range of compounds, and the conversion efficiency of organic matter increases with temperature. Higher temperatures promote the repolymerization of small molecules and improve solids depolymerization and gas aggregation to produce biocrude oils [40].

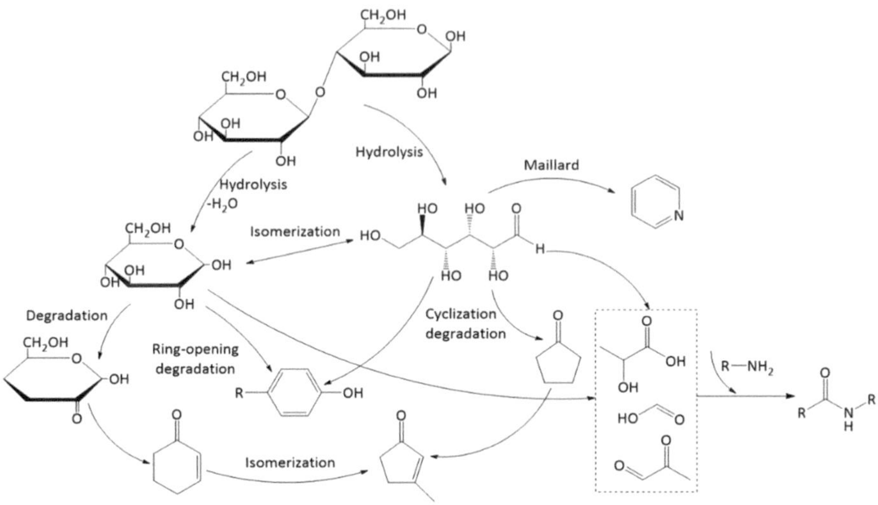

Fig. 4 Potential chemical reactions pathways of carbohydrates in HTL1 [95]

For example, heavier molecules such as carbohydrates, proteins and lipids are cleaved by the rapid increase in temperature. During rapid high-temperature thermoliquefaction, the temperature will increase rapidly, and high-temperature thermoliquefaction can be carried out in a short period of time to convert biomass into biocrude oil.

The catalyst has a selective effect on chemical reactions. Adding Na_2CO_3 can promote carbohydrate decarboxylation [53]. K_2CO_3 can improve the denitrification reaction, but it inhibits sulfur removal and simultaneously increases the sulfur content in biocrude [87]. The addition of HCO_2H has a scavenging effect on reactive species, thereby inhibiting the polymerization process and coke formation. Sulfuric and acetic acids promote decarboxylation and dehydration of intermediates from polysaccharide decomposition, thereby increasing the amount of gaseous products and residues. Srokol et al. [108] investigated the HTL of several monosaccharide model compounds and found that glucose and fructose may undergo both acid-catalyzed and base-catalyzed reactions, resulting in the formation of different products under different catalytic conditions. For example, 5-hydroxymethylfurfural and acetaldehyde were the major products under acid-catalyzed and base-catalyzed conditions, respectively.

Other factors, such as the reducing atmosphere and the ratio of water to feedstock, also significantly affect carbohydrate conversion. Li et al. [75] found that hydrogenation during high-temperature catalytic reduction favors the formation of cyclic nitrogenous compounds and polycyclic aromatic compounds and enhances decarboxylation and dehydroxylation. Increasing the ratio of water to feedstock promotes the formation of more water-soluble substances in sugars [88], which may be related to the increase in water to provide more hydrogen.

1.4 Lignocelluloses

Lignocellulose differs from proteins, non-cellulosic carbohydrates, and lipids and is found mainly in agricultural and forestry wastes. Lignocellulose can be classified as cellulose, hemicellulose and lignin [115]. They are rich in carbon and are good materials for the production of biocrude oil. Cellulose is a polysaccharide consisting of linear chains of d-glucose units with molecular weight greater than 100,000 and is insoluble in water and conventional organic solvents. Hemicellulose is polymerized from mannose, glucose, and xylan. Lignin is a complex amorphous phenolic polymer consisting of three alcohol monomers (i.e., p-coumaroyl alcohol, coniferyl alcohol, and mustelin alcohol) with a structure similar to polyphenols [116, 117].

Figure 5a and b shows plausible reaction routes for cellulose and hemicellulose, respectively. Cellulose and hemicellulose are first degraded to produce C5 or C6 sugars, which are then isomerized between aldoses and alcohol sugars. They further undergo condensation, degradation, hydrogenation, and other reactions to produce unstable products based on C2–C4 intermediates such as 2,3-butanedione and acetone aldehyde. Due to the presence of carbonyl groups, these unstable substances can be hydrogenated to stable C2–C4 fatty alcohols and other substances. These substances form a number of water-soluble biocrude oils, some of which condense to water-insoluble biocrude oils or coke.

Many researchers have studied the conversion of lignocellulosic HTL to biocrude oil using poplar, sludge, and straw as feedstock and have broadly formulated the chemistry of these processes. First, cellulose, hemicellulose, and lignin are hydrolyzed to produce glucose and different alcohol monomers. These are reacted by dehydration, deoxygenation, and decarboxylation to form low-molecular-weight intermediates, which are further converted to biocrude oil by condensation, cyclization and isomerization [120]. Biocrude oils mainly include furfural, esters, alcohols, acids, ketones, phenols, and other oxygenated compounds. These products have relatively few carbon atoms and are mostly C6-C9 light compounds [30, 51, 121]. Due to the structure of lignin, phenolic compounds are mainly derived from the degradation of lignin. Figure 6 shows a simplified schematic of lignin degradation in HTL. The C–O ester bond in lignin is initially broken to form different phenolic compounds and methoxybenzene, and the products can be further hydrolyzed to phenol.

Under hydrothermal conditions, more lignocelluloses are degraded and converted into low-molecular-weight substances [30]. Unlike other compounds, the degradation of lignocellulose is largely influenced by the type of solvent. Commonly used solvents can be categorized into alcohols and phenols, with alcohols including glycerol [123], methanol [124, 125]. The effects of solvents on lignocellulosic HTL are mainly in the mass loss rate, product ratio and liquefaction rate. For example, when using PEG400 as a solvent, more esters are formed through esterification. The use of methanol or ethanol as solvents under supercritical conditions favors lignin degradation, but lignocellulosic conversion and biocrude yield are lower [116]. However, when these two solvents were mixed as co-solvents, the conversion of lignocellulose and the yield of biocrude oil were significantly increased [124]. In addition, phenol as a

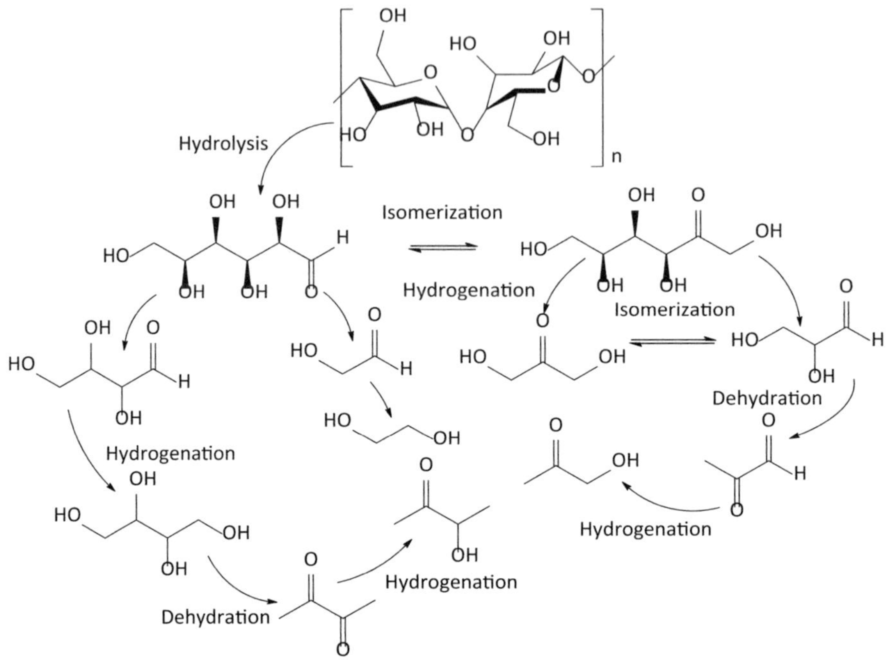

a)

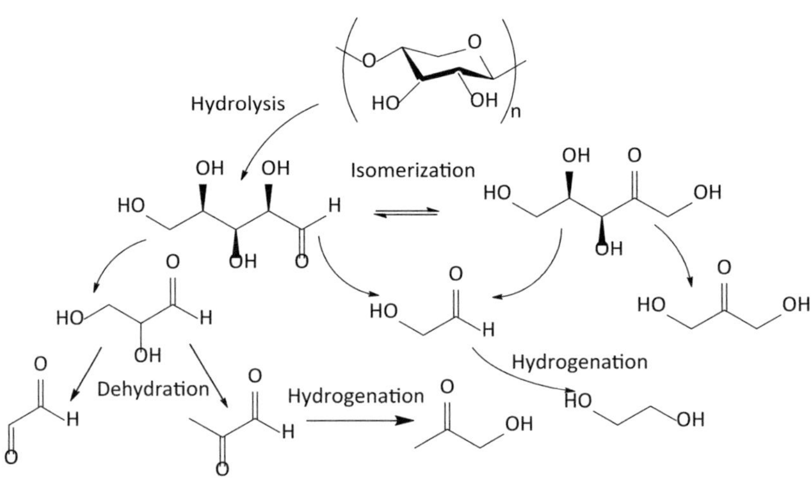

b)

Fig. 5 Plausible reaction pathways of cellulose and hemicellulose in HTL, **a** cellulose, **b** hemicellulose [118, 119]

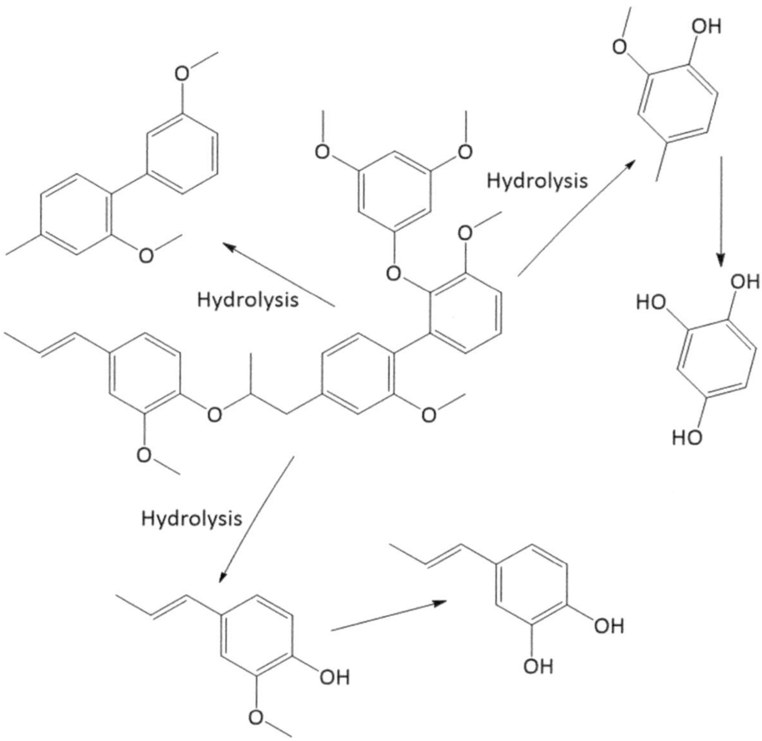

Fig. 6 Simplified schematic diagram of lignin degradation in HTL [122]

solvent has been reported to liquefy faster than methanol or ethanol as a solvent, and phenol has been reported to prevent coke formation [126].

At the initial stage of conversion, the most important reaction is the degradation of biomass. Elevated temperatures can increase the rate of degradation and the degree of conversion of large molecules to small molecules. In particular, higher temperatures can promote lignin decomposition and significantly increase the conversion rate and liquefaction efficiency [127]. At the later stage of the conversion process, it is mainly the repolymerization of intermediate substances. Too low a temperature is detrimental to the polymerization of intermediate substances to produce biocrude oil, while too high a temperature may promote the formation of coke or gas from intermediate substances. Therefore, the temperature has an important influence on the conversion efficiency, conversion rate, and composition of biocrude oil. The yield of biocrude oil from high-temperature catalytic combustion of swine manure reaches a maximum at 340 °C and then decreases with increasing temperature [128]. Of course, the main reactions are different at different temperatures. Due to the structure of the material, in the lower temperature range, the liquefaction rate of cellulose does not change significantly with increasing temperature, and the main degradation is hemicellulose and lignin. As the temperature continues to increase, cellulose degradation begins to dominate, at which point the biocrude yield is higher. However, decomposition or polymerization of organic matter increases gas and coke yields as the temperature increases. In general, high-temperature catalytic combustion of cellulose yields the highest biocrude yields at 250 °C. The temperature at which cellulose degradation occurs is the highest. As the temperature increases to 285 °C, the yield decreases and the gas yield increases, and the major compounds in the biocrude oil vary at four different temperatures (i.e., 200, 250, 300, and 350 °C) [129].

Catalysts can be targeted to promote certain reactions in the thermal liquefaction reaction, which can lead to changes in the composition of biocrude oil and alter the quality of biocrude oil. Therefore, catalysts are widely used in experiments to produce biocrude oil from biomass thermal liquefaction reactors. As alkaline catalysts, KOH, NaOH and K_2CO_3 are beneficial to break the glycosidic bonds in cellulose and hemicellulose as well as the ester bonds in lignin, and promote the decomposition of polymers to form biocrude oil. Especially, lignin is more sensitive to alkaline catalysts, and it will be degraded to form more aldehydes, hydrocarbons, and phenolic compounds under the promotion of alkaline catalysts. At the same time, alkaline catalysts also promote decarboxylation reaction, which reduces the content of oxygenated compounds and improves the quality of biocrude oil [130]. However, the addition of alkaline catalysts to promote biomass decomposition may have negative effects. For example, Yang et al. [127] observed that cellulose was modified to form small soluble molecules when 5 wt% NaOH was added, thus reducing the yield of biocrude oil. Acid catalysts apparently affect dehydration and transesterification reactions. Dehydration and degradation reactions increase the conversion of biomass, while esterification reactions increase the ester content of the product [131]. Chen et al. [128] noted that solid residues can be greatly reduced in acidic environments, and more cellulose and hemicellulose structures can be disrupted and degraded to other substances. In addition, some data suggest that linoleic acid as a catalyst can promote lignin degradation to form anaerobic benzene compounds, which has a positive impact on the deoxygenation of biocrude oil.

1.5 Chlorophyll

The common structural feature of all chlorophylls is the hydrophilic chlorophyll ring, which consists of a light-absorbing network of conjugated double bonds encapsulated in a 4-ion nitrogen ring and hydrophobic ester-linked phytol side chains [132]. Algae commonly contains up to 5 wt% chlorophyll on a dry basis [133]. These chlorophyll-like macromolecules decompose into phytol (< 250 °C), phytane or phytoene (> 300 °C) in the HTL reaction. Similar to chlorophylls, both phytanes and phytoenes contain a chlorine ring attached to phytol and other straight chains. In addition, chlorophylls can be hydrolyzed to nitrogen-containing heterocyclic compounds such as indoles and methylindoles [91]. Changi et al. [132], to our knowledge, first studied the reaction kinetics and pathways of phytol as a model compound for algae chlorophyll in subcritical water. Figure 7 shows the proposed reaction network for HTL of phytol.

1.6 Products from HTL

The products of algal thermal liquefaction reactions are usually categorized into oil-phase products, water-soluble products, gaseous products, and solid products. In this paper, we focus on the oil-phase products and water-soluble products because they are most relevant to the final bio-oil yield. The GC–MS results on the bio-oil produced in previous algae HTL studies are summarized in Table 2. The main oil products in algae HTL can be divided into six main categories by functional group [134]: (1) monoaromatic hydrocarbons and their derivatives, such as benzene, methylbenzene, and phenol; (2) aliphatic compounds; (3) fatty acids, such as hexadecanoic acid; (4) nitrogen-containing compounds, such as amines and amides; (5) polyaromatic hydrocarbons; and (6) other oxygen-containing compounds, such as alcohols, aldehydes, and ketones.

1.6.1 Free Fatty Acids and Alkanes

Free fatty acids are direct products from triglyceride hydrolysis [135, 145]. A number of studies have found that free fatty acids were stable in subcritical water [145, 146]. Algae HTL oil contains a large amount of fatty acids and different long-chain alkanes [135]. As shown in Table 2, the fatty acids mainly consist of palmitic and palmitoleic acids, while the long-chain alkanes contain a number of different species such as pentadecane, nonane, and docosane. The presence of these long-chain fatty acids leads to the high viscosity of the bio-oil, which seriously affects its fluidity.

At temperatures above 310 °C, long-chain fatty acids are decarboxylated to produce the corresponding alkanes [147]. However, at the critical point of water, some unstable fatty acids (C18:1, C18:2, and C20:4, etc.) decompose into a variety of compounds such as glycerol, methanol, acrolein, acetaldehyde, and propionaldehyde [102]. Gasification dominates at the supercritical temperature resulting in further decomposition of these compounds to CO, CO_2, H_2, and some other gas products [148].

1.6.2 Phenols and Aromatic Compounds

HTL converts the carbohydrate content of algae to phenols [149]. Phenols and their alkyl derivatives are produced in HTL of *Nannochloropsis* sp. at higher temperatures than 300 °C [91]. Luijkx et al. [150] proposed the mechanism of glucose transforming into phenols in HTL reactions. Glucose or fructose first converts to a furan derivative

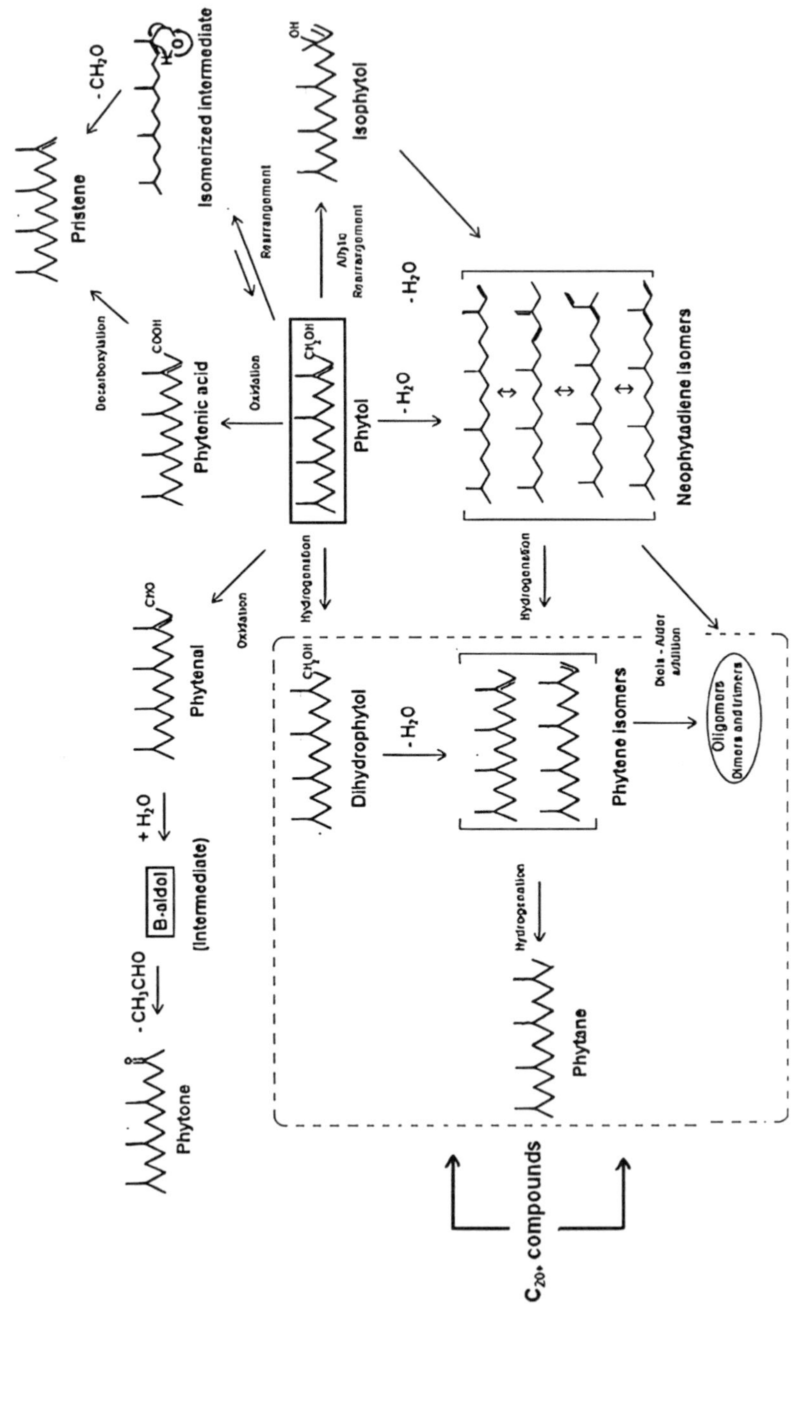

Fig. 7 Reaction network for phytol in subcritical water [132]

Table 2 Identities of compounds in bio-oil produced from HTl of algae in previous literatures

Compounds	References
Monoaromatics	
Phenol	[23, 54, 92, 134–137]
Benzene/Toluene/Styrene	[7, 23, 36, 54, 92, 134, 136, 138]/[36, 136]/[7, 20, 36, 136]
Cholesterol and Cholestene	[91, 135]
Vitamin E	[139, 140]
Fatty acids	
Myristic acid/Myristoleic acid	[91, 135]/[91]
Palmitic acid/Palmitoleic acid	[91, 135, 136, 138, 141]/[91, 135, 138]
Stearic acid/Oleic acid	[54, 91, 138]/[23, 91, 135, 138]
Tetradecanoic and Octanoic acid	[142]
Hexadecanoic acid	[23, 36, 54, 134, 137, 140, 142]
Arachidic acid	[135]
Eicosapentaenoic acid	[91]
Alkane/alkenes	
Hexadecane/Hexadecene	[20, 36, 54, 134, 135, 142]/[36, 134–136, 140]
He ADDIN EN.CITE ptadecane/ Heptadecene	[20, 36, 54, 91, 134–136, 138, 142]/[23, 140]
Pentadecane/Pentadecene	[135, 136]/[7, 23, 36, 54, 91, 92, 134, 139, 140, 142]
Phytane/Phytene	[20, 91, 138, 140]/[91, 138]
Docosane and Cholestane	[135, 138]
Cycloalkane	[7]
Eicosane, Coprostane, Tricontane, Hentriacontane and Dotriacontane	[135]
Polyaromatic compounds	
Naphthalene	[20, 23, 91, 92]
Quinoline	[23, 36, 91]
Indene	[36, 91]
Anthracene and Phenanthrene	[20, 91]
Pyrene and Carbazole	[91]
Fluorene	[20, 23, 91]
Nitrogenated compounds	
Piperidines	[36, 54, 135, 142]
Pyrrols/Pyrrolidines	[23, 36, 54, 134, 136–138, 142]
Indoles	[36, 54, 91, 92, 134, 137, 142]
Pyridines	[137]
Pyrazines	[23, 36]
Pyrrolidinones	[23, 54, 92, 135]

(continued)

Table 2 (continued)

Compounds	References
Amides	[23, 36, 54, 135, 136, 138–140, 142]
Amines and nitriles	[92, 137]
Other oxygenated compounds	
Esters	[23, 36, 139, 140, 143, 144]
Aldehydes	[7, 36, 140]
Ketones	[7, 23]
Alcohols	[54, 92, 138, 142]
Acetic acid	[23, 144]
Furans	[36, 134]

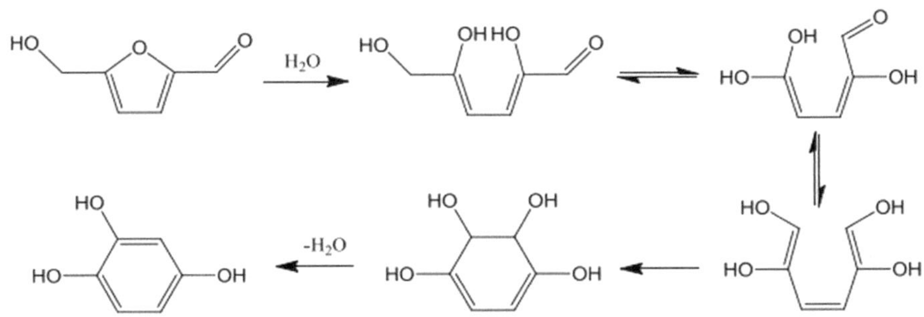

Fig. 8 1,2,4-Benzenetriol formation mechanism of HMF rearrangement in subcritical water [150]

and second to phenols via rearrangement reactions by ring opening and closure as shown in Fig. 8.

Above 300 °C, some polycyclic aromatic hydrocarbons (PAHs) and their derivatives are found in bio-oil [91]. Typical aromatic compounds in heavy oil from HTL are naphthalene, dibenzothiophene, etc. [20]. The yield of PAHs increased slightly with temperature for HTL of *Spirulina* [134].

1.6.3 Heteroatoms Containing Compounds

Algal crude bio-oils typically contain 10–15 wt% of heteroatoms (mainly O and N) [151]. Removal of these heteroatoms from algal bio-oils using conventional refining techniques is a major challenge. Some carboxylic acids may contain additional functional groups such as hydroxyl or alkenyl groups in the alkyl chain, which may lead to polymerization. Hydrolysis, oxidation, and polymerization products can affect fuel quality and may adversely affect engine life [152]. In addition, nitrogen and sulfur in bio-oils need to be removed during the upgrading process as they produce NOx and SOx during combustion [153].

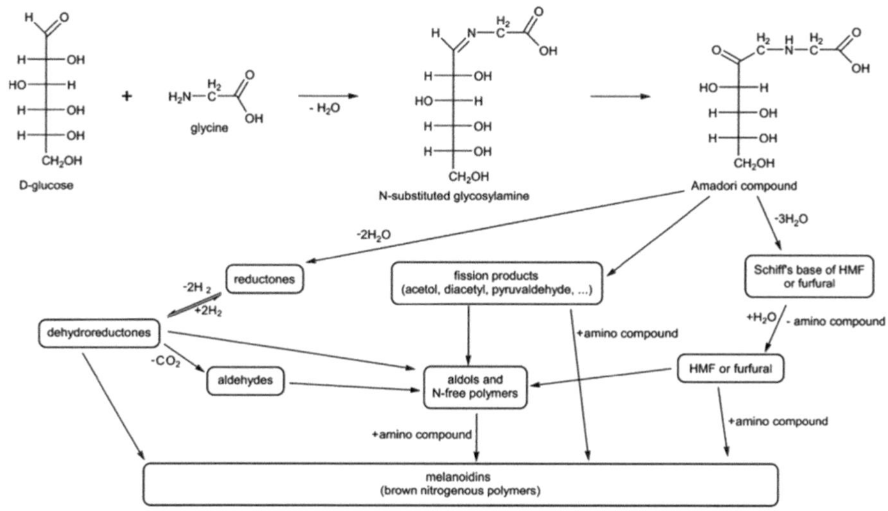

Fig. 9 Maillard reaction network [158]

Nitrogen-containing compounds are produced mainly as amines and amides from the decarboxylation and deamidation of amino acids during protein hydrolysis. In high-temperature catalytic reduction reactions, most of the nitrogen produced by algae is transferred to the aqueous phase, and its content is highly dependent on the conditions [154]. Both longer reaction times and higher temperatures lead to increased nitrogen content in liquid phase [31, 155]. Almost half of the aqueous nitrogen is ammonia, and a small fraction is nitrogen oxides [20]. Some nitrogen-containing rings (e.g., indoles and quinolines) are usually derived from the breakdown of proteins and are often identified as nitrogenous compounds in bio-oils. These compounds are products of the Maillard reaction of amines with carbohydrates [156, 157]. Peterson et al. [158] demonstrated the role of Maillard reactions in HTL of glucose and glycine through kinetics and mechanistic studies as shown in Fig. 9. Once these compounds form, they are usually difficult to decompose, even in the severity of supercritical water [138, 142, 159].

The oxygen content in bio-oil decreases with the increasing temperature and time [31]. This oxygen is mainly present as alcohols, aldehydes, and carboxylic acids. By contrast, the sulfur in bio-oil, primarily dimethyl sulfide, is typically low [138].

1.6.4 Small Molecular and Water-Soluble Compounds

Seaweeds are high in proteins that may be broken down into amino acids, amides, and other short-chain carboxylic acids by cleavage of peptide bonds. These undesirable water-soluble products reduce bio-oil yields and should ideally be minimized. Acetic acid and glycerol are products resulting from the direct hydrolysis of lipid HTL in macroalgae [23]. The aqueous phase of *Spirulina* HTL contains many acetates such as acetic acid and acetamide [136]. Other research [54, 134] found the presence of acetate and formate in HTLs, which may be hydrolyzed products of carbohydrates. Due to the production of carboxylic acids, Duna alga pyrolysis products were acidic [141]. This phenomenon was also shown in other biomass liquefaction research [160, 161].

The aqueous phase by-products from high-temperature pyrolysis of algae usually contains significant amounts of nitrogen, phosphorus, potassium, and other minerals and essential trace elements required for algal growth, which can be recovered and reused. Ammonia is the main component of the aqueous phase because algal proteins are converted to amino acids, which are hydrolyzed to form ammonia [28]. *Spirulina* liquefaction produces a high concentration of ammonia (12,700 mg/L) as a by-product of the aqueous phase [162]. Aqueous phosphorus is mainly in the form of PO_4^{3-} [155]. 80% of the phosphorus in *Chlorella* partitioned into the aqueous phase as phosphates in HTL reactions at 250 °C [31]. The ranges of total nitrogen and phosphorus in HTL products of algae were 998–1157 mg/L and 2.45–5.38 mg/L, respectively [20]. The aqueous phase contains a wide range of nutrients required for algal growth; thus, recycling of nitrogen, phosphorus, and carbohydrates is very beneficial from a techno-economic point of view [163]. Miao et al. utilized a continuous hydrothermal equipment to remove algal polysaccharides before biomass liquefaction, which could increase the yield of bio-oil at lower temperatures and reduce the formation of biochar. They also proved that developing by-products while producing biofuels is an effective method to make algal biofuel production economically feasible [164].

1.6.5 Metal and Solid Residue

In addition to nitrogen, phosphorus, and carbohydrates, algae take up trace elements during growth, so algal bio-oils usually contain trace metals. *Spirulina* HTL oil contains Ca, Fe, Mg, Cu, Zn, Ni, Mo, Na, K, Al, B, Pb, Mn, and Cd [134], while *Chlorella* HTL bio-oil contained Co, Mn, Fe, Cu, Zn, Mo, Cr, Mg, Al, Na, and K [31]. Organic compounds in algae are pyrolyzed and liquefied, while certain inorganic components (e.g., calcium phosphate) form solids and precipitate. The solids can be separated from the liquid products [165]. The solid yield of *Microcystis* HTL is less than 5%. The solid residue of *Phaeodactylum* HTL is mainly calcium phosphate by powder X-ray diffraction [166].

1.6.6 Gas Products

Gases are produced as by-products during pyrolysis of algae. Gasification becomes the main reaction when the temperature rises to or above the supercritical point. CH_4 and C_2 gases were produced in HTL of *Nannochloropsis* sp. from 200 to 500 °C [91]. The content of H_2 and CH_4 is relatively low at low temperatures (200–250 °C) and their formation increases significantly from 300 to 380 °C for *Spirulina platensis*. At 380 °C, yields of C_2–C_5 gases become significant and CH_4 production is lower than at 300–350 °C [134]. The percentage of CH_4 increases with temperature (< 340 °C), indicating that HTL is favorable for methanation [20]. In *Desmodesmus* HTL, CO_2 is the main product in the range of 175–350 °C [155]. The content of light hydrocarbons (CH_4 and C_2–C_3 gases) decreases, while H_2 increases rapidly as the temperature increased. An increase in reaction time (460 min) could lead to a higher H_2 and CO_2 yield but a lower CO yield. CO is barely detected because CO participates and is consumed in water–gas shift and/or methanation reactions [167–169]. In hydrothermal conditions, most of the nitrogen is recovered in the aqueous and oil products [110], so gas phase is nearly void of nitrogen [20, 135].

As the temperature increases, the "Fischer–Tropsch" reaction may occur in hydrothermal process, in which water and short-chain carboxylic acid may act as hydrogen sources. Previous studies [170, 171] discovered that, in the range of 150–250 °C for 2–3 days, CO and H_2 combined to form alkanes in a stainless-steel reactor.

2 Influence of Reaction Conditions on HTL Process

Thermal liquefaction of algal biomass and the physicochemical properties of the resulting bio-oil are clearly influenced by operational parameters such as temperature, reaction time, algal species and loading concentration, reaction atmosphere (hydrotreating), and catalyst [23]. In this section, we focus on the effects of HTL reaction conditions reported in the published literature. The relevant aspects we have sorted out are important for future studies on reaction mechanisms, kinetics, and engineering practices.

2.1 Reaction Temperature

Temperature plays the most important role in the yield of bio-oil produced by high-temperature pyrolysis of microalgae, and the choice of temperature is crucial for the safety and economy of its industrial operation. Several papers have investigated the effect of temperature on bio-oil yield and quality [7, 31, 61, 91, 134, 139, 141, 172, 173]. As demonstrated in Fig. 10, almost all studies show the temperature range that can produce the highest oil yield. In summary, the highest bio-oil yields occurred in the temperature range of 250–375 °C. The highest bio-oil yields were found in the temperature range of 250–375 °C. This phenomenon is mainly attributed to the competitive response that dominates at different temperature ranges.

Figure 11 illustrates the types of reactions in the microalgae high-temperature catalytic reduction reaction under different temperature conditions. The ionic properties of water change with temperature, leading to acceleration of various reactions.

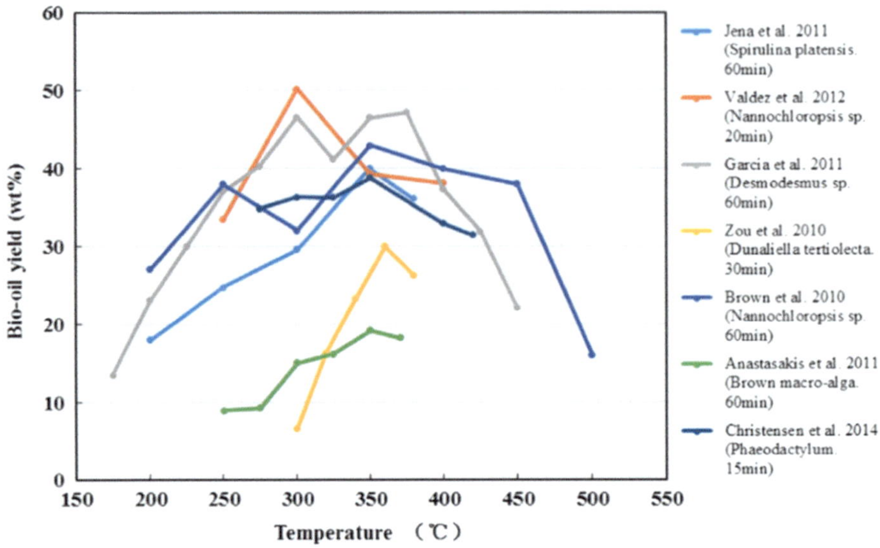

Fig. 10 Influence of reaction temperature on bio-oil yield in published literatures

At low temperatures (< 220 °C), hydrolysis reactions dominate, but as the temperature increases, the repolymerization reactions compete with the hydrolysis reactions [143]. From 220 °C to 250 °C, the hydrolysis products may be more soluble in water, resulting in lower oil production; from 250 °C to the critical point, the oil production may reach a maximum, while the organic content in the aqueous phase decreases [155]. This may be due to the rapid increase of the ionic product of water near the critical point. Due to the higher ionic product, the dissociated H + and OH in the hot liquid water can catalyze acid- and base-catalyzed reactions [174]. Carbohydrates, proteins, and lipids readily undergo isomerization, reforming, depolymerization, and repolymerization reactions under these conditions, ultimately yielding bio-oils. Above the critical point, supercritical conditions favor decarboxylation, cracking, steam reforming, and gasification reactions of intermediates, resulting in higher gas yields [134].

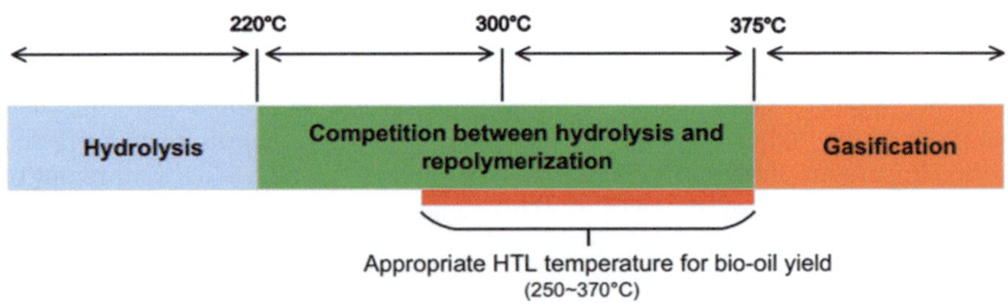

Fig. 11 Reaction types in the temperature ranges of HTL reaction of algae [143]

Fig. 12 Influence of temperature on morphology changing of *Desmodesmus* cells (5 min) [155]

Someone studied the effects of reaction temperature on algae cell morphology [155]. The SEM analysis (Fig. 12) shows that microalgae cell structure was destroyed above 250 °C, resulting in a significant increase in oil production. This may explain the result mentioned above. But when the temperature is between 225 and 250 °C, thermal denaturing of proteins released from the cells can lead to aggregation and adhesion of cells. It turns microalgae into a viscous colloid that has the potential to cause blockages and fouling of equipment.

The higher H/C ratio within *Chlorella* bio-oil was obtained in subcritical reaction conditions than in supercritical conditions [175]. The carbon recovery of bio-oil is less than 5% at temperatures below 160 °C, and the recovery increases to 43.2% as the temperature increases to 240 °C [176]. However, when the temperature further increases to 300 °C, the carbon recovery is basically unchanged. Overall, the C and H content in bio-oil increases with temperature and exceeds the lipid content in the algae feedstock. The oxygen content decreases with temperature, and increasing temperature is favorable for decreasing the O/C ratio in bio-oil while enhancing the H/C ratio, but the sulfur content increases with temperature [176]. There is almost no information available regarding nitrogen distribution with respect to temperature, but it is certain that the N content in bio-oil product is obviously lower than in initial algae biomass [91].

2.2 Reaction Time

To ensure maximum conversion of organic matter from microalgae to bio-oil, sufficient reaction time is necessary. Therefore, reaction time is a key factor in the economic operation of microalgae thermal liquefaction technology. From the results of previous studies, if the reaction time is too short, the oil production will be negatively affected. On the other hand, if the reaction time is too long, a large amount of gas, aqueous products, or coke may be produced, which ultimately leads to lower bio-oil production. When the reaction time is short, microalgae break down and depolymerize into smaller molecules. Over a longer period of time, these compounds may rearrange themselves through condensation or repolymerization to form bio-oil [139]. Some researchers reported that maximum bio-oil yield is obtained in 30–60 min [23, 172]. The trends of bio-oil yields with respect to reaction time are summarized in Fig. 13. Bio-oil yields from HTL of *Dunaliella* increased with reaction time from 0 to 50 min, but the yields plateaued beyond 60 min [141]. Reaction times longer than 60 min negatively impacted bio-oil yields from *Spirulina* [54, 134]. The average bio-oil yield of *Microcystis viridis* at 30 min was higher than that at 60 min [20]. Extending reaction time only had an apparent effect at temperatures below 220 °C, but above 220 °C, the bio-oil yield did not show a clear relationship with reaction time [176]. Some researchers suggested that 30 min is the most appropriate time for treating both *Entermorpha prolifera* [23] and *Dunaliella* [139].

Shorter reaction times are expected to be effective in reducing equipment and operating costs. Several researchers have explored the effect of short reaction times on microalgal HTLs. 15 min is sufficient to convert marine algal HTLs to oil production

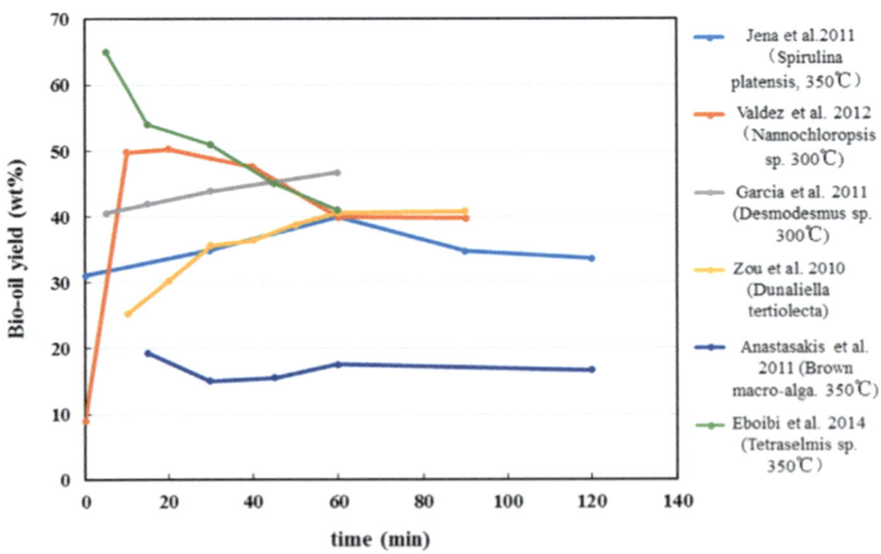

Fig. 13 Influence of reaction time on bio-oil yield in published literatures

(maximum bio-oil yield of 19.3 wt%) [76]. Almost all organic matter can be converted into bio-oil within 10 min [31]. For *Desmodesmus* HTL at temperature of 300 °C, 40.5 wt% bio-oil yield can be achieved within 5 min compared with the maximum yield of 46.6 wt% at 60 min [155]. More recently, rapid liquefaction of *Nannochloropsis* sp. resulted in a bio-oil yield of 66 wt% in 1 min (on a dry and ash-free basis), with a similar carbon content and higher calorific value compared to that obtained by "slow" hydrothermal liquefaction [177]. HTL of *Tetraselmis* sp. at 350 °C resulted in a 65 wt% yield of bio-oil within 5 min [178].

2.3 Water Density or Pressure

The reaction pressure directly reflects the density of water, and liquid water will be characterized by a self-heating vapor pressure. Higher water density accelerates the release of more H + from high-temperature compressed water, which improves the acid-catalyzed reaction [179]. Moderate amount of water (0.5–2.0 g) can effectively facilitate the production of high energy density bio-oil [180]. However, water density has no obvious effect on oil yield from HTL of *Nannochloropsis* sp. (400 °C) [31]. Klingler have studied the HTL of alanine under two different pressures (24 and 34 MPa) at 300 °C and 350 °C, respectively [181]. By comparison, they found that the pressure had no effect on the conversion rate and product distribution. Therefore, the effect of water density on oil production in microalgal HTL remains inconclusive.

2.4 Feedstock

Increasing the algae/water ratio to the extreme may not positively affect the yield and quality of bio-oil, while too low a concentration of algal feedstock may result in poor economics. Based on engineering experience, the target biomass concentration should be 15–20 wt% to achieve practical economic benefits [100]. 20 wt% of biomass loading is the optimum for *S. platensis* liquefaction and greater algal concentrations show no significant effect on product yields [134].

For *Chlorella*, the bio-oil yield increased from 36 to 46 wt% as the algae concentration increased from 5 to 35 wt% [31]. However, bio-oil yield decreased as the ratio of biomass/water increased [139]. While relatively low algal concentrations can increase oil production to a limited extent in this case, the additional amount of water required can lead to increased costs, such as increased energy required for heating and waste water treatment costs. These additional costs associated with increased water content make the process uneconomic. Utilizing algal polycultures grown in waste water may be a viable route to bio-oil production [128, 182]. The highest bio-oil yield (35.7 wt% based on dry matter) and the most energy dense oil (27.5 MJ/kg) came from a swine manure to an algae polyculture with a ratio of 3:1 and 1:3 by mass, respectively.

At present, the research mainly focuses on *Desmodesmus* [155], *Scenedesmus* [183], *Chlorella* [176, 184], *Nannochloropsis* [31, 91, 138, 175, 177], *Dunaliella* [139], *Spirulina* [92, 134, 136, 175, 183], *Phaeodactylum* [166], etc. Differences in the composition of different algae may lead to different optimal treatment conditions, bio-oil yields, composition and distribution of HTL products. Strain-specific parameters (cellular structure, biochemical composition and growth environment) led to significant differences in bio-oil yields at 250 °C (17.6–44.8 wt%), but these differences became less significant at 375 °C (45.6–58.1 wt%) [185]. High-protein content is a remarkable feature of *Spirulina* and *Nannochloropsis* sp [186]. Studies on *Spirulina* have found that the highest yield of bio-oil was 32.6%, but the oil consisted of compounds with low molecular weights [92]. High nitrogen content in *Spirulina* directly results in high N content (5.3–6.6%) in HTL bio-oil [134]. Moreover, at 20 wt% algae concentration, 350 °C and 60 min batch holding time, the S content of the resulting bio-oil reaches 0.5–2.34 wt%. The H/C, O/C, and N/C ratios are 1.44, 0.10, and 0.07, respectively. *Nannochloropsis salina* is more suitable than *Spirulina* for bio-oil production when used as a raw material when considering H/C ratio in the bio-oils [175]. HTL conversion efficiency and oil quality of low-lipid high-protein (*Nannochloropsis* sp.) and high-lipid low-protein (*Chlorella* sp.) microalgae [90]. The highest bio-oil yields were 55% and 82.9% with the highest hydrocarbons of 29.8% and 17.9% for *Nannochloropsis* sp. and *Chlorella* sp. was 82.9%, respectively. Porphyridium is characterized by a high carbohydrate content, which also translates to a high oxygen content. The high oxygen content leads to the lowest bio-oil yield when compared with *Chlorella vulgaris*, *Nannochloropsis occulata*, and *Cyanobacteria spirulina* [142]. *M. viridis*, with 46% carbon, 7.3% hydrogen, and 9.5% nitrogen, produced the highest bio-oil yield at 33 wt%, which corresponds to an energy recovery of 40% [20].

In addition to microalgae, marine macroalgae have also been studied. *E. prolifera* is known to be rich in carbohydrates and proteins and low in lipids and lignin. Although this type of algae is not rich in lipids, it has been reported that higher oil yields can be obtained by the HTL method [23]. In addition, macroalgal cells contain high levels of ash (sea salt), which is released into the aqueous phase during liquefaction and can be recycled [23]. When microalgae (*S. platensis*) and macroalgae (*E. prolifera*) are co-liquefied, a strong synergistic effect occurs, with maximum bio-oil production at a water/algae ratio of 5.8:1 [187]. Recent research on green macroalgae [188] found that marine *Derbesia* [26.2% (dry wt.)], *Ulva* [18.7% (dry wt.)], and freshwater *Oedogonium* [19.7% (dry wt.)] had higher bio-oil yields than marine *Chaetomorpha* and *Cladophora* cultured in both marine and freshwater. Bio-oil with the lowest nitrogen content. The bio-oil also had the lowest nitrogen content and the highest energy recovery.

2.5 Hydrotreating

A reducing atmosphere can promote the generation of bio-oil to an extent [117]. The role of reducing gases or hydrogen donors is to stabilize the fragmented intermediate products obtained from bond breaking and thereby increase oil yield through preventing condensation, cyclization, repolymerization, and reduction of tar and coke [189]. Both H_2 and CO show high reactivity on this stabilization effect, but usage of CO as a reductant poses challenges due to its health hazards [117]. The addition of high pressure H_2 can lead to a higher bio-oil yield with an increased H content and H/C ratio [135]. Hydrogenation improved bio-oil yields in HTL of N. salina from 55.6 to 78.5 wt% [190]. Hydrogen also enhances the removal of hydrophilic functional groups and the stabilization of free radicals, thereby inhibiting mass transfer to aqueous and gaseous products and thus improving the quality of the oil. The use of hydrogen donor solvents is another option for providing a reducing atmosphere for bio-oil production. Formic and acetic acids used as reducing gas sources can be decomposed into H_2 and CO, which then react with hydrolysates from algae to increase bio-oil production. Organic acids are broken down during the HTL process, which is more favorable for bio-oil formation [54]. However, these organic acids increase the sulfur and nitrogen content of bio-oil. Formic acid can increase bio-oil yield from glucose HTL [191]. Another notable feature of the addition of organic acids is the formation of large amounts of phenols, fatty acids, and fatty acid amides in bio-oils. 2-Methylhydrofuran and tetrahydronaphthalene act as hydrogen donors to enhance the hydro-liquefaction of algal biomass, resulting in high oil yields and significant energy values, with tetrahydronaphthalene yielding even more oil [192].

2.6 Catalysts

Catalysts used in HTL of algae can be generalized into two categories: homogeneous and heterogeneous catalysts. The end goals of utilizing catalysts are to improve the yield and quality of bio-oil over that of noncatalyzed processes. Catalytic liquefaction has increased yields by up to 50% in comparison to noncatalytic liquefaction [36]. Catalysts play a role to improve characteristics of bio-oil such as increasing the H/C ratio and the removal of heteroatoms such as O, N, and S, leading to improved characteristics such as lower viscosity. Table 3 demonstrates recent research on catalytic HTL of algae.

2.6.1 Homogeneous Catalyst

Table 3 shows that alkali hydroxides, carbonates, bicarbonates, and alkali formates have been intensively studied as homogeneous catalysts in algal-catalyzed HTL reactions [54]. Previous studies have shown that KOH is capable of increasing bio-oil

Table 3 Recent research on catalytic HIL of microalgae

Catalyst	Algae feedstock	References
Na_2CO_3	*Dunaliella tertiolecta*	[21, 141]
	Cyanobacteria spirulina, Nannochloropsis occulata, Porphyridium cruentum, Chlorella vulgaris	[142]
	Chlorella vulgaris	[54]
	Microcystis viridis	[20]
	Botryococcus braunii	[193]
	Enteromorpha prolifera	[23]
	Chlorella pyrenoidosa	[194]
KOH	*Chlorella vulgaris*	[54]
	Laminaria saccharina	[7]
Pd/C, Pt/C, Ru/C, Ni/ SiO_2–Al_2O_3, CoMo/γ-Al_2O_3 (sulfided)	*Nannochloropsis* sp.	[135]
$Ca_3(PO_4)_2$	*Spirulina platensis*	[36]
Pd/Al_2O_3, Pd/C, Pt/ Al_2O_3, Pt/ C, Raney Ni, NaOH	*Chlorella pyrenoidosa*	[194]
NiO	*Spirulina platensis*	[36]
Ce/HZSM-5	*Chlorella pyrenoidosa*	[195]

yield [142, 196]. Adding base catalysts, formic acid, or cobalt catalysts for the HTL of glucose showed a 5–15% promotion of bio-oil yield [191]. However, the HTL bio-oil yield of brown macro-alga *Laminaria saccharina* was suppressed by increasing KOH, and the same negative effect on gas products was also observed [7]. KOH increases the solubility of organic matter in the aqueous phase, which may account for the lower bio-oil yield. In addition, alkali catalysts promote the formation of aqueous ammonia [54]. The catalytic effect of H_2SO_4 on the liquefaction of *Dunaliella tertiolecta* was tested by Zou et al. [172]. They reported that adding sulfuric acid can significantly reduce the nitrogen below 1% in bio-oil and the highest liquefaction yield (97.05%) was obtained with 2.4% H_2SO_4 and the reaction conditions optimized by response surface method. Although most of these homogeneous catalysts demonstrate significant beneficial effects, if the solvent and catalyst cannot be recycled, the process will not be economically viable.

Related studies on HTL of terrestrial plants [197–201] have shown that sodium carbonate played a positive role on the reaction proceeding and improving bio-oil yield. This observation prompted further work that shows that sodium carbonate also catalytically increases the conversion rate of microalgae HTL and the generation of bio-oil [36, 54, 141]. Sodium carbonate promotes hydrolysis, isomerization, decarboxylation, and hydroxylation of organic, and this generates a large number of $OH^-$and $HCOO^-$, which promotes further decomposition of larger hydrocarbons or macromolecules. These decomposed compounds will rearrange by condensation, cyclization, and polymerization and ultimately form bio-oil, while inhibiting coke formation [198, 200]. However, excessive amounts of sodium carbonate may also adversely affect the reaction. Although the catalytic process can minimize the energy consumption while maximizing the energy recovery of bio-oil, it is selective to monocyclic aromatics, and the resulting bio-oil has high N and O content [36]. HTL with 5% of Na_2CO_3 can get the highest bio-oil yield (25.8 wt%) of *Dunaliella*, but higher loadings of Na_2CO_3 inhibited bio-oil yields [141]. Na_2CO_3 did not improve the HHV of bio-oil because the atomic ratio of C/O of the bio-oil remained unchanged, but the product profile was shifted to contain more acetic acid after adding 5 wt% Na_2CO_3 [23]. Na_2CO_3 can hinder the migration of N-containing compounds into oil phase [154] because Na_2CO_3 promotes protein hydrolysis to ammonia rather than amino acids. Ammonia is water-soluble and has no contribution to bio-oil yield. Na_2CO_3 facilitates saponification of lipids. Na_2CO_3 should not be used as catalyst for high-protein and high-lipid algal feedstocks [142].

2.6.2 Heterogeneous Catalyst

Table 2 shows that heterogeneous catalysts have not been studied as extensively as homogeneous catalysts. When liquefying *Spirulina* and mixed algal feedstocks at 350 °C, 60 min, and 20% solids, the addition of NiO and $Ca_3(PO_4)_2$ negatively affected the production of bio-oil, but the catalysts favored the formation of gaseous products [36]. Addition of $Ca_3(PO_4)_2$ produces more oxygenated compounds, while addition of NiO produces more nitrogenous compounds in bio-oil. NiO promotes further decomposition and reduction of intermediates such as aliphatic compounds and nitrogenous aromatic compounds, thereby reducing the viscosity of the oil product. *Duan et al. studied the catalytic HTL of Nannochloropsis sp* [135]. They found that in the absence of H_2, there was no obvious trend of bio-oil yield with the loading of Pd/C, Pt/C, Ru/C, and CoMo/γ-Al_2O_3, but the viscosity of the bio-oil was lowered significantly. Noble metals (Pt, Pd, Ru) improved H/C ratio of the resultant bio-oil product. Pt, Ni, and CoMo also decreased O/C ratio in the bio-oil significantly (< 0.090 vs. 0.45 in the feedstock). Although Ru/C and CoMo are effective for desulfurization, the effect is not as pronounced in Ni catalysts. Adding Pd/Al_2O_3, Pd/C, Pt/Al_2O_3, Pt/C, or Raney Ni increased the bio-oil yield from HTL of *Chlorella pyrenoidosa* by about 10 wt% at 280 °C [194]. Ce/HZSM-5 significantly enhanced the Lewis acidity, smaller particle size, larger specific surface, and highly dispersed Ce_4O_7 with trivalent and tetravalent cerium in the zeolite skeleton channel that accelerates the catalytic liquefaction of *C. pyrenoidosa* and contributes higher contents of elements C, H, and lower content of N in bio-oil [195].

2.7 Reaction Process

A number of creative experimental methods have improved the yield and economics of algal hydrothermal liquefaction products. The two-step sequential hydrothermal liquefaction method starts with the separation of sugars and proteins at a lower temperature [202]. The highest yields of sugars and proteins and the fewest inhibitory compounds were obtained at 180 °C. The second step was to separate the sugars and proteins at a higher temperature. The second step converts the remaining biomass to bio-oil at a higher temperature, with the highest yield at 240 °C. The second step is to convert the remaining biomass to bio-oil at a higher temperature. Compared to the one-step hydrothermal liquefaction method, this novel process produces higher quality bio-oils with higher fatty acid content and lower nitrogen content. *Nannochloropsis* sp. was thermally liquefied using Pt/C at 350 °C in a two-chamber batch reactor [203]. The catalyst and algae were separated by a porous metal frit. This physical separation produced a hydrocarbon-rich gaseous product with higher bio-oil yields and a larger soluble fraction of hexane than an equivalent system in which the algae and catalyst were in physical contact. One technology evaluation for a novel biorefinery is the production of biofuels from the HTL of *Nannochloropsis gaditana* and *Scenedesmus almeriensis* using enzymatic extraction of proteins [204]. The algal lipids were first extracted by Soxhlet extraction with hexane. Subsequently, proteins (amino acid concentrates) from the algae were extracted by enzymatic hydrolysis. The residual algae are subsequently liquefied into bio-oil in subcritical water. Enzymatic extraction of proteins is promising because it increases the bio-oil yield and reduces the nitrogen content.

Co-HTL is a process in which different organic substances are placed in a reactor to undergo a HTL reaction. Currently, co-HTL reactions of binary substances are the most common. Mixing different substances together to participate in the reaction not only produces a simple mixing effect but also produces a compound effect due to the synergistic or antagonistic effects between these substances. The aim of Co-HTL is to increase the yield and quality of biocrude oil, reduce the harshness of reaction conditions, and reduce the logistic cost through the synergistic effect of different substances, thus improving the economic efficiency of the whole process [187]. The analysis of synergistic or antagonistic effects can reveal co-HTL effects and optimize the type and ratio of feedstock combinations, thus facilitating the efficient conversion of biomass to biocrude oil [205]. Microalgae, sludge, pine, poplar, and cow dung are frequently used in co-HTL experiments. The synergistic and antagonistic mechanisms of these substances in co-HTL are described in detail below.

Under certain conditions, when a small amount of crude glycerin and poplar wood are mixed for Co-HTL experiments, there is a synergistic effect to reduce the formation of coke during the conversion of lignin. It has been speculated that glycerol can undergo dehydrogenation reactions under alkaline conditions and act as a hydrogen donor to stabilize or saturate active substances. On the contrary, it can degrade to produce ethanol and formic acid, which scavenges free radicals and provides hydrogen, thereby reducing the formation of coke. When rice husk (RH) is mixed with microalgae (*C. pyronoidosa*, CP), the content of organic acids in biocrude decreases and the hydrocarbon content increases as the ratio of CP/RH increases. Hence, the quality of biocrude is improved due to the synergistic effect between CP and RH [205]. The co-HTL of lignocellulosic substances and other substances not only produces a synergistic effect but also probably induces an antagonistic effect. Brilman et al. [206] observed an antagonistic effect during co-HTL experiments using *Desmodesmus* sp., beet pulp, and pine wood, and they found that the biocrude yield of mixed feedstocks was always lower than that of a single type of feedstock. They speculated that the intermediate products of microalgae degradation promoted the conversion of beet pulp into aqueous phase products and the conversion of pine wood into solid phase products, thereby decreasing the yield of biocrude.

In co-HTL, most microalgae show a synergistic effect on improving the yield and quality of biocrude. Jin et al. [187] used microalgae and macroalgae as the raw materials to carry out the co-HTL experiments, and they found that fatty acids in the microalgae catalyzed the hydrolysis of proteins and carbohydrates in the macroalgae. Further, a relatively higher yield of biocrude was obtained at 250 °C, thereby making HTL reaction conditions much milder. Similarly, two types of algae (*Cyanidioschyzon merolae*, CM, and *Galdieria sulfuraria*, GS) were mixed for experiments, and the yield of biocrude could be increased by 25.5% in the case of the strongest synergistic effect at 300 °C, 30 min, and CM/GA ratio of 4/1 [207]. Therefore, it is believed that fatty acids produced in the degradation process promote the conversion of organic substances into biocrude and strengthen the synergistic effect. However, the relevant synergistic mechanism in co-HTL is still unclear till now and therefore requires more research.

3 Chemical Reactions in the Catalytic Hydrogenation Upgrading of Biocrude

Although higher oil yields can be obtained through hydrothermal liquefaction, algal derived bio-oils still contain a large number of heteroatoms such as N, S, and O [25]. The long-chain fatty acids present in the oil contribute to a high viscosity. The high viscosity leads to poor cold flow characteristics that pose problems for transportation applications [132]. In addition, N and S in bio-oil will be converted into NO_x and SO_x, respectively, after the combustion and contribute to air pollution. Subsequent upgrading is necessary for bio-oil from microalgae before it is suitable for fuel [90]. Figure 14 shows that deoxygenation and ash removal from bio-oil can be achieved through purely thermal means without catalysts or hydrogen [132]. Although the original seaweed cells contain conventional catalytic metals such as Fe, Mg, Ni, and Zn, these metals make the upgrading process difficult. The extraction of HTL bio-oil from *Spirulina* and *Tetrahymena* by vacuum distillation without catalysts has been carried out, which greatly improved the yield (from 33–58 wt% to 73 wt%) and properties (97% deoxygenation and a significant reduction in metal content) [133].

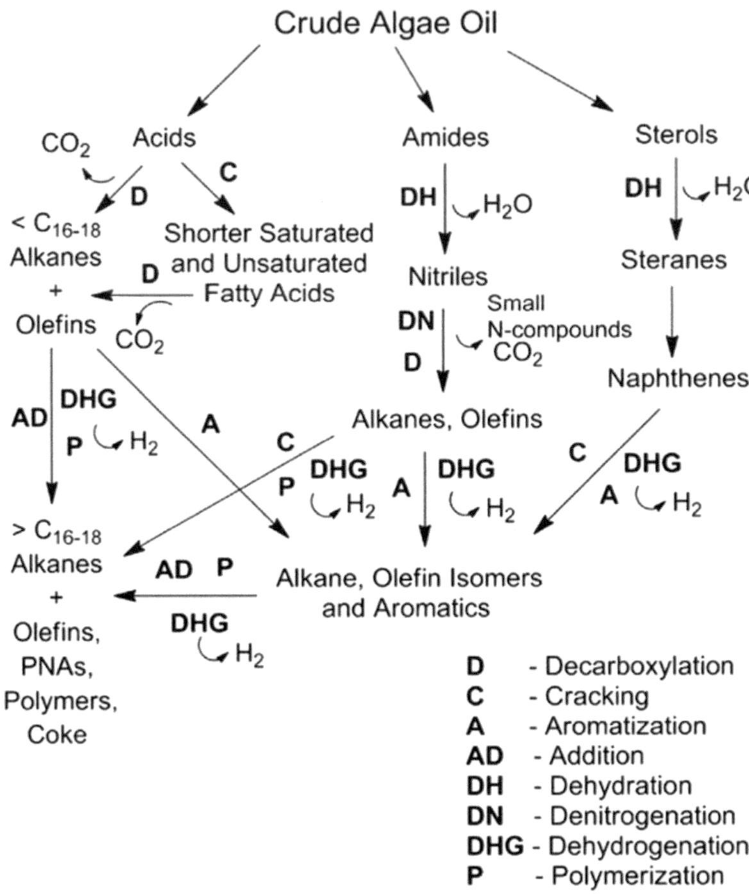

Fig. 14 Scheme for upgrading process of crude algal oil in hydrothermal reaction [208]

Recent studies have shown that the addition of catalysts significantly improves the upgrading process of algal derived bio-oils. Pt/C is very effective for decarboxylation of fatty acids [134, 135] as well as deoxygenation of benzofuran [136]. That Pt/C and high pressure H_2 (3.4 MPa) hydrothermally liquefied *Nannochloropsis* sp. to yield a bio-oil with lower fatty acid, O, and N content and the S was virtually eliminated from the upgraded oil [137]. They followed up with an optimization study on the four operating parameters: temperature (430–530 °C), reaction time (2–6 h), type of catalysts (Pt/C, Mo_2C and HZSM-5), and the catalyst loading (5–20 wt%). They found that liquefaction at 430 °C, 6 h, and 10 wt% of Mo_2C produced the highest content of saturated hydrocarbons [90]. Even without the presence of hydrogen, Pt/C can decarboxylate most of the lipids from *Chlorella* after 120 min at both 330 and 370 °C [135]. A follow-up study on ester model compounds showed that catalytic decarboxylation rate for longer fatty acid esters in water was faster than shorter chain esters. And this catalytic activity retained through its third recycling use.

Similarly, Pd/C is effective for upgrading bio-oil from *Nannochloropsis* sp. Both long reaction times and a high loadings of catalyst were beneficial for increasing the H/C ratio of the upgraded oil while simultaneously lowering the O/C and N/C In hydrothermal catalytic denitrification studies of pyridine (250–400 °C) [20], 138. Pt/γ-Al_2O_3 has the highest catalytic activity when compared to Pt/C, Pd/C, Ru/C, Rh/C, sulfided Pt/C and sulfide Pt/γ-Al_2O_3. HZSM-5 is catalytically active for hydrothermal upgrading of bio-oil from *Nannochloropsis* sp. [88]. The low cost of HZSM potentially makes it more economical than Pt and catalysts. More recently, the first bench scale catalytic hydroprocessing experiments using sulfided CoMo/F-Al_2O_3 (KF-1001, 4% Co, 15% Mo, 1/16 in diameter) catalyst were performed in a continuous-flow reactor by Elliott et al. at Pacific Northwest National Laboratory [99]. The study showed that hydrotreating effectively removed sulfur and nitrogen down to nearly immeasurably low levels. ratios, but gasification and coking were enhanced [90].

3.1 Biocrude Upgrading

Biocrude oils are usually dark brown, viscous, tarry, and have a smoky odor [209], and contain more than 400 organic compounds, typically consisting of phenols, hydrocarbons, fatty acids, carbonyl-containing compounds, nitrogen-containing compounds, and some mixtures of gases and solids [210]. Depending on the type of functional group, biocrude oils include monoaromatic and monoheterocyclic compounds, aliphatic compounds, oxygenated compounds (e.g., 4-chloro-2-nitrobenzyl alcohol and 3-methylbutanol ether), nitrogen-containing compounds (e.g., 2,5-piperazinediones and 3-benzyl-6-isopropyl), and polyaromatic compounds (e.g., 7-ethyl-4,6-heptanediones and 2-(m-trimethylolazolyl) naphthalene). Biocrude oil has the potential to be used as a gas turbine fuel, and its physical and chemical properties depend mainly on the feedstock [211], reaction conditions (e.g., reaction temperature [128], residence time [128, 140, 212], solvent type [213], and catalyst type and amount, and process route [214]. Due to the complexity and variation in the reaction conditions, there are large differences in the yield and quality of biocrude. Table 4 lists the physical and chemical properties of biocrude derived from biomass HTL. It can be seen that the properties of biocrude oil from various biomasses vary considerably. The N content of biocrude oil from microalgae, cow dung or sludge is significantly higher than that of biocrude oil from bark and straw, which is consistent with the protein content of the biomass. The O content does not have a significant relationship with the type of biomass [215], but its average value is about 16.8 wt%, which is significantly higher than that of conventional diesel. The C and H contents of the biocrude oil ranged from 45.7 to 80.2 wt% and 6.0–11.6 wt%, respectively, and their values were lower than those of diesel. In addition, the moisture content,

HHV, density, and other indicators of biocrude oil extracted from biomass are still far from those of transportation fuels.

In fact, the composition and properties of biocrude oil obtained from the thermal liquefaction of biomass can be closer to those of fossil fuels by selecting appropriate thermal liquefaction conditions and catalysts. However, the following problems still exist and limit the use of biocrude oil [215], biocrude oil has high viscosity and is not

Table 4 Physical and chemical properties of biocrude derived from biomass HTL

Biomass	Biocrude							
	Element composition (wt%)					Density (kg/m^3)	HHV (MJ/kg)	References
	C	H	O	N	S			
Microalgae (*Spirulina platensis*)	45.73	5.96	47.04	1.27	–[a]	1380	15.60	[53]
Microalgae (*D. tertiolecta*)	68.53	8.65	19.75	2.89	0.18	–	31.97	[58]
Microalgae (*Chlorella*)	74.70	10.40	5.50	7.70	0.76	–	39.80	[216]
Microalgae (*C. pyrenoidosa*)	56.14	6.49	32.40	4.97	–	1100–1250	26.62	[217]
Microalgae (*Dunaliella tertiolecta*)	78.60	10.40	5.30	4.20	0.50	960	–	[139]
Microalgae	66.01	11.56	12.43	9.54	0.46	1020	36.76	[214]
Microalgae (*Spirulina*)	70.31	8.72	11.27	9.70	–	970	34.33	[218]
Microalgae and lemon peel	69.18	8.17	12.92	9.73	–	–	33.00	[219]
Swine manure	72.58	9.76	13.19	4.47	–	–	36.05	[220]
Swine manure	73.66	11.61	11.48	3.25	–	–	33.98	[221]
Cow manure	73.72	8.13	16.76	1.39	–	–	35.53	[61]
Willow	78.15	8.50	12.40	0.95	–	–	38.57	[222]
Aspen	75.20	8.20	15.80	0.50	0.3	–	34.30	[30]
Bark	80.21	9.42	9.95	0.42	–	–	38.87	[223]
Beech wood	65.95	7.04	27.01	–	–	–	27.53	[224]
Straw	67.89	6.42	23.18	0.75	0.56	–	27.29	[130]
Sludge	75.13	9.41	8.97	5.31	1.17	–	37.35	[115]
Diesel[b]	86.00	13.20	0.80	–	–	817	47.70	[53]

[a] "–" means that the value is not reported
[b] It refers to general diesel

favorable for transportation. Compared to transportation fuels, biocrude oils do not provide sufficient power due to their higher O content and lower C and H content and HHV. The N and S content results in the formation of NOx and SOx after combustion. In addition, the total acid value of biocrude oil is as high as 256 mg KOH g^{-1}, thereby easily inducing a strong corrosion effect. Therefore, the abovementioned information shows the importance of biocrude upgrading after biomass HTL.

Through biocrude upgrading, the content of the desired components can be increased while substances that are not conducive to the quality of biocrude are reduced.

Table 5 lists the comparisons of common biocrude upgrading methods. Upgrading methods can be divided into physical chemistry, biochemistry, and thermochemistry. There is no strict limitation between two thermochemistry methods. Hydrocracking reactions may occur during the hydrotreating process, and hydrogen can be used during supercritical fluid treatment to improve the final effect. The viscosity, density, acid value, and stability of biocrude can be improved through these upgrading methods.

Figure 15 shows a schematic flow diagram for the production of biocrude oil from the thermal liquefaction of biomass and subsequent upgrading treatments. Biomass is converted to a four-phase product by the thermoliquefaction process and then separated by centrifugation to obtain the desired biocrude oil, which is further upgraded to improve its quality by emulsification, electrocatalysis, esterification, hydrotreating, hydrocracking, and/or supercritical or subcritical fluid treatments. The chemical reactions involved are important for revealing the upgrading mechanism of the biocrude. Therefore, the chemical reactions involved in the catalytic hydrotreating upgrading process are discussed in detail below.

3.2 *Hydrogenation*

Apparently, hydrocarbons are the desired components of bio-oil [245]. Hydrotreating can considerably increase the content of hydrocarbons in bio-oil. Figure 16 shows the two main chemical reactions that occur during hydrogenation of components in biocrude oil. One is the hydrogenation of unsaturated substances in the presence of hydrogen and a catalyst, where C=C and C–O bonds are converted to C–C and C–O bonds, respectively. This greatly reduces the proportion of unsaturated substances, increases saturated hydrocarbons and improves the stability of bio-oils [96]. The other is the hydrogen cleavage reaction with the help of hydrogen, which is a process that breaks the C–C bond and cleaves complex macromolecular organic substances into simpler small molecules, thus reducing the viscosity of the biocrude oil and making it easy to flow. For example, Duan et al. [93] confirmed that no olefins were detected in the upgraded bio-oil and the alkane content was considerably increased when Pt/C and H_2 were used for hydrothermal upgrading. This indicates that these olefins are hydrogenated and some of them undergo hydrocracking reactions. Similarly, thermal cracking decomposes larger hydrocarbons into smaller hydrocarbons or even into

Table 5 Comparisons of common biocrude upgrading methods

Methods	Emulsification/ microemulsification	Electrocatalysis	Esterification	Hydrotreating	Hydro-cracking	Sup-or subcritical fluid treatment
Category	Physical chemistry	Physical chemistry	Biochemistry	Thermochemistry	Thermochemistry	Thermochemistry
Content	Using emulsifiers and microemulsifies to mix two liquids into a uniform whole [225]	Hydrogenating organic materials in the cathode compartment [226, 227]	Producing esters via reactions of alcohols with acids, aldehydes, and ketones in biocrude [228]	Acting with hydrogen to separate impurities or saturate C–C bonds [229]	Combining cracking and hydrogenation with bifunctional catalyst [48]	Realize biocrude upgrading in a medium near its critical point [230]
Advantage	Reducing viscosity, increase stability and density [231–233]	No H_2 storage and supply problems, controlling reaction rate [234]	Decreasing viscosity and acid value, improving volatility [48]	Simple process, mild reaction conditions, removing heteroatoms [235]	Obtaining more light components, reducing contents of water and oxygen [229, 236]	Obtaining more light components, and less coke, char and tar, promoting mass and heat transfer [49, 237–239]
Disadvantage	Lowering flash point, reducing HHV [232, 240]	Needing to develop catalyst [241, 242]	Unclear impact of simultaneous esterification and acetalization [243]	Low yield, forming char and tar, and catalyst deactivation [48, 244]	Severe reaction conditions, not economical and energy efficient [235]	Expensive media [244]

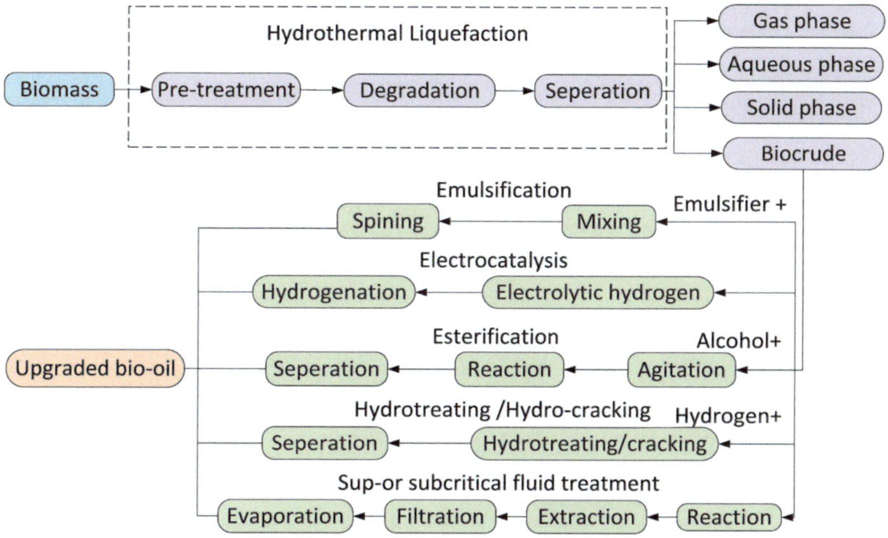

Fig. 15 Schematic diagram for biomass HTL and biocrude upgrading processes

gases (C_xH_y, where $x < 5$ and $y < 12$). These gases are eliminated together with H_2 (not involved in the reactions), CO_2, CO, NH_3, and other gases. The present research shows that bio-oil components after hydrocracking tend to be similar regardless of the initial biomass [12].

Temperature has a great influence on the hydrogenation process, and increasing the temperature can increase the hydrogenation rate and hydrogenation degree of biocrude oil [247]. When the temperature was increased from 340 °C to 400 °C, the hydrogen content in the biocrude oil increased significantly. The number of branched chains and chain lengths of hydrocarbons in the bio-oil was reduced, and the interaction of alkanes was weakened, which decreased the viscosity of the bio-oil. However, Elliott et al. [248] concluded that despite the consumption of hydrogen at higher temperatures, the corresponding bio-oil yield did not show an upward trend due to the cracking of some hydrocarbons into gas. Hydrogenation of biocrude oil has a great impact on the stability, viscosity, and combustion characteristics of biocrude oil. Some components escaped as gases after cracking, thus reducing the bio-oil yield.

Analysts play an important role in the hydrocracking process. It reduces large molecules and increases small and light molecules, thus lowering the boiling point of bio-oil. Catalysts used in the hydrogenation process include zeolite catalysts, metal catalysts, etc., and different catalysts have different effects on the composition of bio-oil [249]. For example, a large amount of phenol can be dehydrated, deoxygenated, and hydrogenated to form cyclopentane, cyclohexane, cyclohexene, and other substances under Ni-Mo/γ-Al catalyst conditions. With the increase of Y-zeolite catalyst dosage, the content of aromatic compounds in biocrude oil decreases while the content of olefinic compounds increases [249]. This is due to the fact that

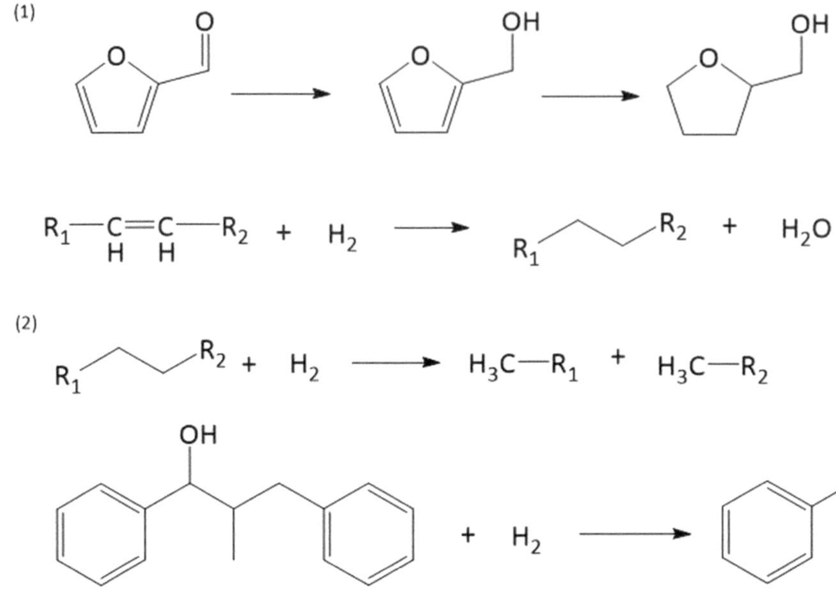

Fig. 16 Significant chemical reactions in the hydrogenation process [11, 12, 246]

more benzene rings are broken down into chain hydrocarbons in the presence of the catalyst, thus increasing the amount of olefins that can be hydrogenated into alkanes. In addition, the most important role of the catalyst is to increase the chemical reaction rate. In the presence of acid and Pt/C, the consumption of H_2 is significantly increased, thus facilitating the hydrogenation process. Hydrogenation upgrading with the addition of ZSM-5 inhibits the formation of solid and gaseous materials and improves the yield and quality of biocrude oil [250]. Nonetheless, the effects of adding metal catalysts (e.g., Pt and Ni) are more remarkable than that of adding ZSM-5.134. A catalyst has different catalytic mechanisms on the reactions. Therefore, there may be synergistic effects among the different catalysts used together. This is an important direction for the study of catalysts for biocrude upgrading.

3.3 *Deoxygenation*

Oxygenated compounds in biocrude are mainly phenols, esters, carboxylic acids, aldehydes/ketones, etc. [251] with a relatively high content of phenols. Phenols, cresols, furfural, dibenzofurans, palmitic acid, and glycerol are often used as model compounds in experiments to explore the deoxygenation process of oxygenated compounds for a better understanding of the corresponding reactions [252]. Kay et al. [253] summarized the deoxygenation process of biocrude oils using model compounds and concluded that deoxygenation can be achieved by hydrodeoxygenation, decarboxylation, and decarbonylation. In this process, oxygen is usually removed in the form of water, carbon monoxide, and carbon dioxide. The various forms of deoxygenation correspond to different types of functional groups, such as carbon dioxide for carboxylic acids, carbon monoxide for ketones, aldehydes and carboxylic acids, and water for alcohols. Organic acids, phenols, and esters are typical oxygenated compounds and are found in relatively high amounts in biocrude oils, so the deoxygenation of these three types of compounds is described below.

The presence of organic acids increases the acid value of biocrude. Organic acids can be hydrodeoxygenated to obtain hydrocarbons so that the acid value can be decreased through decarboxylation. Figure 17a shows the chemical reactions of organic acids in hydrodeoxygenation. Organic acids are mainly derived from lipids and mainly include palmitic acid, stearic acid, and oleic acid [254]. The carboxyl group in the acid is hydrogenated to produce an alkanal (intermediate), which undergoes two main reaction pathways. The main pathway is the sequential hydrogenation, dehydration and hydrogenation reactions, which finally produce alkanes. The most abundant substances in upgraded bio-oils are pentadecane and heptadecane, which may be produced by decarboxylation of n-hexadecanoic acid and 9-octadecenoic acid [255]. Another pathway is the removal of carbonyl groups to produce carbon dioxide and olefins, which are then hydrogenated to produce alkanes. Of course, these olefins can also be cyclized, oligomerized, and aromatized to form aromatic compounds. It has been shown that the hydrodeoxygenation process exists for organic acids with or without a catalyst, but the addition of a metal catalyst or a basic catalyst can facilitate the decarboxylation reaction, converting more of the acid to alkanes. For example, Pt/C favors the hydrothermal decarboxylation reaction of fatty acids, so that fatty acids can be removed from biocrude oil in large quantities and the total acid value can be greatly reduced. [242]. Co/SiO_2, Fe/SiO_2, and $Fe–Co/SiO_2$ catalysts are beneficial to hydrodeoxygenate biocrude, and the biocrude properties can be improved; further, the effect was more remarkable due to the synergy between Fe and Co [256]. Increasing the temperature can promote the deoxygenation reactions. With an increase in temperature, the content of fatty acid amides and other oxygenates can be gradually decreased and the content of hydrocarbons increases in bio-oil [245]. However, an excessively high temperature can cause unstable substances to form coke-like materials [252].

Phenol is commonly used as a model compound for exploring the transformation of phenolic compounds, and its deoxygenation process can be categorized into the following two pathways. The first one is hydrodeoxygenation. Phenol successively produces cyclohexanone and cyclohexanol in the presence of hydrogen, and then cyclohexanol is deoxygenated to water and hydrogenated to produce cyclohexane [261]. Conversion is mainly carried out through this route when Pd/C and H_3PO_4 are used as dual-function catalysts to upgrade biocrude. Pd/C and H_3PO_4 promote both hydrogenation and deoxygenation in reactions [262]. Figure 17b shows the reaction pathway for phenol hydrodeoxygenation. The second one is hydrodeoxygenation. First, the C-O bond in phenol is broken to water to remove the oxygen and then hydrogenated to cyclohexene and cyclohexane. This pathway occurs mainly when

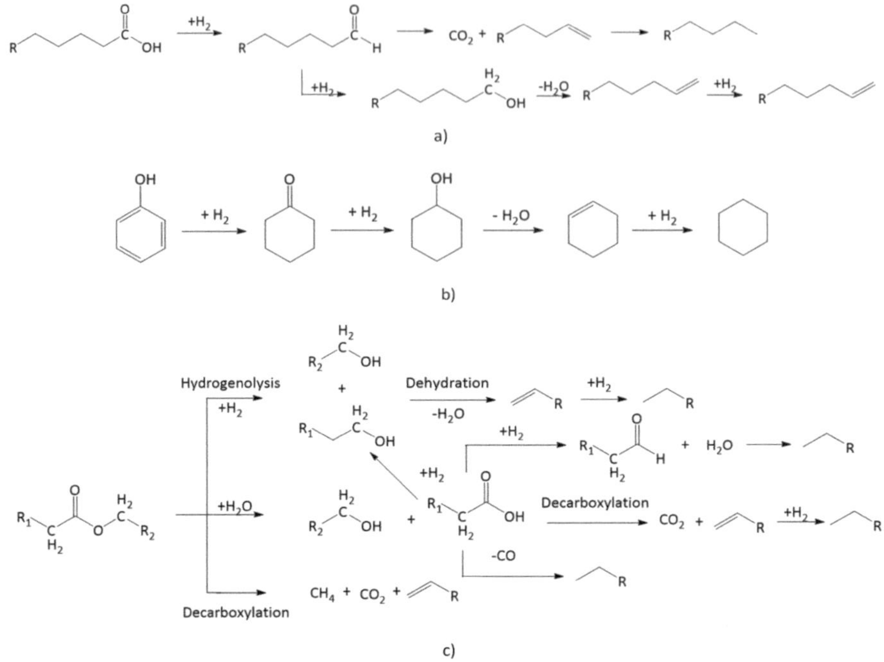

Fig. 17 Chemical reactions in deoxygenation of typical oxygenated compounds **a** organic acids [257], **b** phenol [1, 258], **c** Ester [259, 260]

MoO_3/TiO_2 is used as the catalyst. However, the products are mainly benzene with a relatively low content of non-aromatic compounds such as cyclohexane, cyclohexene, and methylcyclopentane. This may be due to the insufficiently good hydrogenation activity of the catalyst. The deoxygenation of phenol is influenced by the catalyst and temperature. Phenol cannot be completely deoxygenated at 300 °C, but it can be completely deoxygenated when catalytic upgrading is carried out at 350 °C, which increases the content of cyclohexane and benzene in the bio-oil. Zhao et al. [263]. found that cyclohexane was not detected in products at 150 °C, but the cyclohexane yield was considerably improved at 200 and 250 °C. This may be explained by the fact that the content of hydrogen ions in solution increases as the temperature increases, thereby promoting hydrogenolysis and hydrogenation reactions.

Methyl heptanoate and methyl caproate are used as model compounds to study the deoxygenation process of esters. Figure 17c shows the reaction pathways of esters in deoxygenation. Some esters are sequentially hydrogenolyzed, dehydrated, and hydrogenated to produce alkanes. However, some esters are hydrolyzed to form carboxylic acids and alcohols. Some carboxylic acids are reduced to aldehyde, and the aldehyde is dealdehyded to olefins and alkanes. Some carboxylic acids are converted into alcohol and the reactions then proceed according to the first route. The remaining carboxylic acids are decarboxylated into the corresponding olefins, which are further hydrogenated into alkanes. Moreover, some esters can also directly generate olefins, methane, and CO_2.

For aldehydes or ketones, Sanna et al. [264] investigated the catalyst to upgrade water-soluble components in biocrude and stated that carbonyls were hydrogenated to generate alcohols, such as hydroxyacetaldehyde conversion to ethylene glycol, hydroxyacetone conversion to propylene glycol, and polyol conversion to monoalcohol, which can form alkane compounds via secondary hydrogenation.

3.4 Denitrogenation

Nitrogen-containing compounds in biocrude include indole, pyridine, pyrazine, amines, etc. Further, proteins in feedstock are the main source of nitrogen-containing compounds [118]. Compared with biocrude oil, the decrease in N content in treated bio-oil is smaller than the decrease in O and S content. This suggests that during the upgrading process, the hydrogenolysis reaction of C–N bonds is limited and N heteroatoms are difficult to remove. The mechanism of nitrogenation in biocrude oil is mainly saturation, elimination and nucleophilic substitution of intermediates, and the nitrogen is usually removed in the form of ammonia. Figure 18 shows the main chemical reactions involved in the hydrodenitrogenation of pyridine. Pyridine has received much attention as a model compound for nitrogen-containing compounds in biological crude oils. Due to the high energy required to break the C–N bond of the aromatic ring in organic compounds, pyridine is first hydrogenated and ring-opened to form piperidines and pentylamines, and then denitrified to form alkanes and ammonia. Duan et al. [265] performed denitrification reactions during supercritical upgrading to convert piperidines and indoles to olefins and alkanes, but not all substances can be denitrified [266]. These processes are accomplished through a sequence of hydrogenation and hydrogenolysis. Under milder conditions, hydrogenation is the main reaction and the products are mainly piperidines and their alkyl-substituted derivatives. However, at higher reaction temperatures and longer reaction times, hydrogenolysis becomes the main reaction. In addition to pyridine, nitroquinone, pyrrole, indole, and carbazole have been used as model compounds to study the reactions during denitrification of biocrude oil. Due to the different activation energies of different bonds, their reaction order is also hydrogenation followed by hydrolysis [96].

A two-component catalyst is commonly used in denitrogenation, such as transition metals, activated carbon, etc. [96]. Adding Ni to WP/SiO_2 can enhance the active sites and the C–N bond breakage in the hydrogenation process to strengthen the denitrogenation reactions and increase the yields of pentane and pentamine. Adding activated carbon can catalyze the removal of N, O, and S, and this ability is usually better than that of Ru/C [254]. They explained that this was due to the activated carbon immersing ions into the solution to promote the removal of heteroatoms. Experiments have found that treating a catalyst before the reactions and adjusting the acid site of the catalyst can improve the denitrogenation performance of the

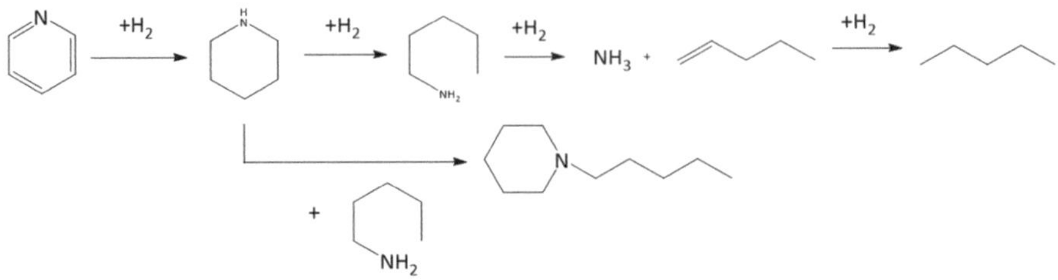

Fig. 18 Chemical reactions in the denitrogenation process of pyridine [254, 267, 268]

catalyst. For example, modifying the NiWF/γ-Al_2O_3 catalyst with an appropriate amount of F—can improve the acid sites and the total number of acid sites in the catalyst, thereby improving the catalyst activity and enhancing the denitrogenation reaction [96].

Temperature affects catalyst activity, bond breaking, and the rate of chemical reactions. An increase in temperature not only accelerates the rate of chemical reaction but also increases the likelihood of certain bonds breaking. For example, the denitrogenation conversion of Pt/Al_2O_3 increases from about 45–80% when the temperature is increased from 300 °C to 340 °C. This is because the C–N bonds are more susceptible to breakage at higher temperatures. This is due to the fact that C–N bonds are more likely to break at higher temperatures.

Hydrogen pressure affects the hydrogenation and hydrogenolysis reactions during denitrification. At lower hydrogen pressures, the denitrification process does not proceed completely and stays mainly in the hydrogenation reaction stage; therefore, the products of pyridine denitrification are mainly alkyl-substituted pyridines [254]. When the hydrogenolysis stage is reached, a large number of alkanes are produced. The amount of hydrogen consumed during denitrification of nitrogen-containing compounds depends on the amount of chemical reactions occurring [269]. Therefore, in order to ensure the efficiency of denitrification of nitrogen-containing compounds, the hydrogen pressure must be maintained at a relatively high level.

3.5 Desulfurization

Sulfur in biocrude oil is mainly derived from proteins. Desulfurization is the process of converting sulfur-containing substances in biocrude oil into hydrocarbons [270]. Due to the low content of sulfur in the compounds, the values are sometimes below the detection limit of the instrument. The desulfurization process is relatively simple, so some special studies have been carried out on the desulfurization reactions during the purification of biocrude oils. In biocrude oil, most of the S atoms are present in the form of thermally unstable S-containing compounds such as thiols and thioethers [271]. Most sulfur-containing compounds can be decomposed and removed in the form of H_2S as the temperature rises in the hydrogenation process. Therefore,

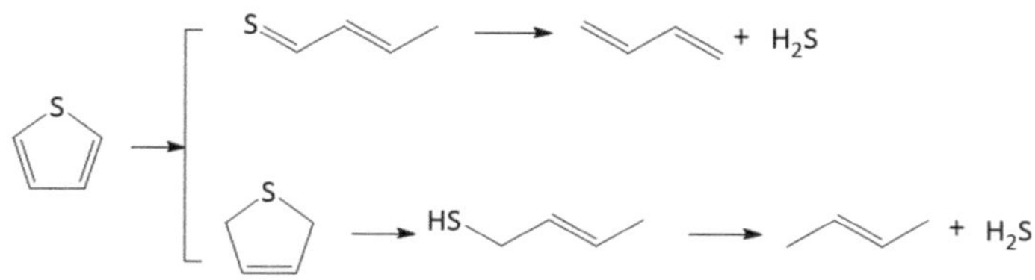

Fig. 19 Main chemical reactions in the desulfurization process of thiophene [273]

biocrude can usually be completely desulfurized only via the catalytic hydrogenation treatment [95].

Thiophene is a sulfide with high thermal stability. Its desulfurization efficiency is only 1% without any catalyst and at 300 °C for 5 h [272]. Therefore, thiophene is used as a model compound to explain the chemical reactions in the desulfurization process of biocrude. Figure 19 shows the main chemical reactions in the desulfurization of thiophene. Some of these processes are the same as those for nitrogen-containing compounds. First hydrogenation takes place, followed by hydrogenation ring opening and desulfurization reactions to remove S as H_2S. Of course, the other route is direct desulfurization, where the C–S bonds in thiophene are broken and the ring is opened in the absence of hydrogen [273]. This step is much more effective in terms of desulfurization, as its intermediates are desulfurized to produce H_2S and olefins. It is worth noting that there is a competition between these two desulfurization methods. In the desulfurization process, nitrogen-containing substances and sulfur-containing substances also have competitive relationships for denitrification and desulfurization, respectively. For example, the presence of nitrogen-containing compounds inhibits the desulfurization reaction in biocrude oil when pyridine and thiophene are simultaneously denitrified and desulfurized [272].

Catalysts are known to have a significant impact on chemical reactions. Catalysts also affect the desulfurization process of sulfur-containing compounds. Platinum and palladium are the two catalysts usually chosen for desulfurization processes and they have good catalytic properties [274, 275]. Pt and Pd have different selectivities in the two desulfurization processes mentioned above. In general, Pt has higher selectivity for direct desulfurization, while Pd has higher selectivity for hydrodesulfurization [275]. The removal efficiency of sulfur increases with the amount of catalyst used [276]. Pretreatment of Pd with montmorillonite or early chlorination of Pd and Pt can increase the efficiency of desulfurization and remove more S in the form of H_2S.In addition, nickel, molybdenum, cobalt, magnesium, and aluminum can be used in the desulfurization process, and all of them can increase the desulfurization efficiency of sulfur-containing compounds to a certain extent [277].

4 Prospects and Challenges of Biomass HTL and Biocrude Upgrading

High-temperature liquefaction technology can produce liquid biofuels for energy consumption and environmental protection. It is promising due to its high yield, low equipment requirements and high energy density [278]. As mentioned above, a lot of research has been carried out on the chemical reaction pathways in the thermal liquefaction of biomass and catalytic biocrude upgrading, which can help to improve its operating conditions. Catalysts can significantly improve the yield and quality of biocrude oil, and multi-component catalysts are more effective than single-component catalysts due to the synergistic effect of different components. Currently, there are many types of single-component catalysts, such as acid catalysts, base catalysts, and metal catalysts. However, there are few studies related to multi-component catalysts, so it is valuable to further explore multi-component catalysts. Catalytic thermal liquefaction of biomass and catalytic upgrading of biocrude oil have some limitations, such as catalyst poisoning, coking, plugging, sintering, low stability, and short service time. These issues need to be taken into account when developing catalysts for bio-oil production through thermal catalytic hydrogenation of biomass. In addition, the whole high-temperature liquefaction and catalytic upgrading process is complex, and one catalyst usually catalyzes or even negatively affects some reactions. Therefore, it is necessary to consider the catalysts with multiple indicators and explore their effects in a comprehensive manner.

Biomass thermal liquefaction technology can be improved to reduce energy input and increase conversion efficiency. In addition to isothermal thermoliquefaction, rapid thermoliquefaction and multistage thermoliquefaction can also be used to convert biomass to biocrude oil. Gollakota et al. [77] observed that rapid thermoliquefaction for 1 min at 500 °C resulted in higher biocrude oil yields than isothermal thermoliquefaction for 100 min at 350 °C. This rapid thermoliquefaction method has a low heat input and is therefore promising. This rapid HTL method is promising because of its low heat input, high biomass loading, and economy. He et al. [279] reported that the biocrude oil obtained from multistage HTL was similar to that obtained from optimal isothermal HTL with the right combination of reaction temperature and residence time. However, these two high-temperature liquefaction methods have been less studied, and their suitable reaction conditions and specific conversion processes are still unknown, so further detailed studies are still needed.

Currently, most biomass thermal liquefaction and catalytic biocrude upgrading experiments are still at the laboratory stage, on a relatively small scale and in batch processing. This is not suitable for future large-scale continuous commercial production. Therefore, it is necessary to promote continuous large-scale experiments in subsequent studies until actual industrial production is realized. Several research groups have designed continuous reaction systems for thermal liquefaction of biomass [280]. Through their experiments, some issues need to be investigated in subsequent work, such as the pumpability and behavior of biomass slurry, salt precipitation, secondary reactions, and catalyst recovery and replacement. Meanwhile, biocrude purification has similarities with crude oil processing. Therefore, the operational flow and management system of crude oil refining can be studied, which is conducive to the rapid diffusion of the process.

References

1. D. Xu, G. Lin, S. Guo, S. Wang, Y. Guo, Z. Jing, Catalytic hydrothermal liquefaction of algae and upgrading of biocrude: a critical review. Renew. Sustain. Energy Rev. **97**, 103–118 (2018)
2. Y. Chisti, Biodiesel from microalgae. Biotechnol. Adv. **25**, 294–306 (2007)
3. C. Posten, G. Schaub, Microalgae and terrestrial biomass as source for fuels—a process view. J. Biotechnol. **142**, 64–69 (2009)
4. F. Behrendt, Y. Neubauer, M. Oevermann, B. Wilmes, N. Zobel, Direct liquefaction of biomass. Chem. Eng. Technol. **31**, 667–677 (2008)
5. M. Kishimoto, T. Okakura, H. Nagashima, T. Minowa, S.-Y. Yokoyama, K. Yamaberi, CO_2 fixation and oil production using micro-algae. J. Ferment. Bioeng. **78**, 479–482 (1994)
6. L. Brennan, P. Owende, Biofuels from microalgae—a review of technologies for production, processing, and extractions of biofuels and co-products. Renew. Sustain. Energy Rev. **14**, 557–577 (2010)
7. K. Anastasakis, A.B. Ross, Hydrothermal liquefaction of the brown macro-alga *Laminaria saccharina*: effect of reaction conditions on product distribution and composition. Biores. Technol. **102**, 4876–4883 (2011)
8. V. Patil, K.Q. Tran, H.R. Giselrød, Towards sustainable production of biofuels from microalgae. Int. J. Mol. Sci. **9**, 1188–1195 (2008)
9. K. Tsukahara, S. Sawayama, Liquid fuel production using microalgae. J. Jpn. Petrol. Inst. **48**, 251–259 (2005)
10. S. Amin, Review on biofuel oil and gas production processes from microalgae. Energy Convers. Manage. **50**, 1834–1840 (2009)
11. M. Saber, B. Nakhshiniev, K. Yoshikawa, A review of production and upgrading of algal bio-oil. Renew. Sustain. Energy Rev. **58**, 918–930 (2016)
12. S. Arvindnarayan, K.K. Sivagnana Prabhu, S. Shobana, G. Kumar, J. Dharmaraja, Upgrading of micro algal derived bio-fuels in thermochemical liquefaction path and its perspectives: a review. Int. Biodeteriorat. Biodegrad. **119**, 260–272 (2017)
13. D. Xu, P.E. Savage, Characterization of biocrudes recovered with and without solvent after hydrothermal liquefaction of algae. Algal Res. **6**, 1–7 (2014)
14. M. Kumar, A. Olajire Oyedun, A. Kumar, A review on the current status of various hydrothermal technologies on biomass feedstock. Renew. Sustain. Energy Rev. **81**, 1742–1770 (2018)
15. X.-Q. Ma, J.-J. Liao, D.-B. Chen, Z.-X. Xu, Hydrothermal carbonization of sewage sludge: catalytic effect of Cl^- on hydrochars physicochemical properties. Mol. Catal. **513**, 111789 (2021)
16. D. López Barreiro, W. Prins, F. Ronsse, W. Brilman, Hydrothermal liquefaction (HTL) of microalgae for biofuel production: state of the art review and future prospects. Biomass Bioenergy **53**, 113–127 (2013)
17. R. Cherad, J.A. Onwudili, P. Biller, P.T. Williams, A.B. Ross, Hydrogen production from the catalytic supercritical water gasification of process water generated from hydrothermal liquefaction of microalgae. Fuel **166**, 24–28 (2016)
18. H. Li, Z. Ai, L. Yang, W. Zhang, Z. Yang, H. Peng, L. Leng, Machine learning assisted predicting and engineering specific surface area and total pore volume of biochar. Biores. Technol. **369**, 128417 (2023)

19. S. Sawayama, T. Minowa, S.Y. Yokoyama, Possibility of renewable energy production and CO_2 mitigation by thermochemical liquefaction of microalgae. Biomass Bioenergy **17**, 33–39 (1999)
20. Y.F. Yang, C.P. Feng, Y. Inamori, T. Maekawa, Analysis of energy conversion characteristics in liquefaction of algae. Resour. Conserv. Recycl. **43**, 21–33 (2004)
21. T. Minowa, S.-Y. Yokoyama, M. Kishimoto, T. Okakura, Oil production from algal cells of *Dunaliella tertiolecta* by direct thermochemical liquefaction. Fuel **74**, 1735–1738 (1995)
22. Y. Guo, T. Yeh, W. Song, D. Xu, S. Wang, A review of bio-oil production from hydrothermal liquefaction of algae. Renew. Sustain. Energy Rev. **48**, 776–790 (2015)
23. D. Zhou, L. Zhang, S. Zhang, H. Fu, J. Chen, Hydrothermal liquefaction of macroalgae *Enteromorpha prolifera* to bio-oil. Energy Fuels **24**, 4054–4061 (2010)
24. P. Kiran Kumar, S. Vijaya Krishna, K. Verma, K. Pooja, D. Bhagawan, K. Srilatha, V. Himabindu, Bio oil production from microalgae via hydrothermal liquefaction technology under subcritical water conditions. J. Microbiol. Methods **153**, 108–117 (2018)
25. S.N. Hansen, A. Mirkouei, L.A. Diaz, A comprehensive state-of-technology review for upgrading bio-oil to renewable or blended hydrocarbon fuels. Renew. Sustain. Energy Rev. **118**, 109548 (2020)
26. D. Klingler, J. Berg, H. Vogel, Hydrothermal reactions of alanine and glycine in sub- and supercritical water. J. Supercrit. Fluids **43**, 112–119 (2007)
27. C. Gai, Y. Zhang, W.-T. Chen, P. Zhang, Y. Dong, Thermogravimetric and kinetic analysis of thermal decomposition characteristics of low-lipid microalgae. Biores. Technol. **150**, 139–148 (2013)
28. N. Sato, A.T. Quitain, K. Kang, H. Daimon, K. Fujie, Reaction kinetics of amino acid decomposition in high-temperature and high-pressure water. Ind. Eng. Chem. Res. **43**, 3217–3222 (2004)
29. L. Zhou, D. Liu, J. Wang, Z.L. Wang, Triboelectric nanogenerators: fundamental physics and potential applications. Friction **8**, 481–506 (2020)
30. T.H. Pedersen, I.F. Grigoras, J. Hoffmann, S.S. Toor, I.M. Daraban, C.U. Jensen, S.B. Iversen, R.B. Madsen, M. Glasius, K.R. Arturi, R.P. Nielsen, E.G. Søgaard, L.A. Rosendahl, Continuous hydrothermal co-liquefaction of aspen wood and glycerol with water phase recirculation. Appl. Energy **162**, 1034–1041 (2016)
31. P.J. Valdez, M.C. Nelson, H.Y. Wang, X.N. Lin, P.E. Savage, Hydrothermal liquefaction of *Nannochloropsis* sp.: systematic study of process variables and analysis of the product fractions. Biomass Bioenergy **46**, 317–331 (2012)
32. D.K. Alargov, S. Deguchi, K. Tsujii, K. Horikoshi, Reaction behaviors of glycine under super- and subcritical water conditions. Orig. Life Evol. Biosph. **32**, 1–12 (2002)
33. O.İ Şenol, E.M. Ryymin, T.R. Viljava, A.O.I. Krause, Reactions of methyl heptanoate hydrodeoxygenation on sulphided catalysts. J. Mol. Catal. A: Chem. **268**, 1–8 (2007)
34. L. Zhang, C. Xu, P. Champagne, Overview of recent advances in thermo-chemical conversion of biomass. Energy Convers. Manage. **51**, 969–982 (2010)
35. H.-J. Huang, X.-Z. Yuan, Recent progress in the direct liquefaction of typical biomass. Prog. Energy Combust. Sci. **49**, 59–80 (2015)
36. U. Jena, K.C. Das, J.R. Kastner, Comparison of the effects of Na_2CO_3, $Ca_3(PO_4)_2$, and NiO catalysts on the thermochemical liquefaction of microalga *Spirulina platensis*. Appl. Energy **98**, 368–375 (2012)
37. D.M. Gao, T. Kobayashi, S. Adachi, Kinetic analysis for the isomerization of glucose, fructose, and mannose in subcritical aqueous ethanol. Biosci. Biotechnol. Biochem. **79**, 1005–1010 (2015)
38. P. Ranganathan, S. Savithri, Computational fluid dynamics simulation of hydrothermal liquefaction of microalgae in a continuous plug-flow reactor. Biores. Technol. **258**, 151–157 (2018)
39. P. Manara, A. Zabaniotou, Towards sewage sludge based biofuels via thermochemical conversion—a review. Renew. Sustain. Energy Rev. **16**, 2566–2582 (2012)

40. J. Yu, M. Audu, M.T. Myint, F. Cheng, J.M. Jarvis, U. Jena, N. Nirmalakhandan, C.E. Brewer, H. Luo, Bio-crude oil production and valorization of hydrochar as anode material from hydrothermal liquefaction of algae grown on brackish dairy wastewater. Fuel Process. Technol. **227**, 107119 (2022)
41. P.J. Valdez, V.J. Tocco, P.E. Savage, A general kinetic model for the hydrothermal liquefaction of microalgae. Bioresour. Technol. **163**, 123–127 (2014)
42. D. López Barreiro, S. Riede, U. Hornung, A. Kruse, W. Prins, Hydrothermal liquefaction of microalgae: effect on the product yields of the addition of an organic solvent to separate the aqueous phase and the biocrude oil. Algal Res. **12**, 206–212 (2015)
43. M. Tabatabaei, M. Aghbashlo, E. Valijanian, H. Kazemi Shariat Panahi, A.-S. Nizami, H. Ghanavati, A. Sulaiman, S. Mirmohamadsadeghi, K. Karimi, A comprehensive review on recent biological innovations to improve biogas production, Part 1: Upstream strategies. Renew. Energy **146**, 1204–1220 (2020)
44. L. Qian, S. Wang, P.E. Savage, Hydrothermal liquefaction of sewage sludge under isothermal and fast conditions. Biores. Technol. **232**, 27–34 (2017)
45. L. Yan, Y. Wang, J. Li, Y. Zhang, L. Ma, F. Fu, B. Chen, H. Liu, Hydrothermal liquefaction of *Ulva prolifera* macroalgae and the influence of base catalysts on products. Biores. Technol. **292**, 121286 (2019)
46. W. Wang, Y. Xu, X. Wang, B. Zhang, W. Tian, J. Zhang, Hydrothermal liquefaction of microalgae over transition metal supported TiO_2 catalyst. Bioresour. Technol. **250**, 474–480 (2018)
47. W. Ding, H. Jin, Q. Zhao, B. Chen, W. Cao, Dissolution of polycyclic aromatic hydrocarbons in supercritical water in hydrogen production process: a molecular dynamics simulation study. Int. J. Hydrogen Energy **45**, 28062–28069 (2020)
48. H. KazemiShariatPanahi, M. Tabatabaei, M. Aghbashlo, M. Dehhaghi, M. Rehan, A.-S. Nizami, Recent updates on the production and upgrading of bio-crude oil from microalgae. Bioresour. Technol. Rep. **7**, 100216 (2019)
49. W. Li, C. Pan, L. Sheng, Z. Liu, P. Chen, H. Lou, X. Zheng, Upgrading of high-boiling fraction of bio-oil in supercritical methanol. Bioresour. Technol. **102**, 9223–9228 (2011)
50. G. Teri, L. Luo, P.E. Savage, Hydrothermal treatment of protein, polysaccharide, and lipids alone and in mixtures. Energy Fuels **28**, 7501–7509 (2014)
51. Y. Kang, R. Li, X. Shen, B. Jin, Y. Zhan, A. Wang, B. Li, Y. Song, G. Xiao, Q. Dong, Stability-enhanced perovskite heterointerfaces and solar cells via strongly anchored and sterically hindered ligands. Nano Energy **120**, 109178 (2024)
52. W.H. Chen, B.J. Lin, M.Y. Huang, J.S. Chang, Thermochemical conversion of microalgal biomass into biofuels: a review. Bioresour. Technol. **184**, 314–327 (2015)
53. Y. Xu, Y. Hu, Y. Peng, L. Yao, Y. Dong, B. Yang, R. Song, Catalytic pyrolysis and liquefaction behavior of microalgae for bio-oil production. Bioresour. Technol. **300**, 122665 (2019)
54. A.B. Ross, P. Biller, M.L. Kubacki, H. Li, A. Lea-Langton, J.M. Jones, Hydrothermal processing of microalgae using alkali and organic acids. Fuel **89**, 2234–2243 (2010)
55. S. Kandasamy, B. Zhang, Z. He, H. Chen, H. Feng, Q. Wang, B. Wang, V. Ashokkumar, S. Siva, N. Bhuvanendran, M. Krishnamoorthi, Effect of low-temperature catalytic hydrothermal liquefaction of *Spirulina platensis*. Energy **190**, 116236 (2020)
56. Y. Chen, R. Mu, M. Yang, L. Fang, Y. Wu, K. Wu, Y. Liu, J. Gong, Catalytic hydrothermal liquefaction for bio-oil production over CNTs supported metal catalysts. Chem. Eng. Sci. **161**, 299–307 (2017)
57. Y. Chen, K. Wu, Q. Lin, Y. Wu, Y. Tang, M. Yang, Catalytic hydrothermal liquefaction of microalgae over metal incorporated mesoporous SBA-15 with high hydrothermal stability. Carbon Resour. Convers. **1**, 251–259 (2018)
58. Q. Lin, Y. Chen, Y. Tang, K. Wu, M. Yang, H. Hu, Y. Wu, Catalytic hydrothermal liquefaction of *D. tertiolecta* over multifunctional mesoporous silica-based catalysts with high stability. Micropor. Mesopor. Mater. **250**, 120–127 (2017)
59. Z. Zhu, L. Rosendahl, S.S. Toor, D. Yu, G. Chen, Hydrothermal liquefaction of barley straw to bio-crude oil: effects of reaction temperature and aqueous phase recirculation. Appl. Energy **137**, 183–192 (2015)

60. J. Lu, J. Watson, J. Zeng, H. Li, Z. Zhu, M. Wang, Y. Zhang, Z. Liu, Biocrude production and heavy metal migration during hydrothermal liquefaction of swine manure. Process Saf. Environ. Prot. **115**, 108–115 (2018)
61. S. Yin, R. Dolan, M. Harris, Z. Tan, Subcritical hydrothermal liquefaction of cattle manure to bio-oil: effects of conversion parameters on bio-oil yield and characterization of bio-oil. Biores. Technol. **101**, 3657–3664 (2010)
62. L. Zhang, P. Champagne, C. Xu, Bio-crude production from secondary pulp/paper-mill sludge and waste newspaper via co-liquefaction in hot-compressed water. Energy **36**, 2142–2150 (2011)
63. H. Li, S. Hurley, C. Xu, Liquefactions of peat in supercritical water with a novel iron catalyst. Fuel **90**, 412–420 (2011)
64. A.A. Shah, S.S. Toor, F. Conti, A.H. Nielsen, L.A. Rosendahl, Hydrothermal liquefaction of high ash containing sewage sludge at sub and supercritical conditions. Biomass Bioenergy **135**, 105504 (2020)
65. D. Xu, G. Lin, L. Liu, Y. Wang, Z. Jing, S. Wang, Comprehensive evaluation on product characteristics of fast hydrothermal liquefaction of sewage sludge at different temperatures. Energy **159**, 686–695 (2018)
66. L. Qian, S. Wang, P.E. Savage, Hydrothermal liquefaction of sewage sludge under isothermal and fast conditions. Bioresour. Technol. **232**, 27–34 (2017)
67. L. Nazari, Z. Yuan, S. Souzanchi, M.B. Ray, C. Xu, Hydrothermal liquefaction of woody biomass in hot-compressed water: catalyst screening and comprehensive characterization of bio-crude oils. Fuel **162**, 74–83 (2015)
68. P. Sun, M. Heng, S. Sun, J. Chen, Direct liquefaction of paulownia in hot compressed water: influence of catalysts. Energy **35**, 5421–5429 (2010)
69. Y. Wang, H. Wang, H. Lin, Y. Zheng, J. Zhao, A. Pelletier, K. Li, Effects of solvents and catalysts in liquefaction of pinewood sawdust for the production of bio-oils. Biomass Bioenergy **59**, 158–167 (2013)
70. H. Mazaheri, K.T. Lee, S. Bhatia, A.R. Mohamed, Subcritical water liquefaction of oil palm fruit press fiber for the production of bio-oil: effect of catalysts. Bioresour. Technol. **101**, 745–751 (2010)
71. C. Perego, D. Bianchi, Biomass upgrading through acid–base catalysis. Chem. Eng. J. **161**, 314–322 (2010)
72. S. Nagappan, R.R. Bhosale, D.D. Nguyen, N.T.L. Chi, V.K. Ponnusamy, C.S. Woong, G. Kumar, Catalytic hydrothermal liquefaction of biomass into bio-oils and other value-added products—a review. Fuel **285**, 119053 (2021)
73. J. Zhang, Catalytic transfer hydrogenolysis as an efficient route in cleavage of lignin and model compounds. Green Energy Environ. **3**, 328–334 (2018)
74. W. Tian, R. Liu, W. Wang, Z. Yin, X. Yi, Effect of operating conditions on hydrothermal liquefaction of *Spirulina* over Ni/TiO_2 catalyst. Biores. Technol. **263**, 569–575 (2018)
75. D. Li, L. Chen, D. Xu, X. Zhang, N. Ye, F. Chen, S. Chen, Preparation and characteristics of bio-oil from the marine brown alga *Sargassum patens* C. Agardh. Bioresour. Technol. **104**, 737–742 (2012)
76. W. Song, S. Wang, Y. Guo, D. Xu, Bio-oil production from hydrothermal liquefaction of waste *Cyanophyta* biomass: influence of process variables and their interactions on the product distributions. Int. J. Hydrog. Energy **42**(31), 20361–20374 (2017)
77. A. Gollakota, P.E. Savage, Biocrude production from fast and isothermal hydrothermal liquefaction of chitin. Energy Fuels **33**, 11328–11338 (2019)
78. K. Anastasakis, A.B. Ross, Hydrothermal liquefaction of the brown macro-alga *Laminaria saccharina*: effect of reaction conditions on product distribution and composition. Bioresour. Technol. **102**, 4876–4883 (2011)
79. C. Tian, Z. Liu, Y. Zhang, B. Li, W. Cao, H. Lu, N. Duan, L. Zhang, T. Zhang, Hydrothermal liquefaction of harvested high-ash low-lipid algal biomass from Dianchi Lake: effects of operational parameters and relations of products. Bioresour. Technol. **184**, 336–343 (2015)

80. S. Yin, Z. Tan, Hydrothermal liquefaction of cellulose to bio-oil under acidic, neutral and alkaline conditions. Appl. Energy **92**, 234–239 (2012)
81. R. Younas, S. Hao, L. Zhang, S. Zhang, Hydrothermal liquefaction of rice straw with NiO nanocatalyst for bio-oil production. Renew. Energy **113**, 532–545 (2017)
82. H. Liu, Y. Chen, H. Yang, F.G. Gentili, U. Söderlind, X. Wang, W. Zhang, H. Chen, Hydrothermal carbonization of natural microalgae containing a high ash content. Fuel **249**, 441–448 (2019)
83. F. Cheng, Z. Cui, K. Mallick, N. Nirmalakhandan, C.E. Brewer, Hydrothermal liquefaction of high- and low-lipid algae: mass and energy balances. Bioresour. Technol. **258**, 158–167 (2018)
84. Y. Han, S.K. Hoekman, Z. Cui, U. Jena, P. Das, Hydrothermal liquefaction of marine microalgae biomass using co-solvents. Algal Res. **38**, 101421 (2019)
85. R. Singh, B. Balagurumurthy, A. Prakash, T. Bhaskar, Catalytic hydrothermal liquefaction of water hyacinth. Biores. Technol. **178**, 157–165 (2015)
86. L. Jiang, X. Yuan, G. Zeng, J. Liang, X. Chen, H. Yu, H. Wang, Z. Wu, J. Zhang, T. Xiong, In-situ synthesis of direct solid-state dual Z-scheme $WO_3/g\text{-}C_3N_4/Bi_2O_3$ photocatalyst for the degradation of refractory pollutant. Appl. Catal. B **227**, 376–385 (2018)
87. H. Chen, J. Liu, X. Chang, D. Chen, Y. Xue, P. Liu, H. Lin, S. Han, A review on the pretreatment of lignocellulose for high-value chemicals. Fuel Process. Technol. **160**, 196–206 (2017)
88. J. Yang, Q. He, K. Corscadden, H. Niu, J. Lin, T. Astatkie, Advanced models for the prediction of product yield in hydrothermal liquefaction via a mixture design of biomass model components coupled with process variables. Appl. Energy **233–234**, 906–915 (2019)
89. M. Pecchi, A. Cascioli, A.R. Maag, J.L. Goldfarb, M. Baratieri, Uncovering the transition between hydrothermal carbonization and liquefaction using differential scanning calorimetry. Fuel Process. Technol. **235**, 107349 (2022)
90. H. Li, Z. Liu, Y. Zhang, B. Li, H. Lu, N. Duan, M. Liu, Z. Zhu, B. Si, Conversion efficiency and oil quality of low-lipid high-protein and high-lipid low-protein microalgae via hydrothermal liquefaction. Biores. Technol. **154**, 322–329 (2014)
91. T.M. Brown, P. Duan, P.E. Savage, Hydrothermal liquefaction and gasification of *Nannochloropsis* sp. Energy Fuels **24**, 3639–3646 (2010)
92. D.R. Vardon, B.K. Sharma, J. Scott, G. Yu, Z. Wang, L. Schideman, Y. Zhang, T.J. Strathmann, Chemical properties of biocrude oil from the hydrothermal liquefaction of *Spirulina* algae, swine manure, and digested anaerobic sludge. Biores. Technol. **102**, 8295–8303 (2011)
93. P. Duan, P.E. Savage, Upgrading of crude algal bio-oil in supercritical water. Bioresour. Technol. **102**, 1899–1906 (2011)
94. C. Gai, Y. Zhang, W.-T. Chen, P. Zhang, Y. Dong, An investigation of reaction pathways of hydrothermal liquefaction using *Chlorella pyrenoidosa* and *Spirulina platensis*. Energy Convers. Manage. **96**, 330–339 (2015)
95. W. Yang, X. Li, Z. Li, C. Tong, L. Feng, Understanding low-lipid algae hydrothermal liquefaction characteristics and pathways through hydrothermal liquefaction of algal major components: crude polysaccharides, crude proteins and their binary mixtures. Bioresour. Technol. **196**, 99–108 (2015)
96. C.-X. Zhao, J.-N. Liu, B.-Q. Li, D. Ren, X. Chen, J. Yu, Q. Zhang, Multiscale construction of bifunctional electrocatalysts for long-lifespan rechargeable zinc-air batteries. Adv. Func. Mater. **30**, 2003619 (2020)
97. Z. Wang, X. You, C. Yang, W. Li, Y. Li, Y. Li, J. Shen, R. Zhang, Y. Su, Z. Jiang, Ultrathin polyamide nanofiltration membranes with tunable chargeability for multivalent cation removal. J. Membr. Sci. **642**, 119971 (2022)
98. X. Ruan, X. Cui, G. Jia, J. Wu, J. Zhao, D.J. Singh, Y. Liu, H. Zhang, L. Zhang, W. Zheng, Intramolecular heterostructured carbon nitride with heptazine-triazine for enhanced photocatalytic hydrogen evolution. Chem. Eng. J. **428**, 132579 (2022)
99. R. Obeid, D.M. Lewis, N. Smith, T. Hall, P. van Eyk, Reaction kinetics and characterisation of species in renewable crude from hydrothermal liquefaction of monomers to represent organic fractions of biomass feedstocks. Chem. Eng. J. **389**, 124397 (2020)

100. A.A. Peterson, F. Vogel, R.P. Lachance, M. Froeling, M.J. Antal Jr., J.W. Tester, Thermochemical biofuel production in hydrothermal media: a review of sub- and supercritical water technologies. Energy Environ. Sci. **1**, 32–65 (2008)
101. H.L. Barnebey, Continuous fat splitting plants using the colgate-emery process. J. Am. Oil Chem. Soc. **25**, 95–99 (1948)
102. W. Bühler, E. Dinjus, H.J. Ederer, A. Kruse, C. Mas, Ionic reactions and pyrolysis of glycerol as competing reaction pathways in near- and supercritical water. J. Supercrit. Fluids **22**, 37–53 (2002)
103. Z. Shen, F. Jin, Y. Zhang, A. Kishita, K. Tohji, Effect of Base Catalysts on Hydrothermal Conversion of Glycerin, in: 2009 3rd International Conference on Bioinformatics and Biomedical Engineering, 2009, pp. 1–4
104. S. Xiu, A. Shahbazi, V.B. Shirley, L. Wang, Swine manure/crude glycerol co-liquefaction: physical properties and chemical analysis of bio-oil product. Biores. Technol. **102**, 1928–1932 (2011)
105. M.P. Caporgno, J. Pruvost, J. Legrand, O. Lepine, M. Tazerout, C. Bengoa, Hydrothermal liquefaction of *Nannochloropsis oceanica* in different solvents. Biores. Technol. **214**, 404–410 (2016)
106. X. Zhou, Y. Hao, X. He, D. Zhou, L. Xie, S. Liu, B. Qiao, Protean morphology of waterborne polyurethane dispersion: an overview of nanoparticles from sphere to irregular elongated shape. Prog. Org. Coat. **146**, 105742 (2020)
107. J.-Y. Li, Y.-H. Li, F. Zhang, Z.-R. Tang, Y.-J. Xu, Visible-light-driven integrated organic synthesis and hydrogen evolution over 1D/2D CdS-Ti_3C_2Tx MXene composites. Appl. Catal. B **269**, 118783 (2020)
108. Z. Srokol, A.-G. Bouche, A. van Estrik, R.C.J. Strik, T. Maschmeyer, J.A. Peters, Hydrothermal upgrading of biomass to biofuel; studies on some monosaccharide model compounds. Carbohyd. Res. **339**, 1717–1726 (2004)
109. B. Zhang, Q. Lin, Q. Zhang, K. Wu, W. Pu, M. Yang, Y. Wu, Catalytic hydrothermal liquefaction of Euglena sp. microalgae over zeolite catalysts for the production of bio-oil. RSC Adv. **7**, 8944–8951 (2017)
110. Y. Dote, S. Inoue, T. Ogi, S. Yokoyama, Distribution of nitrogen to oil products from liquefaction of amino acids. Biores. Technol. **64**, 157–160 (1998)
111. J. Li, T.B. Brill, Spectroscopy of hydrothermal reactions 20: Experimental and DFT computational comparison of decarboxylation of dicarboxylic acids connected by single, double, and triple bonds. J. Phys. Chem. A **106**, 9491–9498 (2002)
112. M.J. Stablein, A. Aierzhati, J. Watson, B. Si, Y. Zhang, Characterization and bioremediation potential of byproducts from hydrothermal liquefaction of food wastes. Biores. Technol. Rep. **12**, 100555 (2020)
113. T. Rogalinski, K. Liu, T. Albrecht, G. Brunner, Hydrolysis kinetics of biopolymers in subcritical water. J. Supercrit. Fluids **46**, 335–341 (2008)
114. Z. Fang, R.L. Smith, J.A. Kozinski, T. Minowa, K. Arai, Reaction of d-glucose in water at high temperatures (410 °C) and pressures (180 MPa) for the production of dyes and nano-particles. J. Supercrit. Fluids **56**, 41–47 (2011)
115. K. Kapusta, Effect of ultrasound pretreatment of municipal sewage sludge on characteristics of bio-oil from hydrothermal liquefaction process. Waste Manage. **78**, 183–190 (2018)
116. M.P. Pandey, C.S. Kim, Lignin depolymerization and conversion: a review of thermochemical methods. Chem. Eng. Technol. **34**, 29–41 (2011)
117. J. Akhtar, N.A.S. Amin, A review on process conditions for optimum bio-oil yield in hydrothermal liquefaction of biomass. Renew. Sustain. Energy Rev. **15**, 1615–1624 (2011)
118. Y. Miyata, K. Sagata, Y. Yamazaki, H. Teramura, Y. Hirano, C. Ogino, Y. Kita, Mechanism of the Fe-assisted hydrothermal liquefaction of lignocellulosic biomass. Ind. Eng. Chem. Res. **57**, 14870–14877 (2018)
119. M. Scarsella, B. de Caprariis, M. Damizia, P. De Filippis, Heterogeneous catalysts for hydrothermal liquefaction of lignocellulosic biomass: a review. Biomass Bioenergy **140**, 105662 (2020)

120. D.R. Vardon, B.K. Sharma, J. Scott, G. Yu, Z. Wang, L. Schideman, Y. Zhang, T.J. Strathmann, Chemical properties of biocrude oil from the hydrothermal liquefaction of Spirulina algae, swine manure, and digested anaerobic sludge. Bioresour. Technol. **102**, 8295–8303 (2011)
121. L. Jasiūnas, T.H. Pedersen, S.S. Toor, L.A. Rosendahl, Biocrude production via supercritical hydrothermal co-liquefaction of spent mushroom compost and aspen wood sawdust. Renew. Energy **111**, 392–398 (2017)
122. A. Liu, Y. Park, Z. Huang, B. Wang, R.O. Ankumah, P.K. Biswas, Product identification and distribution from hydrothermal conversion of walnut shells. Energy Fuels **20**(2), 446–454 (2006)
123. T.H. Pedersen, N.H. Hansen, O.M. Pérez, D.E.V. Cabezas, L.A. Rosendahl, Renewable hydrocarbon fuels from hydrothermal liquefaction: a techno-economic analysis. Biofuels Bioprod. Biorefin. **12**, 213–223 (2018)
124. H. Li, S. Feng, Z. Yuan, Q. Wei, C.C. Xu, Highly efficient liquefaction of wheat straw for the production of bio-polyols and bio-based polyurethane foams. Ind. Crops Prod. **109**, 426–433 (2017)
125. C. Zhou, D. Huang, P. Xu, G. Zeng, J. Huang, T. Shi, C. Lai, C. Zhang, M. Cheng, Y. Lu, A. Duan, W. Xiong, M. Zhou, Efficient visible light driven degradation of sulfamethazine and tetracycline by salicylic acid modified polymeric carbon nitride via charge transfer. Chem. Eng. J. **370**, 1077–1086 (2019)
126. M. Takada, E. Minami, S. Saka, Decomposition behaviors of the lignocellulosics as treated by semi-flow hot-compressed water. J. Supercrit. Fluids **133**, 566–572 (2018)
127. L. Yang, Q.S. He, P. Havard, K. Corscadden, C.C. Xu, X. Wang, Co-liquefaction of spent coffee grounds and lignocellulosic feedstocks. Bioresour. Technol. **237**, 108–121 (2017)
128. W.-T. Chen, Y. Zhang, J. Zhang, L. Schideman, G. Yu, P. Zhang, M. Minarick, Co-liquefaction of swine manure and mixed-culture algal biomass from a wastewater treatment system to produce bio-crude oil. Appl. Energy **128**, 209–216 (2014)
129. Y. Gao, X.-H. Wang, H.-P. Yang, H.-P. Chen, Characterization of products from hydrothermal treatments of cellulose. Energy **42**, 457–465 (2012)
130. Z. Zhu, S.S. Toor, L. Rosendahl, D. Yu, G. Chen, Influence of alkali catalyst on product yield and properties via hydrothermal liquefaction of barley straw. Energy **80**, 284–292 (2015)
131. T.H. Pedersen, L. Jasiūnas, L. Casamassima, S. Singh, T. Jensen, L.A. Rosendahl, Synergetic hydrothermal co-liquefaction of crude glycerol and aspen wood. Energy Convers. Manage. **106**, 886–891 (2015)
132. S. Changi, T.M. Brown, P.E. Savage, Reaction kinetics and pathways for phytol in high-temperature water. Chem. Eng. J. **189**, 336–345 (2012)
133. M. Piorreck, K.H. Baasch, P. Pohl, Biomass production, total protein, chlorophylls, lipids and fatty acids of freshwater green and blue-green algae under different nitrogen regimes. Phytochemistry **23**, 207–216 (1984)
134. U. Jena, K.C. Das, J.R. Kastner, Effect of operating conditions of thermochemical liquefaction on biocrude production from *Spirulina platensis*. Biores. Technol. **102**, 6221–6229 (2011)
135. P. Duan, P.E. Savage, Hydrothermal liquefaction of a microalga with heterogeneous catalysts. Ind. Eng. Chem. Res. **50**, 52–61 (2011)
136. U. Jena, K.C. Das, Comparative evaluation of thermochemical liquefaction and pyrolysis for bio-oil production from microalgae. Energy Fuels **25**, 5472–5482 (2011)
137. X. Yuan, J. Wang, G. Zeng, H. Huang, X. Pei, Z. Liu, M. Cong, Comparative studies of thermochemical liquefaction characteristics of microalgae using different organic solvents. Energy **36**, 6406–6412 (2011)
138. P.J. Valdez, J.G. Dickinson, P.E. Savage, Characterization of product fractions from hydrothermal liquefaction of *Nannochloropsis* sp. and the influence of solvents. Energy Fuels **25**, 3235–3243 (2011)
139. S. Zou, Y. Wu, M. Yang, C. Li, J. Tong, Bio-oil production from sub- and supercritical water liquefaction of microalgae *Dunaliella tertiolecta* and related properties. Energy Environ. Sci. **3**, 1073–1078 (2010)

140. Z. Shuping, W. Yulong, Y. Mingde, I. Kaleem, L. Chun, J. Tong, Production and characterization of bio-oil from hydrothermal liquefaction of microalgae *Dunaliella tertiolecta* cake. Energy **35**, 5406–5411 (2010)
141. S. Zou, Y. Wu, M. Yang, I. Kaleem, L. Chun, J. Tong, Production and characterization of bio-oil from hydrothermal liquefaction of microalgae *Dunaliella tertiolecta* cake. Energy **35**, 5406–5411 (2010)
142. P. Biller, A.B. Ross, Potential yields and properties of oil from the hydrothermal liquefaction of microalgae with different biochemical content. Biores. Technol. **102**, 215–225 (2011)
143. X.Z. Yuan, J.Y. Tong, G.M. Zeng, H. Li, W. Xie, Comparative studies of products obtained at different temperatures during straw liquefaction by hot compressed water. Energy Fuels **23**, 3262–3267 (2009)
144. C. Yang, L. Jia, C. Chen, G. Liu, W. Fang, Bio-oil from hydro-liquefaction of *Dunaliella sauna* over Ni/REHY catalyst. Biores. Technol. **102**, 4580–4584 (2011)
145. R.L. Holliday, J.W. King, G.R. List, Hydrolysis of vegetable oils in sub- and supercritical water. Ind. Eng. Chem. Res. **36**, 932–935 (1997)
146. J.W. King, R.L. Holliday, G.R. List, Hydrolysis of soybean oil in a subcritical water flow reactor. Green Chem. **1**, 261–264 (1999)
147. M. Watanabe, T. Iida, H. Inomata, Decomposition of a long chain saturated fatty acid with some additives in hot compressed water. Energy Convers. Manage. **47**, 3344–3350 (2006)
148. A. Kruse, Supercritical water gasification. Biofuels Bioprod. Bioref. **2**, 415–437 (2008)
149. D. Knezevic, W.P.M. van Swaaij, S.R.A. Kersten, Hydrothermal conversion of biomass: I, glucose conversion in hot compressed water. Ind. Eng. Chem. Res. **48**, 4731–4743 (2009)
150. G.C.A. Luijkx, F. Vanrantwijk, H. Vanbekkum, Hydrothermal formation of 1,2,4-benzenetriol from 5-hydroxymethyl-2-furaldehyde and d-fructose. Carbohyd. Res. **242**, 131–139 (1993)
151. Z. Li, P.E. Savage, Feedstocks for fuels and chemicals from algae: treatment of crude bio-oil over HZSM-5. Algal Res. Biomass Biofuels Bioprod. **2**, 154–163 (2013)
152. N.H. Tran, J.R. Bartlett, G.S.K. Kannangara, A.S. Milev, H. Volk, M.A. Wilson, Catalytic upgrading of biorefinery oil from micro-algae. Fuel **89**, 265–274 (2010)
153. P. Duan, P.E. Savage, Catalytic hydrotreatment of crude algal bio-oil in supercritical water. Appl. Catal. B Environ. **104**, 136–143 (2011)
154. Y. Dote, S. Inoue, T. Ogi, S. Yokoyama, Studies on the direct liquefaction of protein-contained biomass: the distribution of nitrogen in the products. Biomass Bioenerg. **11**, 491–498 (1996)
155. L.G. Alba, C. Torri, C. Samori, J. van der Spek, D. Fabbri, S.R.A. Kersten, D.W.F. Brilman, Hydrothermal treatment (HIT) of microalgae: evaluation of the process as conversion method in an algae biorefinery concept. Energy Fuels **26**, 642–657 (2012)
156. A. Kruse, A. Krupka, V. Schwarzkopf, C. Gamard, T. Henningsen, Influence of proteins on the hydrothermal gasification and liquefaction of biomass. 1. Comparison of different feedstocks. Indus. Eng. Chem. Res. **44**, 3013–3020 (2005)
157. A. Kruse, P. Maniam, F. Spieler, Influence of proteins on the hydrothermal gasification and liquefaction of biomass. 2. Model compounds. Indus. Eng. Chem. Res. **46**, 87–96 (2007)
158. A.A. Peterson, R.P. Lachance, J.W. Tester, Kinetic evidence of the maillard reaction in hydrothermal biomass processing: glucose–glycine interactions in high-temperature, high-pressure water. Ind. Eng. Chem. Res. **49**, 2107–2117 (2010)
159. Y. Guo, S. Wang, C.M. Huelsman, P.E. Savage, Products, pathways, and kinetics for reactions of indole under supercritical water gasification conditions. J. Supercrit. Fluids **73**, 161–170 (2013)
160. C.L. Zhong, X.M. Wei, A comparative experimental study on the liquefaction of wood. Energy **29**, 1731–1741 (2004)
161. Z. Fang, T. Sato, R.L. Smith Jr., H. Inomata, K. Arai, J.A. Kozinski, Reaction chemistry and phase behavior of lignin in high-temperature and supercritical water. Biores. Technol. **99**, 3424–3430 (2008)
162. U. Jena, N. Vaidyanathan, S. Chinnasamy, K.C. Das, Evaluation of microalgae cultivation using recovered aqueous co-product from thermochemical liquefaction of algal biomass. Biores. Technol. **102**, 3380–3387 (2011)

163. P. Biller, A.B. Ross, S.C. Skill, A. Lea-Langton, B. Balasundaram, C. Hall, R. Riley, C.A. Llewellyn, Nutrient recycling of aqueous phase for microalgae cultivation from the hydrothermal liquefaction process. Algal Res. Biomass Biofuels Bioprod. **1**, 70–76 (2012)
164. C. Miao, M. Chakraborty, S. Chen, Impact of reaction conditions on the simultaneous production of polysaccharides and bio-oil from heterotrophically grown *Chlorella sorokiniana* by a unique sequential hydrothermal liquefaction process. Biores. Technol. **110**, 617–627 (2012)
165. D.C. Elliott, T.R. Hart, A.J. Schmidt, G.G. Neuenschwander, L.J. Rotness, M.V. Olarte, A.H. Zacher, K.O. Albrecht, R.T. Hallen, J.E. Holladay, Process development for hydrothermal liquefaction of algae feedstocks in a continuous-flow reactor. Algal Res. Biomass Biofuels Bioprod. **2**, 445–454 (2013)
166. P.S. Christensen, G. Peng, F. Vogel, B.B. Iversen, Hydrothermal liquefaction of the microalgae *Phaeodactylum tricornutum*: impact of reaction conditions on product and elemental distribution. Energy Fuels **28**, 5792–5803 (2014)
167. T. Sato, S. Kurosawa, R.L. Smith, T. Adschiri, K. Arai, Water gas shift reaction kinetics under noncatalytic conditions in supercritical water. J. Supercrit. Fluids **29**, 113–119 (2004)
168. F.L.P. Resende, P.E. Savage, Expanded and updated results for supercritical water gasification of cellulose and lignin in metal-free reactors. Energy Fuels **23**, 6213–6221 (2009)
169. I.G. Lee, M.S. Kim, S.K. Ihm, Gasification of glucose in supercritical water. Ind. Eng. Chem. Res. **41**, 1182–1188 (2002)
170. T.M. McCollom, G. Ritter, B.R.T. Simoneit, Lipid synthesis under hydrothermal conditions by Fischer-Tropsch-type reactions. Orig. Life Evol. Biosph. **29**, 153–166 (1999)
171. A.I. Rushdi, B.R.T. Simoneit, Lipid formation by aqueous Fischer-Tropsch-type synthesis over a temperature range of 100 to 400 °C. Orig. Life Evol. Biosph. **31**, 103–118 (2001)
172. S. Zou, Y. Wu, M. Yang, C. Li, J. Tong, Thermochemical catalytic liquefaction of the marine microalgae *Dunaliella tertiolecta* and characterization of bio-oils. Energy Fuels **23**, 3753–3758 (2009)
173. M. Aresta, A. Dibenedetto, M. Carone, T. Colonna, C. Fragale, Production of biodiesel from macroalgae by supercritical CO_2 extraction and thermochemical liquefaction. Environ. Chem. Lett. **3**, 136–139 (2005)
174. N. Akiya, P.E. Savage, Roles of water for chemical reactions in high-temperature water. Chem. Rev. **102**, 2725–2750 (2002)
175. S.S. Toor, H. Reddy, S. Deng, J. Hoffmann, D. Spangsmark, L.B. Madsen, J.B. Holm-Nielsen, L.A. Rosendahl, Hydrothermal liquefaction of *Spirulina* and *Nannochloropsis salina* under subcritical and supercritical water conditions. Biores. Technol. **131**, 413–419 (2013)
176. G. Yu, Y. Zhang, L. Schideman, T. Funk, Z. Wang, Distributions of carbon and nitrogen in the products from hydrothermal liquefaction of low-lipid microalgae. Energy Environ. Sci. **4**, 4587–4595 (2011)
177. J.L. Faeth, P.J. Valdez, P.E. Savage, Fast hydrothermal liquefaction of *Nannochloropsis* sp. to produce biocrude. Energy Fuels **27**, 1391–1398 (2013)
178. B.E. Eboibi, D.M. Lewis, P.J. Ashman, S. Chinnasamy, Effect of operating conditions on yield and quality of biocrude during hydrothermal liquefaction of halophytic microalga *Tetraselmis* sp. Biores. Technol. **170**, 20–29 (2014)
179. N. Akiya, P.E. Savage, Kinetics and mechanism of cyclohexanol dehydration in high-temperature water. Ind. Eng. Chem. Res. **40**, 1822–1831 (2001)
180. T.O. Matsui, A. Nishihara, C. Ueda, M. Ohtsuki, N.O. Ikenaga, T. Suzuki, Liquefaction of micro-algae with iron catalyst. Fuel **76**, 1043–1048 (1997)
181. D. Klingler, J. Berg, H. Vogel, Hydrothermal reactions of alanine and glycine in sub- and supercritical water. J. Supercrit. Fluids **43**, 112–119 (2007)
182. W.-T. Chen, Y. Zhang, J. Zhang, G. Yu, L.C. Schideman, P. Zhang, M. Minarick, Hydrothermal liquefaction of mixed-culture algal biomass from wastewater treatment system into bio-crude oil. Biores. Technol. **152**, 130–139 (2014)
183. D.R. Vardon, B.K. Sharma, G.V. Blazina, K. Rajagopalan, T.J. Strathmann, Thermochemical conversion of raw and defatted algal biomass via hydrothermal liquefaction and slow pyrolysis. Biores. Technol. **109**, 178–187 (2012)

184. G. Yu, Y. Zhang, L. Schideman, T.L. Funk, Z. Wang, Hydrothermal liquefaction of low lipid content microalgae into bio-crude oil. Trans. ASABE **54**, 239–246 (2011)
185. D.L. Barreiro, W. Prins, F. Ronsse, W. Brilman, Hydrothermal liquefaction (HTL) of microalgae for biofuel production: State of the art review and future prospects. Biomass Bioenerg. **53**, 113–127 (2013)
186. L. Rodolfi, G.C. Zittelli, N. Bassi, G. Padovani, N. Biondi, G. Bonini, M.R. Tredici, Microalgae for oil: strain selection, induction of lipid synthesis and outdoor mass cultivation in a low-cost photobioreactor. Biotechnol. Bioeng. **102**, 100–112 (2009)
187. B. Jin, P. Duan, Y. Xu, F. Wang, Y. Fan, Co-liquefaction of micro- and macroalgae in subcritical water. Biores. Technol. **149**, 103–110 (2013)
188. N. Neveux, A.K.L. Yuen, C. Jazrawi, M. Magnusson, B.S. Haynes, A.F. Masters, A. Montoya, N.A. Paul, T. Maschmeyer, R. de Nys, Biocrude yield and productivity from the hydrothermal liquefaction of marine and freshwater green macroalgae. Biores. Technol. **155**, 334–341 (2014)
189. C. Xu, T. Etcheverry, Hydro-liquefaction of woody biomass in sub- and super-critical ethanol with iron-based catalysts. Fuel **87**, 335–345 (2008)
190. H. Li, J. Hu, Z. Zhang, H. Wang, F. Ping, C. Zheng, H. Zhang, Q. He, Insight into the effect of hydrogenation on efficiency of hydrothermal liquefaction and physico-chemical properties of biocrude oil. Biores. Technol. **163**, 143–151 (2014)
191. M. Watanabe, F. Bayer, A. Kruse, Oil formation from glucose with formic acid and cobalt catalyst in hot-compressed water. Carbohyd. Res. **341**, 2891–2900 (2006)
192. F. Lemoine, I. Maupin, L. Lemee, J.M. Lavoie, J.L. Lemberton, Y. Pouilloux, L. Pinard, Alternative fuel production by catalytic hydroliquefaction of solid municipal wastes, primary sludges and microalgae. Biores. Technol. **142**, 1–8 (2013)
193. S. Inoue, S. Sawayama, T. Ogi, Components of oil derived from liquefaction of hydrocarbon-rich microalgae, Abstracts of Papers of the American Chemical Society (vol. 211, pp. 140, 1996)
194. G. Yu, Y. Zhang, B. Guo, T. Funk, L. Schideman, Nutrient flows and quality of bio-crude oil produced via catalytic hydrothermal liquefaction of low-lipid microalgae. Bioenergy Res. **7**, 1317–1328 (2014)
195. Y. Xu, X. Zheng, H. Yu, X. Hu, Hydrothermal liquefactionof *Chlorella pyrenoidosa* for bio-oil production over Ce/HZSM-5. Biores. Technol. **156**, 1–5 (2014)
196. T.M. Yeh, J.G. Dickinson, A. Franck, S. Linic, L.T. Thompson Jr., P.E. Savage, Hydrothermal catalytic production of fuels and chemicals from aquatic biomass. J. Chem. Technol. Biotechnol. **88**, 13–24 (2013)
197. T. Minowa, F. Zhen, T. Ogi, Cellulose decomposition in hot-compressed water with alkali or nickel catalyst. J. Supercrit. Fluids **13**, 253–259 (1998)
198. C. Xu, N. Lad, Production of heavy oils with high caloric values by direct liquefactionof woody biomass in sub/near-critical water. Energy Fuels **22**, 635–642 (2008)
199. Z. Fang, T. Minowa, R.L. Smith, T. Ogi, J.A. Kozinski, Liquefaction and gasification of cellulose with Na_2CO_3 and Ni in subcritical water at 350 °C. Ind. Eng. Chem. Res. **43**, 2454–2463 (2004)
200. C.C. Song, H.Q. Hu, S.W. Zhu, G. Wang, G.H. Chen, Nonisothermal catalytic liquefaction of corn stalk in subcritical and supercritical water. Energy Fuels **18**, 90–96 (2004)
201. S. Karagöz, T. Bhaskar, A. Muto, Y. Sakata, T. Oshiki, T. Kishimoto, Low-temperature catalytic hydrothermal treatment of wood biomass: analysis of liquid products. Chem. Eng. J. **108**, 127–137 (2005)
202. C. Miao, M. Chakraborty, T. Dong, X. Yu, Z. Chi, S. Chen, Sequential hydrothermal fractionation of yeast *Cryptococcus curvatus* biomass. Biores. Technol. **164**, 106–112 (2014)
203. L. Yang, Y. Li, P.E. Savage, Catalytic hydrothermal liquefaction of a microalga in a two-chamber reactor. Ind. Eng. Chem. Res. **53**, 11939–11944 (2014)
204. D.L. Barreiro, C. Samori, G. Terranella, U. Hornung, A. Kruse, W. Prins, Assessing microalgae biorefinery routes for the production of biofuels via hydrothermal liquefaction. Biores. Technol. **174**, 256–265 (2014)

205. J. Yang, Q. He, L. Yang, A review on hydrothermal co-liquefaction of biomass. Appl. Energy **250**, 926–945 (2019)
206. D.W.F. Brilman, N. Drabik, M. Wądrzyk, Hydrothermal co-liquefaction of microalgae, wood, and sugar beet pulp. Biomass Convers. Bioref. **7**, 445–454 (2017)
207. K.P.R. Dandamudi, M. Mathew, T. Selvaratnam, T. Muppaneni, M. Seger, P. Lammers, S. Deng, Recycle of nitrogen and phosphorus in hydrothermal liquefaction biochar from *Galdieria sulphuraria* to cultivate microalgae. Resour. Conserv. Recycl. **171**, 105644 (2021)
208. S.G. Roussis, R. Cranford, N. Sytkovetskiy, Thermal treatment of crude algae oils prepared under hydrothermal extraction conditions. Energy Fuels **26**, 5294–5299 (2012)
209. F. Wang, Z. Chang, P. Duan, W. Yan, Y. Xu, L. Zhang, J. Miao, Y. Fan, Hydrothermal liquefaction of Litsea cubeba seed to produce bio-oils. Biores. Technol. **149**, 509–515 (2013)
210. C. Yang, S. Wang, Z. Jiang, J. Li, C. He, T. Xu, D. Xu, Catalytic hydrotreatment upgrading of biocrude oil derived from hydrothermal liquefaction of animal carcass. Fuel **317**, 123528 (2022)
211. J. Lehto, A. Oasmaa, Y. Solantausta, M. Kytö, D. Chiaramonti, Review of fuel oil quality and combustion of fast pyrolysis bio-oils from lignocellulosic biomass. Appl. Energy **116**, 178–190 (2014)
212. H.K. Reddy, T. Muppaneni, S. Ponnusamy, N. Sudasinghe, A.K. Pegallapati, T. Selvaratnam, M. Seger, B.N. Dungan, N. Nirmalakhandan, T.M. Schaub, F.O. Holguin, P.J. Lammers, W.A.V. Voorhies, S. Deng, Temperature effect on hydrothermal liquefaction of *Nannochloropsis gaditana* and *Chlorella* sp. Appl. Energy **165**, 943–951 (2016)
213. H. Zhang, H. Yang, H. Guo, C. Huang, L. Xiong, X. Chen, Kinetic study on the liquefaction of wood and its three cell wall component in polyhydric alcohols. Appl. Energy **113**, 1596–1600 (2014)
214. D.C. Elliott, T.R. Hart, A.J. Schmidt, G.G. Neuenschwander, L.J. Rotness, M.V. Olarte, A.H. Zacher, K.O. Albrecht, R.T. Hallen, J.E. Holladay, Process development for hydrothermal liquefaction of algae feedstocks in a continuous-flow reactor. Algal Res. **2**, 445–454 (2013)
215. X. Shen, H. Hua, H. Li, R. Li, T. Hu, D. Wu, P. Zhang, J. Zhao, Synthesis and molecular dynamic simulation of a novel single ion conducting gel polymer electrolyte for lithium-ion batteries. Polymer **201**, 122568 (2020)
216. P. Biller, A.B. Ross, Potential yields and properties of oil from the hydrothermal liquefaction of microalgae with different biochemical content. Bioresour. Technol. **102**, 215–225 (2011)
217. Y. Xu, P. Duan, B. Wang, Catalytic upgrading of pretreated algal oil with a two-component catalyst mixture in supercritical water. Algal Res. **9**, 186–193 (2015)
218. H.-J. Huang, X.-Z. Yuan, H.-N. Zhu, H. Li, Y. Liu, X.-L. Wang, G.-M. Zeng, Comparative studies of thermochemical liquefaction characteristics of microalgae, lignocellulosic biomass and sewage sludge. Energy **56**, 52–60 (2013)
219. B. Zhang, J. Chen, S. Kandasamy, Z. He, Hydrothermal liquefaction of fresh lemon-peel and *Spirulina platensis* blending-operation parameter and biocrude chemistry investigation. Energy **193**, 116645 (2020)
220. S. Xiu, A. Shahbazi, V. Shirley, D. Cheng, Hydrothermal pyrolysis of swine manure to bio-oil: effects of operating parameters on products yield and characterization of bio-oil. J. Anal. Appl. Pyrol. **88**, 73–79 (2010)
221. S. Xiu, A. Shahbazi, C.W. Wallace, L. Wang, D. Cheng, Enhanced bio-oil production from swine manure co-liquefaction with crude glycerol. Energy Convers. Manage. **52**, 1004–1009 (2011)
222. F. Conti, S.S. Toor, T.H. Pedersen, A.H. Nielsen, L.A. Rosendahl, Biocrude production and nutrients recovery through hydrothermal liquefaction of wastewater irrigated willow. Biomass Bioenerg. **118**, 24–31 (2018)
223. S. Feng, Z. Yuan, M. Leitch, C.C. Xu, Hydrothermal liquefaction of barks into bio-crude—effects of species and ash content/composition. Fuel **116**, 214–220 (2014)
224. K. Tekin, S. Karagöz, S. Bektaş, Hydrothermal liquefaction of beech wood using a natural calcium borate mineral. J. Supercrit. Fluids **72**, 134–139 (2012)

225. Z. Guo, S. Wang, X. Wang, Stability mechanism investigation of emulsion fuels from biomass pyrolysis oil and diesel. Energy **66**, 250–255 (2014)
226. Z. Li, M. Garedew, C.H. Lam, J.E. Jackson, D.J. Miller, C.M. Saffron, Mild electrocatalytic hydrogenation and hydrodeoxygenation of bio-oil derived phenolic compounds using ruthenium supported on activated carbon cloth. Green Chem. **14**, 2540–2549 (2012)
227. C.H. Lam, C.B. Lowe, Z. Li, K.N. Longe, J.T. Rayburn, M.A. Caldwell, C.E. Houdek, J.B. Maguire, C.M. Saffron, D.J. Miller, J.E. Jackson, Electrocatalytic upgrading of model lignin monomers with earth abundant metal electrodes. Green Chem. **17**, 601–609 (2015)
228. T. Schulzke, S. Conrad, S. Kaluza, T. Van Loo, Upgrading of fast pyrolysis condensates via esterification with higher alcohols. Biomass Bioenerg. **103**, 11–20 (2017)
229. S. Hansen, A. Mirkouei, L.A. Diaz, A comprehensive state-of-technology review for upgrading bio-oil to renewable or blended hydrocarbon fuels. Renew. Sustain. Energy Rev. **118**, 109548 (2020)
230. A.A. Al-Muntaser, M.A. Varfolomeev, M.A. Suwaid, C. Yuan, A.E. Chemodanov, D.A. Feoktistov, I.Z. Rakhmatullin, M. Abbas, E. Domínguez-Álvarez, A.A. Akhmadiyarov, V.V. Klochkov, M.I. Amerkhanov, Hydrothermal upgrading of heavy oil in the presence of water at sub-critical, near-critical and supercritical conditions. J. Pet. Sci. Eng. **184**, 106592 (2020)
231. A. Farooq, H. Shafaghat, J. Jae, S.C. Jung, Y.K. Park, Enhanced stability of bio-oil and diesel fuel emulsion using Span 80 and Tween 60 emulsifiers. J. Environ. Manage. **231**, 694–700 (2019)
232. S.S. Reham, H.H. Masjuki, M.A. Kalam, I. Shancita, I.M. Rizwanul Fattah, A.M. Ruhul, Study on stability, fuel properties, engine combustion, performance and emission characteristics of biofuel emulsion. Renew. Sustain. Energy Rev. **52**, 1566–1579 (2015)
233. M. Ikura, M. Stanciulescu, E. Hogan, Emulsification of pyrolysis derived bio-oil in diesel fuel. Biomass Bioenerg. **24**(3), 221–232 (2003)
234. C.H. Lam, S. Das, N.C. Erickson, C.D. Hyzer, M. Garedew, J.E. Anderson, T.J. Wallington, M.A. Tamor, J.E. Jackson, C.M. Saffron, Towards sustainable hydrocarbon fuels with biomass fast pyrolysis oil and electrocatalytic upgrading. Sustain. Energy Fuels **1**, 258–266 (2017)
235. H.A. Baloch, S. Nizamuddin, M.T.H. Siddiqui, S. Riaz, A.S. Jatoi, D.K. Dumbre, N.M. Mubarak, M.P. Srinivasan, G.J. Griffin, Recent advances in production and upgrading of bio-oil from biomass: a critical overview. J. Environ. Chem. Eng. **6**, 5101–5118 (2018)
236. M. Bertero, G. de la Puente, U. Sedran, Fuels from bio-oils: bio-oil production from different residual sources, characterization and thermal conditioning. Fuel **95**, 263–271 (2012)
237. H. Kondoh, Y. Nakasaka, T. Kitaguchi, T. Yoshikawa, T. Tago, T. Masuda, Upgrading of oil sand bitumen over an iron oxide catalyst using sub- and super-critical water. Fuel Process. Technol. **145**, 96–101 (2016)
238. J. Peng, P. Chen, H. Lou, X. Zheng, Catalytic upgrading of bio-oil by HZSM-5 in sub- and super-critical ethanol. Bioresour. Technol. **100**, 3415–3418 (2009)
239. S. Cheng, L. Wei, J. Julson, K. Muthukumarappan, P.R. Kharel, Upgrading pyrolysis bio-oil to biofuel over bifunctional Co-Zn/HZSM-5 catalyst in supercritical methanol. Energy Convers. Manage. **147**, 19–28 (2017)
240. L. Leng, H. Li, X. Yuan, W. Zhou, H. Huang, Bio-oil upgrading by emulsification/microemulsification: a review. Energy **161**, 214–232 (2018)
241. Y. Song, U. Sanyal, D. Pangotra, J.D. Holladay, D.M. Camaioni, O.Y. Gutiérrez, J.A. Lercher, Hydrogenation of benzaldehyde via electrocatalysis and thermal catalysis on carbon-supported metals. J. Catal. **359**, 68–75 (2018)
242. Y. Song, S.H. Chia, U. Sanyal, O.Y. Gutiérrez, J.A. Lercher, Integrated catalytic and electrocatalytic conversion of substituted phenols and diaryl ethers. J. Catal. **344**, 263–272 (2016)
243. X. Li, R. Gunawan, C. Lievens, Y. Wang, D. Mourant, S. Wang, H. Wu, M. Garcia-Perez, C.-Z. Li, Simultaneous catalytic esterification of carboxylic acids and acetalisation of aldehydes in a fast pyrolysis bio-oil from mallee biomass. Fuel **90**, 2530–2537 (2011)
244. S. Xiu, A. Shahbazi, Bio-oil production and upgrading research: a review. Renew. Sustain. Energy Rev. **16**, 4406–4414 (2012)

245. X. Bai, P. Duan, Y. Xu, A. Zhang, P.E. Savage, Hydrothermal catalytic processing of pretreated algal oil: a catalyst screening study. Fuel **120**, 141–149 (2014)
246. W.-M. Xiong, Y. Fu, F.-X. Zeng, Q.-X. Guo, An in situ reduction approach for bio-oil hydroprocessing. Fuel Process. Technol. **92**, 1599–1605 (2011)
247. T.R. Brown, R. Thilakaratne, R.C. Brown, G. Hu, Techno-economic analysis of biomass to transportation fuels and electricity via fast pyrolysis and hydroprocessing. Fuel **106**, 463–469 (2013)
248. D. Elliott, T. Hart, G. Neuenschwander, L. Rotness, A. Zacher, Catalytic hydroprocessing of biomass fast pyrolysis bio-oil to produce hydrocarbon products. Environ. Prog. Sustain. Energy **28**, 441–449 (2009)
249. M. Bertero, J.R. García, M. Falco, U. Sedran, Equilibrium FCC catalysts to improve liquid products from biomass pyrolysis. Renew. Energy **132**, 11–18 (2019)
250. O.D. Mante, F.A. Agblevor, S.T. Oyama, R. McClung, Catalytic pyrolysis with ZSM-5 based additive as co-catalyst to Y-zeolite in two reactor configurations. Fuel **117**, 649–659 (2014)
251. D. Xu, P.E. Savage, Supercritical water upgrading of water-insoluble and water-soluble biocrudes from hydrothermal liquefaction of *Nannochloropsis* microalgae. J. Supercrit. Fluids **133**, 683–689 (2018)
252. N. Joshi, A. Lawal, Hydrodeoxygenation of pyrolysis oil in a microreactor. Chem. Eng. Sci. **74**, 1–8 (2012)
253. A.N.K. Lup, F. Abnisa, W.M.W. Daud, M.K.T. Aroua, A review on reactivity and stability of heterogeneous metal catalysts for deoxygenation of bio-oil model compounds. J. Ind. Eng. Chem. **56**, 1–34 (2017)
254. P. Duan, P.E. Savage, Catalytic hydrothermal hydrodenitrogenation of pyridine. Appl. Catal. B **108–109**, 54–60 (2011)
255. C. Prestigiacomo, P. Costa, F. Pinto, B. Schiavo, A. Siragusa, O. Scialdone, A. Galia, Sewage sludge as cheap alternative to microalgae as feedstock of catalytic hydrothermal liquefaction processes. J. Supercrit. Fluids **143**, 251–258 (2019)
256. S. Cheng, L. Wei, J. Julson, M. Rabnawaz, Upgrading pyrolysis bio-oil through hydrodeoxygenation (HDO) using non-sulfided Fe-Co/SiO_2 catalyst. Energy Convers. Manage. **150**, 331–342 (2017)
257. J. Liang, R. Ding, Y. Wu, Y. Chen, K. Wu, Y. Meng, M. Yang, Y. Wang, Effective conversion of heteroatomic model compounds in microalgae-based bio-oils to hydrocarbons over β-Mo_2C/CNTs catalyst. J. Mol. Catal. A Chem. **411**, 95–102 (2016)
258. C. Zhao, J. He, A.A. Lemonidou, X. Li, J.A. Lercher, Aqueous-phase hydrodeoxygenation of bio-derived phenols to cycloalkanes. J. Catal. **280**, 8–16 (2011)
259. O.İ Şenol, T.R. Viljava, A.O.I. Krause, Hydrodeoxygenation of methyl esters on sulphided NiMo/γ-Al_2O_3 and CoMo/γ-Al_2O_3 catalysts. Catal. Today **100**, 331–335 (2005)
260. A. Gutierrez, E.-M. Turpeinen, T.-R. Viljava, O. Krause, Hydrodeoxygenation of model compounds on sulfided CoMo/γ-Al_2O_3 and NiMo/γ-Al_2O_3 catalysts. Role of sulfur-containing groups in reaction networks. Catal. Today **285**, 125–134 (2017)
261. S. Boullosa-Eiras, R. Lødeng, H. Bergem, M. Stöcker, L. Hannevold, E.A. Blekkan, Catalytic hydrodeoxygenation (HDO) of phenol over supported molybdenum carbide, nitride, phosphide and oxide catalysts. Catal. Today **223**, 44–53 (2014)
262. B. Pawelec, C.V. Loricera, C. Geantet, N. Mota, J.L.G. Fierro, R.M. Navarro, Factors influencing selectivity in the liquid-phase phenol hydrodeoxygenation over ZSM-5 supported Pt/Ir and Pt+Ir catalysts. Mol. Catal. **482**, 110669 (2020)
263. J. Xing, L. Song, C. Zhang, M. Zhou, L. Yue, X. Li, Effect of acidity and porosity of alkali-treated ZSM-5 zeolite on eugenol hydrodeoxygenation. Catal. Today **258**, 90–95 (2015)
264. A. Sanna, T.P. Vispute, G.W. Huber, Hydrodeoxygenation of the aqueous fraction of bio-oil with Ru/C and Pt/C catalysts. Appl. Catal. B **165**, 446–456 (2015)
265. P. Duan, P.E. Savage, Catalytic treatment of crude algal bio-oil in supercritical water: optimization studies. Energy Environ. Sci. **4**, 1447–1456 (2011)
266. A. Galadima, O. Muraza, Hydrothermal liquefaction of algae and bio-oil upgrading into liquid fuels: role of heterogeneous catalysts. Renew. Sustain. Energy Rev. **81**, 1037–1048 (2018)

267. C.S. Raghuveer, J.W. Thybaut, G.B. Marin, Pyridine hydrodenitrogenation kinetics over a sulphided NiMo/γ-Al_2O_3 catalyst. Fuel **171**, 253–262 (2016)
268. J. Kopyscinski, J. Choi, J.M. Hill, Comprehensive kinetic study for pyridine hydrodenitrogenation on (Ni)WP/SiO_2 catalysts. Appl. Catal. A **445–446**, 50–60 (2012)
269. E. Furimsky, F.E. Massoth, Hydrodenitrogenation of petroleum. Catal. Rev. **47**, 297–489 (2005)
270. H.A. Miran, M.K. Altarawneh, Z.N. Jaf, M.M. Rahman, M.H. Almatarneh, Z.T. Jiang, Influence of the variation in the Hubbard parameter (U) on activation energies of CeO_2-catalysed reactions. Can. J. Phys. **98**(4), 385–389 (2020)
271. S.N. Sahu, N.K. Sahoo, S.N. Naik, D.M. Mahapatra, Chapter 12—Advancements in hydrothermal liquefaction reactors: overview and prospects, in *Bioreactors*. ed. by L. Singh, A. Yousuf, D.M. Mahapatra (Elsevier, Amsterdam, 2020), pp.195–213
272. K. Guo, Y. Ding, Z. Yu, One-step synthesis of ultrafine MoNiS and MoCoS monolayers as high-performance catalysts for hydrodesulfurization and hydrodenitrogenation. Appl. Catal. B **239**, 433–440 (2018)
273. R.H.H. Smits, K. Seshan, J.R.H. Ross, A.P.M. Kentgens, Investigation of V_2O_5/Nb_2O_5 catalysts by 51V solid-state NMR. J. Phys. Chem. **99**, 9169–9175 (1995)
274. S.A. Yashnik, G.A. Urzhuntsev, A.I. Stadnichenko, D.A. Svintsitskiy, A.V. Ishchenko, A.I. Boronin, Z.R. Ismagilov, Effect of Pd-precursor and support acid properties on the Pd electronic state and the hydrodesulfurization activity of Pd-zeolite catalysts. Catal. Today **323**, 257–270 (2019)
275. V.G. Baldovino-Medrano, P. Eloy, E.M. Gaigneaux, S.A. Giraldo, A. Centeno, Factors controlling the development of the HYD route of desulfurization of DBT over γ-alumina supported Pt and Pd catalysts. Catal. Today **150**, 186–195 (2010)
276. P.N. Kuznetsov, E.S. Kamenskiy, L.I. Kuznetsova, Comparative study of the properties of the coal extractive and commercial pitches. Energy Fuels **31**, 5402–5410 (2017)
277. M. Dorneles de Mello, F. de Almeida Braggio, B. da Costa Magalhães, J.L. Zotin, M.A.P. da Silva, Kinetic modeling of deep hydrodesulfurization of dibenzothiophenes on NiMo/alumina catalysts modified by phosphorus. Fuel Process. Technol. **177**, 66–74 (2018)
278. C. Zhang, X. Gong, J. Zeng, Z. Peng, X. Li, L. Lin, Y. Peng, S. Wang, Effects of solvent phase recycling on microalgae liquefaction in ethanol: Bio-oil production and nitrogen transformation. Sci. Total Environ. **902**, 166069 (2023)
279. L. Wang, Z. Zhang, Y. Liu, B. Wang, L. Fang, J. Qiu, K. Zhang, S. Wang, Exceptional thermoelectric properties of flexible organic−inorganic hybrids with monodispersed and periodic nanophase. Nat. Commun. **9**, 3817 (2018)
280. C. Jazrawi, P. Biller, A.B. Ross, A. Montoya, T. Maschmeyer, B.S. Haynes, Pilot plant testing of continuous hydrothermal liquefaction of microalgae. Algal Res. **2**, 268–277 (2013)

A Review on Recent Advances in Clean Microalgal Bio-oil Production via Catalytic Hydrothermal Deoxygenation

1 Introduction

The rapid development of modern economic society comes at the cost of massive fossil energy consumption, which, in turn, restricts the sustainable improvement of global society due to the limited reserves of conventional energy and the severe ecological and environmental issues caused by energy overuse [1, 2]. Therefore, humanity must seek clean alternatives to support continuous economic growth and alleviate the ongoing energy crisis. Biomass, as the only carbon-containing renewable energy and the fourth largest energy reserve after coal, petroleum, and natural gas, shows great potential and has attracted extensive attention over the past decade [3, 4]. Microalgae, as the third-generation biomass for energy production, stands out due to its various advantages, including high photosynthetic efficiency, fast growth rate, strong environmental adaptability, and minimal competition with agricultural land [5, 6]. Furthermore, due to the high moisture content of microalgae, HTL technology, which operates in sub-/supercritical water (240–380 °C, 5–30 MPa), is an efficient process for converting "wet" microalgae without the need for drying [7, 8].

Compared to various technologies for converting microalgae into bio-oil, such as pyrolysis [9, 10], transesterification [11, 12], and fermentation [13, 14], HTL stands out due to its high efficiency in utilizing all components of microalgae [15, 16]. During the HTL process, the main components of microalgae—protein, carbohydrate, and lipid, which have calorific value—undergo hydrolysis, dehydration, decarboxylation, and repolymerization to form oil products [17, 18]. These biochemical components contain significant amounts of oxygen, leading to the formation of 5–20% aliphatic and aromatic oxygen-containing compounds, such as fatty acids, aldehydes, ketones, alcohols, esters, ethers, and phenols, in the biocrude. However, these oxygenated compounds negatively affect bio-oil properties, resulting in low calorific value and high viscosity [19, 20]. Therefore, it is crucial to remove oxygen heteroatoms from the bio-oil phase. Significant efforts have been made to upgrade

Y. Guo et al., *Hydrothermal Processing of Biomass for Hydrogen and Bio-oil Production*, https://doi.org/10.1007/978-981-96-6336-1_5

microalgae biocrude into cleaner biofuels with reduced oxygen content for commercial use. The presence of a large number of heteroatoms in hydrophilic functional groups makes oil and water separation challenging, leading to bio-oil produced from HTL being in an oil/water emulsion state [21]. Therefore, directly upgrading this biocrude in hydrothermal environment is a natural technology option with engineering advantages compared with deoxygenation in traditional petrochemical refining.

Fatty acids are the predominant oxygen-containing compounds in HTL microalgae bio-oil [22] and their deoxygenation has been extensively studied, with three common reactions typically occurring during the process [23, 24].

Decarboxylation (DCX) removes two oxygen atoms while producing alkanes, though it results in the loss of one carbon atom as CO_2 (Eq. 1). Decarbonylation (DCN) occurs in the presence of hydrogen, expelling oxygen as CO and H_2O (Eq. 2). In contrast, hydrodeoxygenation (HDO) consumes significant hydrogen, removing oxygen as H_2O molecules and producing alkanes without any carbon loss (Eq. 3).

DCX:

$$\text{Rn-COOH} \rightarrow \text{Rn-H} + CO_2 \tag{1}$$

DCN:

$$\text{Rn-COOH} + H_2 \rightarrow \text{Rn-H} + CO + H_2O \tag{2}$$

HDO:

$$\text{Rn-COOH} + 3H_2 \rightarrow \text{Rn-CH}_3 + 2H_2O \tag{3}$$

In addition to chain-like oxygenated compounds, phenols or phenolic substances—compounds with at least one hydroxyl group (–OH) attached to a benzene ring—are commonly found in HTL microalgae bio-oil [25]. Hydrodeoxygenation (HDO) is the primary method for removing oxygen from the aromatic rings of phenolic compounds, and it occurs through three distinct pathways: direct-deoxygenation (DDO) [26], hydrogenation-dehydration (HYD) [27], and tautomerization-deoxygenation [28]. In the DDO pathway, the $C(sp^2)$–OH bond is cleaved via hydrogenolysis, producing deoxygenated aromatics. The HYD pathway involves initial hydrogenation of the aromatic ring, followed by dehydration of the hydroxyl group from the $C(sp^3)$–OH bond, resulting in cyclic hydrocarbons. Furthermore, density functional theory (DFT) studies on phenolic tautomerization have shown that tautomerization can transform the $C(sp^3)$–O bond into a $C(sp^2)$–O bond, reducing the activation energy barrier for C–O bond cleavage [29].

Significant progress has been made in the deoxygenation of bio-oil, with key research advancements including the role of bifunctional materials serving as hydrogenation and acid–base centers in HDO reactions [30], the influence of varying Si/

Al ratios in acid-activated zeolite catalysts on bio-oil HDO [31], the use of molecular sieve-loaded noble and transition metal catalysts for HDO of lignin-derived bio-oils [32], and investigations into carbon-loaded catalysts for bio-oil HDO [33]. Additional studies have explored atmospheric HDO of bio-oils or bio-derived oxygenates in pyrolysis oils [34], as well as the kinetics and pathways of HDO reactions over heterogeneous catalysts [35]. It is worth noting that most existing HDO research has focused on lignin-based bio-oils from pyrolysis. Although some studies have examined HDO of microalgae bio-oil under hydrothermal conditions, to the best of our knowledge, no review paper has been published specifically on catalytic deoxygenation of microalgae biocrude under such conditions.

To address this gap, we specifically focus on the deoxygenation of microalgae bio-oil and its oxygen-containing compounds under hydrothermal conditions. This paper reviews the effects of parameters such as temperature, batch holding time, and solvents on the deoxygenation of microalgae biocrude during hydrothermal conversion. Additionally, it examines the deoxygenation mechanisms of metal and acid-based catalysts for typical oxygen-containing compounds, including fatty acids, furans, and phenols. We also analyze both experimental and computational results to explore the role of water, particularly HTW, in HDO reactions. This article aims to provide comprehensive insights into the hydrothermal deoxygenation of microalgae biocrude and oxygen-containing model compounds, enriching the foundational knowledge needed for further research and industrial application in upgrading microalgae biocrude.

2 Effect of HTW on Deoxygenation

Water is typically utilized as a green solvent in aqueous reactions, but the physical and chemical properties of HTW are highly influenced by the temperature–pressure regime, particularly temperature [36]. This dependence gives HTW multiple functions in hydrothermal reactions and affects deoxygenation activity.

2.1 Unique Properties of HTW

HTW demonstrates good solubility for organic compounds and gases, similar to many nonpolar organic solvents, due to its low dielectric constant (ε), which decreases from around 80 at ambient conditions to 5 at the critical point (374 °C and 22.1 MPa) [6]. This property allows organic compounds to dissolve in water, forming a homogeneous phase under supercritical conditions. Furthermore, the viscosity of water at room temperature is approximately 0.001 Pa s, while the viscosity of SCW drops significantly to about 0.298×10^{-2} Pa s at 450 °C, 27 MPa. The diffusion rate of SCW also decreases to nearly 1/100 of that of liquid water [37]. Additionally, the mass transfer resistance between interfaces vanishes in the SCW phase due to the weakened hydrogen bonding among water molecules. As a result, the high diffusion rate and low viscosity of SCW enhance efficient heat transfer and rapid mass diffusion, which accelerates reaction rates, particularly in heterogeneous catalysis [38].

In a subcritical state, another major alteration is observed in the ionic product of water, $K_w = [H^+] \cdot [OH^-]$, where $[H^+] = [OH^-]$. For instance, the value of K_w increases significantly with temperature, rising from 10^{-14} $(mol/kg)^2$ under ambient conditions to a maximum of 10^{-11} at around 250 °C [39]. As a result, the high concentration of H^+ ions produced by HTW can promote acid-catalyzed reactions [40]. Furthermore, due to the high density and elevated K_w in subcritical HTW, ionic reactions such as the dehydration of carbohydrates and alcohols are also accelerated [41]. During hydrothermal reductions, hydrogen can be generated in situ through the splitting of HTW, which can further participate in hydrogenation reactions [42]. Isotope labeling experiments confirmed that acetic acid was formed from the hydrothermal reduction of HCO_3^- with the involvement of deuterium in D_2O [43].

2.2 *Mass Transfer Effect of HTW*

The mass transfer characteristics of HTW are a crucial factor influencing heterogeneous reactions and have been extensively studied [44]. It has been reported that the rate of proton diffusion increases in the presence of water. In such cases, water can donate one of its protons to a metal surface and accept another proton from a different site, facilitating hydrogen diffusion. In catalytic HDO of phenol over Nb_2O_5-MC (micro-mesoporous carbon), research found that water in a biphasic decalin/water solvent system could suppress benzene hydrogenation, thereby enhancing selectivity toward benzene. Additionally, the formation of an emulsion increases the interfacial surface area for the HDO reaction, improving mass transfer between the phases [45]. In another study, vanillin conversion and p-creosol selectivity reached 98.0% and 100%, respectively, over Au/CNTs in a water/oil system at 150 °C and 1 MPa H_2 [46]. This demonstrates that adding water to an organic solvent can improve mass transfer in the HDO process by creating an emulsified state.

In addition, the solvent cage effect has been used to explain the increased HDO activity, where oxygen-containing compounds are confined by the catalyst at the water/oil interface of emulsion droplets, enhancing collisions between the catalyst and oxygenated molecules [47]. A mechanism study also showed that, because hydrogen has a higher solubility in the organic phase compared to water, the formation of a Pickering emulsion can boost the local hydrogen concentration near the interface, thereby promoting hydrogenation [48]. In reaction kinetics research, the Weisz–Prater criterion was employed to assess mass transfer resistance in the hydrothermal deoxygenation of palmitic acid. The Weisz–Prater parameter was found to be only 0.13, much lower than the threshold value of 1, indicating that diffusion and mass transfer limitations can be disregarded under supercritical water conditions [49].

Additionally, both the Mears and Weisz–Prater criteria confirmed that mass transport has minimal influence on the hydrothermal conversion of pyridine [50].

2.3 *HTW as a Reactant for Deoxygenation*

Under hydrothermal conditions, HTW can act as a reactant, influencing deoxygenation activity. This enhancement is achieved through hydrogen transfer, as water takes part in the hydrogenation process during HDO reactions. For example, the conversion of guaiacol and the selectivity for cyclohexanol increased significantly to 92% and 73.1%, respectively, in the presence of water, compared to 25.5% and 2.7% in the organic solvent decalin [51]. Furthermore, in the catalytic HDO of phenol to benzene, isotopic labeling experiments demonstrated that water directly participated in the reaction, reducing the activation energy for breaking the C–O bond by donating a proton [52]. The study revealed that on the surface of a Ni catalyst, adsorbed D_2O easily accepts protons derived from H_2, aiding in the cleavage of the C–O bond, as illustrated in Fig. 1.

With the support of DFT calculations, it was found that phenol molecules are more easily adsorbed on the Pd surface due to the reduced work function when water is present [53]. Additionally, compared to a vapor-phase environment, the activation barrier for adding surface-bound H adatoms to phenol is lowered by 10–20 kJ/mol, owing to the enhanced proton transfer effect in an aqueous medium. More specifically, based on computational studies of the hydrothermal conversion of nine different oxygen-containing compounds—furfural, tetrahydrofuran, xylose, phenol, guaiacol, ferulic acid, acetic acid, 2-hydroxybenzaldehyde, and hydroxyacetone—Agrawal and Kishore [54] defined a descriptor to illustrate the influence of the supercritical phase on HDO reactions. The findings revealed that the Gibbs free energy change (ΔG) serves as a useful indicator of the effects of solvation, temperature, and pressure in SCW. The ΔG for all nine reactions consistently decreased as the temperature increased under hydrothermal conditions. Furthermore, the HDO of aromatic compounds such as phenol and 2-hydroxybenzaldehyde was enhanced in

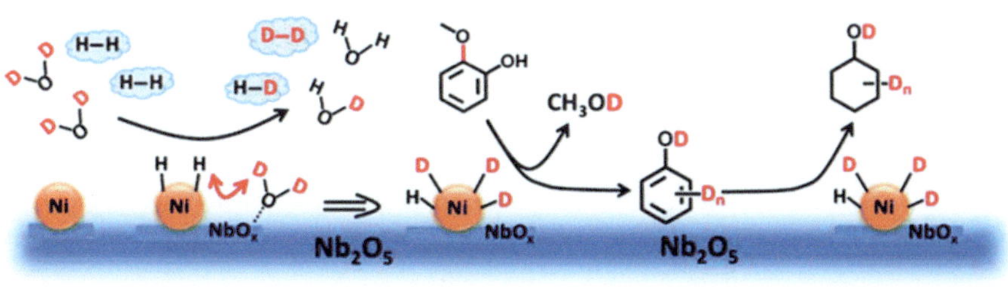

Fig. 1 Schematic of the direct HDO reaction mechanism of cyclohexanol formation through H_2O-promoted hydrogenation via H–D exchange [51]

SCW, whereas smaller molecules like acetic acid and hydroxyacetone showed little change.

2.4 Catalyst Regulation and Deactivation in HTW

On the flip side, water acting as a reactant may also have detrimental effects. Competitive adsorption between water and oxygen-containing compounds can occur during the catalytic process. Based on DFT calculations and experimental data, Engelhardt et al. [55] proposed a two-step deactivation mechanism for the Mo_2C catalyst. Initially, reversible deactivation happens due to competition between water and phenol for the active sites. As the water partial pressure rises, the active sites become fully covered by water, causing phenol to interact with water molecules rather than the Mo active sites on the catalyst. The second step involves irreversible deactivation caused by the formation of MoO_2 under hydrothermal conditions [56]. Therefore, conducting deoxygenation reactions below the critical water partial pressure can help preserve catalytic performance and stability.

Besides competing for active sites, water can also alter catalyst structure. Typically, HTW can cause leaching of active metals and degradation of catalyst supports. For instance, during hydrothermal deoxygenation of palmitic acid, metal particles were observed to aggregate due to the reduced surface area of metal oxide supports [57]. Interestingly, structural changes induced by water may sometimes enhance catalytic activity. In the HDO of p-cresol over MoS_2, water facilitated sulfur-oxygen exchange, converting MoS_2 had lost S with formation of Mo oxide at the edge of MoS_2 at the edges into Mo oxide, which increased both the number and strength of acid sites for C–O bond cleavage [58]. At 275 °C and 4.0 MPa H_2 pressure, p-cresol conversion and toluene selectivity were 83.4% and 74.6%, respectively, in dodecane solvent. When water was added, these values increased to 98.0% and 86.7%, respectively [59]. This effect may be attributed to the formation of Mo–OH and SH^- groups from the interaction between water and the active sites of MoS_2. The Mo–OH group then acts as a Lewis acid site, adsorbing p-cresol by accepting electron pairs from oxygen atoms, facilitating C–O bond rupture. After adsorption, H^+ from Mo–OH attacks the carbon in the C–O bond, producing a phenoxide ion. The C–O bond in the phenoxide ion then breaks, forming toluene and generating $-OH_2^+$ species on the catalyst surface. Lastly, the vacancies on the MoS_2 catalyst are restored when water is removed in the presence of H_2.

2.5 Catalyst Stability Under Hydrothermal Conditions

The durability of a catalyst is another important criterion for evaluating its overall performance, especially in the context of future industrial applications. This section focuses on the stability of metal-based catalysts for deoxygenation under hydrothermal conditions.

Al_2O_3 is commonly used as a catalyst support in HDO reactions because its acidic sites can promote the dehydration step [60–62], However, when Al_2O_3-supported catalysts are employed in hydrothermal conditions, Al_2O_3 can undergo rehydration, which significantly reduces its strength and surface area [57]. Additionally, in the presence of hot water, active metals like Ni can interact with Al_2O_3 to form low-activity Ni–Al compounds, leading to a permanent reduction in catalytic activity [63, 64]. For instance, in the HDO of o-cresol at 380 °C under supercritical water with a continuous flow system [65], the Ni/SiO_2–Al_2O_3 catalyst initially showed moderate conversion and yield of deoxygenated products, but these values dropped substantially over time due to Al_2O_3 degradation and sintering. In contrast, activated carbon-supported Pt/C exhibited the highest turnover frequency (TOF) among the catalysts tested (Pt/C, Ni–SiO_2–Al_2O_3, Raney-Ni, and Raney NiCu). Furthermore, in the hydrothermal HDO of palmitic acid, TEM analysis showed that fresh Pt nanoparticles on activated carbon were mostly in the range of 1.5–2 nm and increased slightly to 2–2.5 nm after 90 min at 340 °C [66]. Additionally, Pt0.5Cu0.5/C and pure Cu demonstrated high stability in hydrothermal CO_2 reduction, with less than a 5% decrease in activity after five reuses [42].

Apart from Al_2O_3, zeolites are another type of catalyst with acidic sites widely used in HDO, but they also suffer from instability under hydrothermal conditions. For example, pure SBA-15 lost 88.5% of its BET surface area and its pore diameter increased by 326% after 4 h in liquid water at 220 °C. In contrast, Zr-modified silicas retained most of their BET surface area, pore size, and pore volume after 16 h of hydrothermal treatment. Moreover, researchers found that using a carbon precursor is an effective method for enhancing the hydrothermal stability of SBA-15-based HDO catalysts. Polyfurfuryl alcohol was identified as the optimal carbon source, as it helped preserve the pore size and surface area during incipient wetness impregnation and subsequent drying processes [67].

3 Deoxygenation Research on HTL Algae Bio-oil

Although HTL is a promising approach for converting microalgae into biocrude oil, several challenges remain. This section will explore the distribution of oxygenated compounds in HTL algae bio-oil, the hydrothermal deoxygenation pathways of several key oxygen-containing model compounds, and the catalytic performance for deoxygenation when processing real algae feedstock under hydrothermal conditions.

3.1 O-Containing Compounds in HTL Algae Bio-oil

During the HTL process, microalgae are hydrolyzed into monomers, which are subsequently converted into bio-oil through various reactions, including DCX, dehydration, cyclization, condensation, and repolymerization [68].

Figure 2 illustrates the overall conversion pathways for the three main components (lipids, proteins, and carbohydrates) in microalgae under hydrothermal conditions [69]. At lower temperatures (0–100 °C), proteins are primarily hydrolyzed into amino acids and some fatty acids [68]. Meanwhile, the other two major components—lipids and carbohydrates—are also hydrolyzed, yielding oxygen-containing compounds such as glycerol, long-chain fatty acids, and sugars. As the temperature rises to 100–200 °C, the resulting amino acids, fatty acids, and sugars undergo further decomposition [70]. Specifically, amino compounds arise from the DCX of amino acids, with carbon dioxide produced from the dissociation of carboxyl groups, facilitating oxygen removal. Additionally, some hydrocarbons, including alkanes and alkenes, are formed through the DCX and HDO of free fatty acids, while cyclic oxygenates like furans are generated from sugar dehydration at this temperature range. When the temperature exceeds 200 °C, hydroxyl groups in long-chain fatty acids are substituted by ammonia from the deamination of amino acids, resulting in the formation of aliphatic amines. Carbohydrates can decompose into water-soluble aldehydes, alcohols, and ketones [71, 72].

The distribution of oxygen-containing compounds in microalgae bio-oil produced through HTL has been thoroughly studied. Statistical analysis by [74] indicates that the yield of oxygenated compounds in bio-oil from the HTL of *Desmodesmus* sp. typically ranges from 32.2 to 46.1%. The specific oxygenated compounds identified include phenol (5–10%), p-cresol (0.5–9%), 2-ethyl phenol (0.5–5%), isophytol (1–10%), 2-undecanone (7–13%), 6,10-dimethyl (0–3%), 1-dodecanol (0–3%), and cyclododecanol (0.7–3%). These compounds primarily originate from the decomposition of carbohydrates and lipids [75], with a small quantity of ketones formed through the hydrolysis and dehydration of polysaccharides [76]. Component analysis, summarized in Table 1, shows that macroalgae have a higher carbohydrate content, leading to increased oxygen levels in both the raw algal feedstock and the corresponding HTL bio-oil [73]. A machine learning study further supports the notion that carbohydrates play a crucial role in generating oxygen-containing substances in HTL algal bio-oil [77].

Among the oxygen-containing compounds, the concentration of ketones is significantly higher—often several times or even dozens of times greater—than that of acids, esters, and alcohols. This is likely due to ketones being formed from the hydrolysis and dehydration of polysaccharides [76]. Notably, a considerable amount of nitrogen and oxygen-containing compounds from amino acid decomposition was also detected in the algae bio-oil. Compounds such as tryptophol and indole-3-propionic acid are particularly resistant to decomposition, even under severe SCW conditions (400 °C for 90 min) [78]. Oxygenated products like phenol and its derivatives show higher concentrations under hydrothermal conditions compared to other

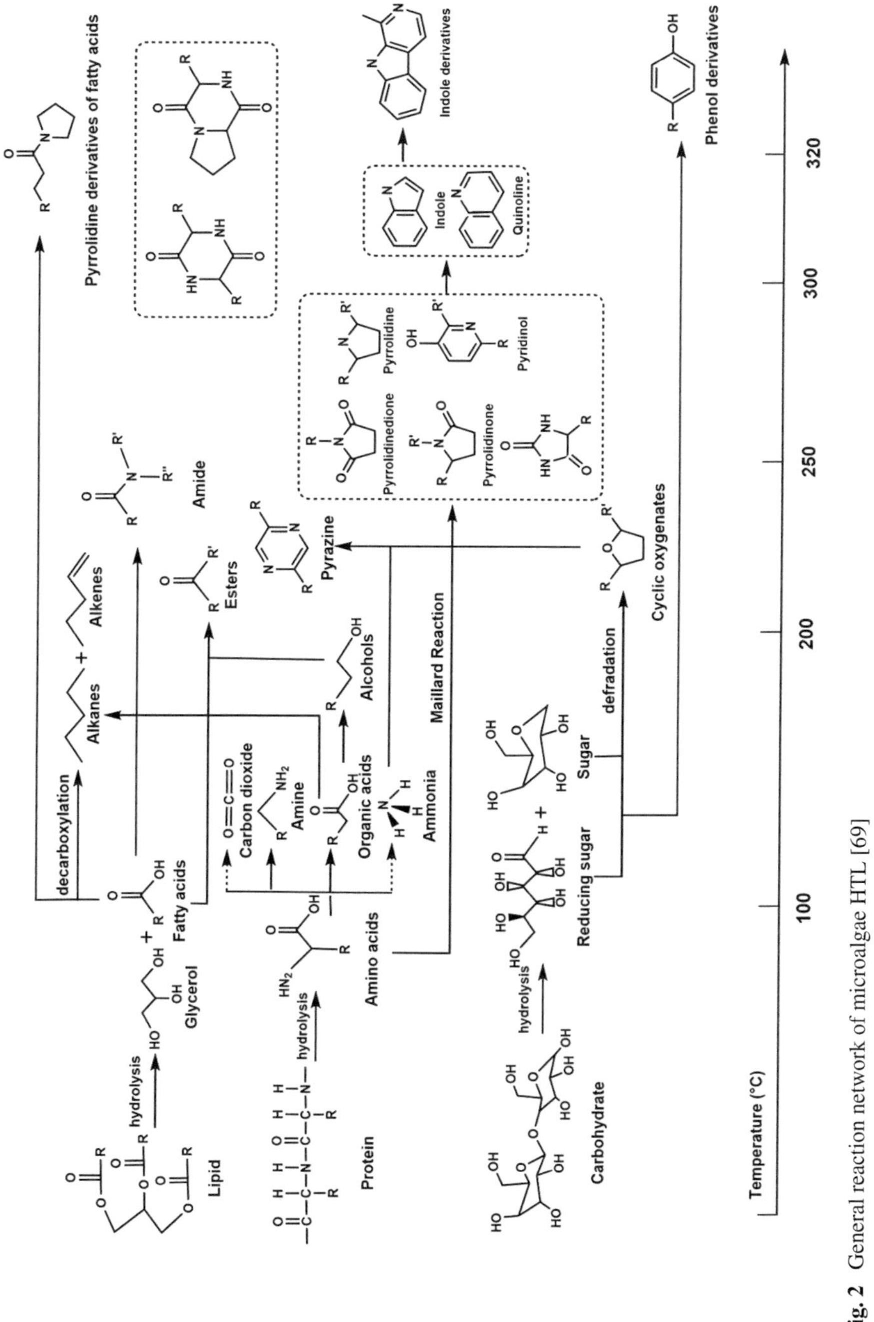

Fig. 2 General reaction network of microalgae HTL [69]

Table 1 O content of raw algal biomass and O-containing compounds distribution of algal bio-oil [73]

Algae	AuP	ArP	SL	NO	UP	SJ	ZM	GHE
O content (%)	22.61	22.48	16.96	21.70	23.55	18.52	28.81	33.35
Acids (%)	2.68	1.50	20.22	0.92	0.09	1.07	1.87	1.09
Alcohols (%)	1.06	1.74	2.00	2.71	2.14	2.64	2.45	0.98
Ester (%)	1.53	1.59	3.15	2.37	0.92	1.03	1.35	1.63
Ketones (%)	8.31	7.78	6.76	12.26	21.21	23.03	36.59	17.10
Amides (%)	30.09	19.75	35.75	26.71	4.59	4.45	3.06	14.29
O-compounds (%)	5.44	3.70	3.86	7.03	9.31	11.79	11.97	15.76
N,O-compounds (%)	14.76	19.92	4.90	12.69	15.29	14.51	8.71	4.79

Abbreviations: Aup, *Auxenochlorella pyrenoidosa*; ArP, *Arthrospira platensis*; SL, *Schizochytrium limacinum*; NO, *Nannochloropsis oceanica*; UP, *Ulva prolifera*; SJ, *Saccharina japonica*; ZM, *Zostera marina*; GHE: *Gracilaria eucheumatoides* Harvey

organic solvents, such as alcohols, indicating the active participation of hot water in the reactions [79]. DFT studies reveal that the energy barrier for the interaction between methyl aniline and hydroxyl groups derived from water is relatively high, at 3.52 eV, with nearly 2.5% of o-cresol detectable during the hydrothermal conversion of indole at 400 °C for 60 min [80]. This further underscores the involvement of hot water with nitrogen-containing substances.

3.2 *Hydrothermal Deoxygenation of O-Containing Model Compounds*

As discussed earlier, microalgae bio-oil primarily comprises oxygen-containing substances such as fatty acids, aldehydes, ketones, alcohols, esters, ethers, and phenols. This section will provide a detailed examination of the deoxygenation mechanisms of these typical oxygenates under hydrothermal conditions to enhance our understanding of the reaction processes involved.

3.2.1 Fatty Acids

Previous research on the deoxygenation of fatty acids primarily focused on organic phases, such as *n*-dodecane, *n*-heptane, *n*-decane, and decahydronaphthalene, where hydrogen was essential for C–O bond hydrogenation [81, 82]. Recently, both subcritical and supercritical water mediums have been employed for the deoxygenation of fatty acids. As mentioned in Sect. 2, in situ hydrogen can be generated from hot water splitting in a subcritical state [83, 84]. For instance, the yield of hydrocarbons from the deoxygenation of palmitic acid over Ni/ZrO_2 increased sharply from

1.8% to 35.5% due to the presence of hot water [85]. In the deoxygenation of stearic acid, significant amounts of CO and carbonyl compounds were produced in organic solvents, indicating that direct catalytic DCN was favored over direct catalytic DCX in that scenario. Conversely, the hydrothermal environment helped stabilize carboxyl groups and suppress cracking reactions, leading to acetic acid (CH_3COOH) and C_{16} alkenes being the major products [86].

Additionally, hydrogen donors such as formic acid, glycerol, and methanol can significantly influence reaction pathways and product distribution under hydrothermal conditions [85, 87, 88]. The presence of H_2 can activate HDO, promoting the formation of hydrocarbons without carbon loss [89]. The hydrothermal deoxygenation of palmitic acid, both with and without formic acid as an additional hydrogen donor, is illustrated in Fig. 3. In this process, pentadecane and hexadecane emerged as the primary deoxygenation products, indicating that direct catalytic DCX and HDO occurred simultaneously in the presence of formic acid. Additionally, kinetic modeling results revealed significant competition between the DCX and HDO reactions. The adsorption equilibrium constant for the hydrogenation intermediate, hexadecanol, reached 1991 L/mol at 310 °C, suggesting that hexadecanol occupied some of the active sites, thereby inhibiting the HDO pathway.

Moreover, hydrogen partial pressure significantly influences product distribution in the hydrothermal deoxygenation of fatty acids. Aromatic hydrocarbons, such as toluene and xylene, were produced from the hydrothermal deoxygenation of palmitic acid over HZSM-5 only when hydrogen pressure exceeded 50 bar. This is because high hydrogen pressure inhibits the dehydrocyclization of the resulting aromatic products [65].

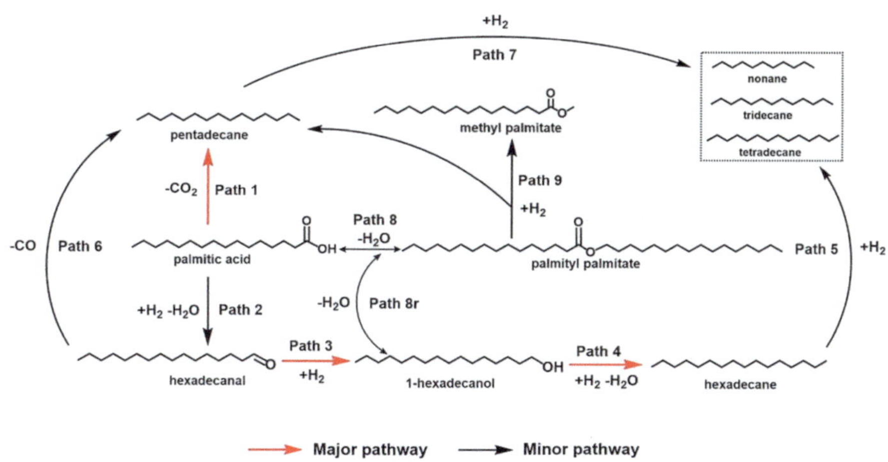

Fig. 3 Possible reaction network for palmitic acid hydrothermal deoxygenation [66]

3.2.2 Furans

Furans are heterocyclic organic compounds characterized by a five-membered aromatic ring that contains four carbon atoms and at least one oxygen atom. These compounds are typically derived from the dehydration and decarboxylation of glucose and galactose found in the bio-oil of microalgae subjected to HTL [92, 93]. The HDO of benzofuran in SCW, illustrated in Fig. 4 [90], reveals a reaction network for benzofuran [94, 95]. The primary distinction is that ethylcyclohexanone was exclusively detected under hydrothermal conditions. Kinetic modeling indicates that the ring-opening product, phenol, inhibits the hydrogenation of ethylbenzene and the dehydrogenation of ethylcyclohexane, suggesting that a pure hydrothermal environment lacks sufficient hydrogen to break C–O bonds in aromatic compounds [90]. This finding aligns with observations regarding the deoxygenation of fatty acids, where competitive adsorption for metal sites between hydrogenated intermediates and oxygenated reactants can significantly reduce HDO activity under hydrothermal conditions [49]. In a DFT calculation for furan HDO, the formation of pentane from furan is both thermodynamically and kinetically favored among several potential pathways for furfural HDO [96]. This supports the observation that heptane is the predominant long-chain alkane produced from the hydrothermal deoxygenation of benzofuran [90].

In contrast to the hydrothermal reaction mechanism, research indicates that the HDO of benzofuran occurs through β-hydrogen elimination (E_2 elimination) or nucleophilic substitution (S_N2) in organic phases, particularly over sulfurized Ni–Mo/γ-Al_2O_3 [97]. In the conversion of 2,3-dihydrobenzofuran to 2-ethylphenol, the cleavage of the C(sp^3)–O bond in the furan ring results from the nucleophilic substitution of SH^-, leading to the eventual opening of the furan ring. The 2-ethylphenol intermediate subsequently undergoes HDO via either direct DDO or HYD pathways to eliminate the hydroxyl group (–OH), yielding ethylbenzene or ethylcyclohexane as the major products in a toluene medium [95]. Conversely, catalytic HDO

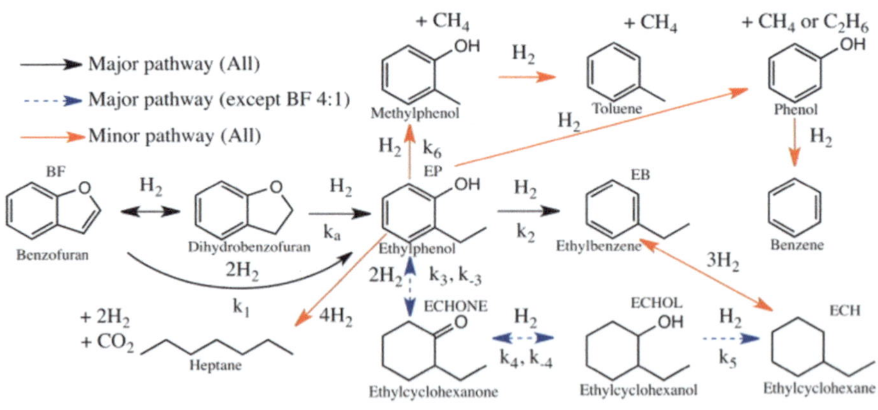

Fig. 4 Proposed reaction network for benzofuran deoxygenation in SCW [90]

of benzofuran in vapor phase does not conform to first-order kinetics; as the reaction rate constant (k_{HDO}) increases, the partial pressure of oxygenated products (Pox) decreases, indicating a self-inhibiting effect of these compounds [98].

3.2.3 Phenols

Phenol is a notable aromatic oxygen-containing compound found in bio-oil produced from the decomposition of cellulose in microalgae and other terrestrial biomass [99, 100]. A study has modeled the reaction kinetics of phenol's hydrothermal conversion, detailing a network of potential reactions leading to key liquid and gaseous products, as illustrated in Fig. 5 [91]. Under SCW conditions, cyclohexanone does not arise from the direct hydrogenation of phenol. Instead, phenol is believed to undergo partial hydrogenation to form an enol intermediate, which rapidly converts to cyclohexanone through keto-enol tautomerism. Other products formed from phenol include cyclopentanone and 3-cyclohexene-1-one, also via partial hydrogenation and keto-enol tautomerism. Various alkylphenols, such as 2-methylphenol, 3-methylphenol, 2,3-dimethylphenol, and 3,4-dimethylphenol, are generated from the alkylation of phenol. Additionally, the formation of 2-methylcyclohexanone can be attributed to the alkylation of cyclohexanone. Aromatic hydrocarbons, including toluene and xylenes (p-xylene and m-xylene), are produced through the alkylation of the hydroxyl group on phenol. In contrast to bio-oil derived from pyrolysis [101, 102], only trace amounts of biphenyl were detected as a by-product from toluene pyrolysis in a submerged carbon arc process. Kinetic studies on phenol hydrothermal HDO indicate that reaction rates follow this order: phenol hydrogenation < cyclohexanone hydrogenation < cyclohexanol dehydration < cyclohexene hydrogenation, establishing phenol hydrogenation as the rate-determining step in the overall reaction sequence [103].

A high density of acid sites in the support material enhances the oxygen removal from phenol through dehydration. For instance, an HZSM-5 supported Ni–Co catalyst demonstrated the highest HDO activity for phenol compared to other supports like HBeta, HY, and ZrO_2 at a hydrothermal temperature of 250 °C. Additionally, phenol conversion and benzene selectivity were significantly influenced by reaction temperature. Infrared spectroscopic measurements of adsorbed pyridine (pyr-IR) revealed that phenol hydrogenation on Ni–Co metal sites occurs around 200 °C, with a high activation energy for breaking the C–O bond. Dehydration steps necessitate temperatures exceeding 225 °C. When temperatures surpass 250 °C, undesired dehydrogenation toward benzene and isomerization become more prominent, leading to chemical equilibrium controlling product distribution.

In a comparative study of the HDO of 2-phenoxy-1-phenylethanol (PPE) in organic versus aqueous systems over Ni/HZSM-5, it was found that PPE primarily yields ethylbenzene and phenol as the major products (35% and 23%, respectively) under hydrothermal conditions at 250 °C, with relatively low amounts of ring-saturation products such as cyclohexanone and cyclohexanol, particularly in comparison to reactions conducted in nonpolar organic solvents [104].

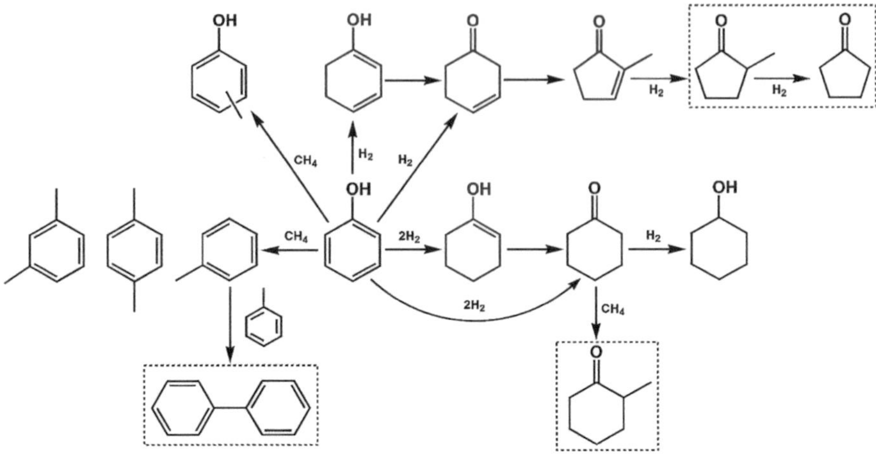

Fig. 5 Potential reaction pathways for liquid products for phenol hydrothermal conversion [91]

3.2.4 Amino Acids

Microalgae contain varying amounts of proteins, ranging from 6 wt% to as high as 71 wt%, depending on the species. During hydrothermal treatment, amino acids such as leucine, valine, lysine, phenylalanine, arginine, aspartic acid, glutamine, alanine, and glycine are quickly generated through peptide bond hydrolysis of algal proteins [105].

The DCX of short-chain amino acids like alanine and glycine follows first-order kinetics in subcritical water [106]. Several other short-chain amino acids, including alanine, leucine, phenylalanine, serine, and aspartic acid, can be fully decomposed within 180 s at 200–340 °C through two main pathways: DCX to amines and deamination to organic acids [107]. When the temperature is raised to 450 °C under a pressure of 34 MPa, glycine dimerizes to diketopiperazine [108], and trimers from glycine have been detected in SCW [109, 110]. Furthermore, a kinetic study found that second-order kinetics better describes glycine oligomerization in HTW [111, 112]. Beyond temperature and pressure, pH also significantly affects the dimerization rate of glycine in hydrothermal conditions. Most amino acids are unstable in acidic or near-neutral environments, while a basic pH of 10 has been identified as the most stable condition [113]. The DCX rate of α-alanine is notably reduced when potassium chloride (KCl) is added to the system. This may be due to the less polar transition state of the DCX process compared to the zwitterion form of α-alanine, as well as the reduced reaction activity caused by the complexity of ionic sites formed by interactions between α-alanine and K^+ or Cl^- ions [112].

In addition to linear amino acids like alanine, leucine, serine, and aspartic acid, hydrothermal conversion of amino acids with aromatic rings has also been studied. Research on the products, pathways, kinetics, and mechanisms of phenylalanine under hydrothermal conditions showed that phenylethylamine is easily formed

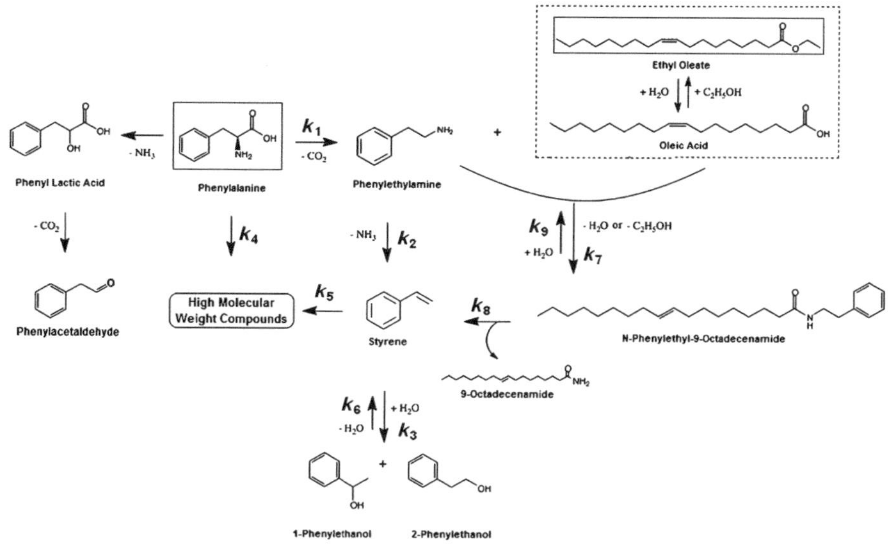

Fig. 6 Possible reaction pathways for phenylalanine under hydrothermal conditions [114]

through the DCX of phenylalanine [114]. As shown in Fig. 6, at higher temperatures (350 °C), phenylethylamine undergoes deamination to produce styrene, and further reactions lead to the formation of phenyl ethanol and oligomers via styrene hydration. In a binary mixture, new products such as 9-octadecenamide, N-ethyl-9-octadecenamide, and N-phenylethyl-9-octadecenamide are formed due to the interaction between phenylethylamine/ammonia and ethyloleate/oleic acid. For phenylalanine, the removal of the amino group is more challenging than the removal of the carboxyl group. This phenomenon has also been observed in the hydrothermal reaction of tryptophan, where indole derivatives, such as 3-methyl-indole and 2,3-dimethyl-indole, are predominantly produced from the hydrothermal deoxygenation of tryptophan over a Ni–Ru bimetallic catalyst [78].

4 Effects of Reaction Variables on O Content in Algae Bio-oil

In general, operational conditions such as temperature, reaction time, and solvent significantly affect the hydrothermal process. Therefore, in this section, the influence of these experimental variables on deoxygenation is discussed individually.

4.1 Temperature and Reaction Time

Temperature significantly influences the product distribution of microalgae bio-oil during the HTL process. Hydrolysis primarily occurs at lower temperatures (below 220 °C), with lipids being the main reactants during this phase. As the temperature rises to 250–300 °C, depolymerization reactions become more prominent, leading to a reduction in bio-oil yield [115]. When the temperature exceeds 300 °C, protein decomposition and the polymerization of intermediates increase, resulting in a higher yield of hydrocarbons and a decline in organic acid production [69]. However, at temperatures approaching the critical point of water (374.15 °C), N,O-containing heterocyclic compounds are formed through the recombination and repolymerization of intermediates [116].

A study examining the effect of temperature on oxygenated compounds in microalgae bio-oil using Response Surface Methodology found that, within the temperature range of 320–370 °C, the content of oxygenated compounds increased almost linearly as the temperature rose, while maintaining a fixed catalyst loading [74]. At temperatures above 400 °C, even refractory substances like phenols can decompose [117], and bio-oil tends to become gasified under the more extreme conditions of SCW [118]. A machine learning model based on HTL data from 42 species of microalgae and 21 species of macroalgae highlighted the relative importance of different variables in achieving specific target outcomes.

Among four variable categories—elemental composition, biochemical composition, elemental composition ratios, and operating conditions—operating conditions were identified as the most influential factors [77]. Specifically, as shown in Fig. 7a, temperature and reaction time account for 61% of the total importance in producing high oxygen content algae bio-oil (O_oil). Increasing the temperature to 350–450 °C and extending the reaction time to 30–60 min can significantly reduce the oxygen content in the oil phase, as depicted in Fig. 7b. Compared to operating conditions, biomass composition has a relatively lower impact on the oxygen content of HTL microalgae bio-oil. Among biochemical components, carbohydrates contribute more to oxygen yield than lipids or proteins due to their higher hydroxyl and ether content. As for elemental composition, oxygen unsurprisingly ranks first in importance, as it directly generates oxygen-containing chemicals in microalgae.

4.2 Solvents

Compared to water, organic solvents like methanol, ethanol, and acetone have lower critical points, requiring milder reaction conditions when used in HTL for bio-oil production. In recent years, the use of organic and water co-solvents in hydrothermal deoxygenation for upgrading microalgae bio-oil has been extensively studied. Research has shown that adding ethanol as a co-solvent in biomass HTL promotes deoxygenation due to enhanced hydrogenation activity [119–121]. For

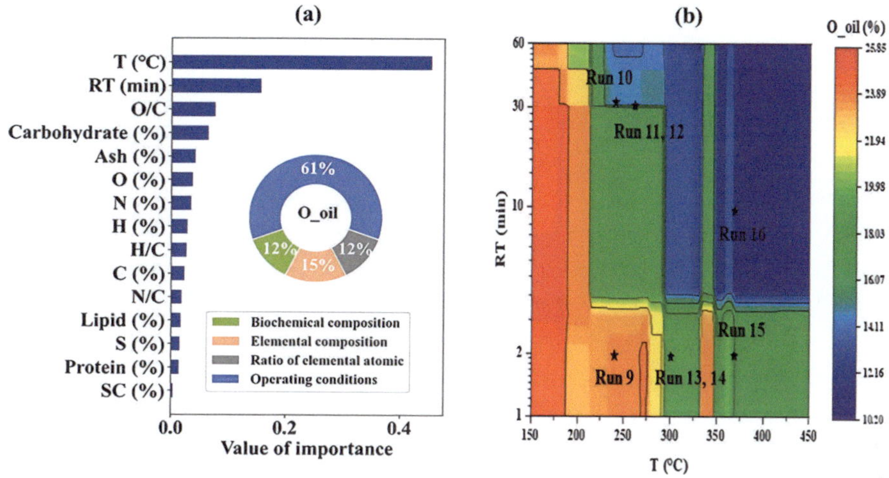

Fig. 7 **a** Feature importance plots on algae O_oil and **b** The optimal operating conditions for bio-oil production with *Chlorella* of minimized O_oil [77]

example, in the HTL of rice husk, a water–ethanol co-solvent showed a synergistic effect on deoxygenation, reducing oxygen content to 25.99% compared to 30.24% in pure water and 26.23% in pure ethanol [122]. Similarly, for the HTL of Dunaliella tertiolecta, the water–ethanol mixture exhibited higher deoxygenation activity compared to sub- or supercritical water alone [123]. This suggests that ethanol not only acts as a hydrogen donor but also reacts with amides and acids to form ethyl esters during the liquefaction process. Water–ethanol co-solvents have also demonstrated better oxygen removal capabilities in the HTL of single components like starch and protein [124].

The ratio of water to ethanol plays a crucial role in determining the composition of the resulting algal biocrude oil. When the ethanol ratio exceeds 40%, n-hexadecanoic acid is not detected, likely due to esterification between n-hexadecanoic acid and ethyl ester, which indicates that a sub-/supercritical ethanol–water system can facilitate ester formation [125]. Additionally, a higher ethanol ratio enhances the breakdown of oxygen-containing heterocyclic and aromatic compounds through hydrogen free radical attack, promoting the elimination of oxygen heteroatoms [121]. Figure 8a illustrates this mechanism, while Fig. 8b explains how in mixed-solvent phases, hydrophilic reactants can enhance the formation of localized solvent domains with higher local water density near the reactant compared to bulk solvent. This facilitates two steps in acid-catalyzed reactions: (1) proton transfer from the bulk solvent to the reactant and (2) the formation of a carbocation-like transition state [126].

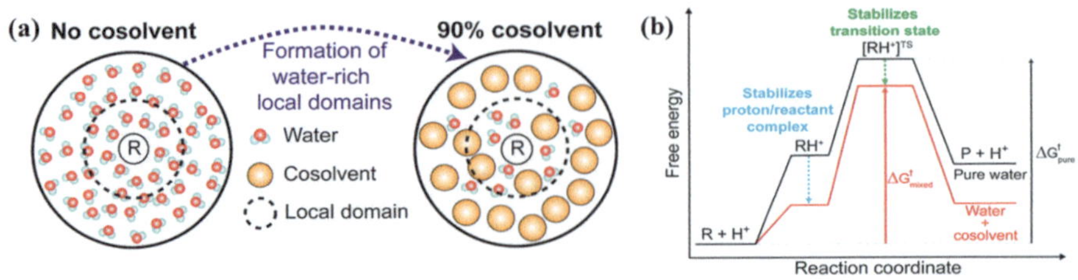

Fig. 8 **a** Role of co-solvent molecules on the distribution of solvent molecules. **b** Proposed effect of co-solvent molecules on a reaction free energy landscape [126]

4.3 Other Avenues Enhancing Deoxygenation

By optimizing the HTL process and algae cultivation methods, the oxygen content in algae-derived bio-oil can be reduced. Differences in the chemical composition of various algal species result in distinct reaction mechanisms due to interactions among the components. Therefore, interactions between micro- and macroalgae may facilitate deoxygenation under hydrothermal conditions.

Studies have shown that the oxygen content in the oil product from the co-liquefaction of *Spirulina platensis* and *Entermorpha prolifera* (5.0%) was lower than that from the liquefaction of *S. platensis* (6.1%) and *E. prolifera* (6.2%) individually [127]. In the co-liquefaction of microalgae *Chlorella pyrenoidosa* (CP) and lignocellulosic biomass rice husk (RH) under subcritical water conditions at 300 °C, the yield of hydrocarbons was higher than that obtained from CP or RH alone, demonstrating the synergistic effect between different biomass species [128]. Additionally, the yields of straight and branched amides increased linearly with the CP ratio, while the concentration of cyclic oxygenates decreased. Considering that each component of microalgae has an optimal temperature for hydrothermal conversion, a multi-temperature HTL process for *Chlorella* was explored to assess its impact on oxygen content in bio-oil. Compared to single-step HTL, the oxygen content decreased from 16.07% to 10.49% after a two-step treatment at 300–350 °C without using catalysts [129]. Higher H_2 pressure (3.4 MPa) during the upgrading step further reduced the oxygen content in crude *Nannochloropsis* biofuel [130]. The percentage of deoxygenation also correlated with catalyst loading. In a study on cattail biomass hydrothermal deoxygenation, the highest yield of HDO bio-oil (24.1 wt%) was obtained with 3 wt% Nb-AC catalyst at 300 °C for 90 min, producing 5.1 wt% biochar and 37.2 wt% aqueous phase [131]. Both lower (1 wt%) and higher (5 wt%) catalyst loadings reduced the HDO bio-oil yield, as insufficient catalyst loading could not effectively break the C–O bond, while excessive loading enhanced gasification of hydrocarbons. Additionally, increasing the hydrogen atmosphere from 100 to 200 bar further decreased the oxygen content from 11 to 5% [131].

5 Catalysts for Hydrothermal Deoxygenation

Converting bio-oil into long-chain hydrocarbons generally requires catalysts with both deoxygenation and hydrogenation sites to break C–O bonds [136, 142, 143]. Transition and noble metals, with their unfilled outer d orbitals, typically serve as active sites for deoxygenation, while zeolites are commonly used as acid centers for hydrogenation [144, 145]. This section separately examines the deoxygenation activity of metal and acid-based catalysts. Table 2 summarizes the deoxygenation activity of these catalysts under HTL conditions.

5.1 *Metal-Based Catalysts*

Metal-based catalysts have been extensively used in deoxygenation research and have proven effective in reducing the oxygen content in algal biocrude. Among monometallic noble metal catalysts (such as Ru/γ-Al_2O_3, Pt/γ-Al_2O_3, Pd/γ-Al_2O_3, Pt/C, Ru/C, and Pd/C), Pt/γ-Al_2O_3 exhibited the highest deoxygenation activity for upgrading *Nannochloropsis* biocrude oil, reducing the oxygen content from 12.8 to 1.6% at 400 °C for 20 min under 50 bar H_2 pressure after hydrothermal treatment [146]. Compared to noble metals, transition metals were less active. Among CNT-supported magnetic metal catalysts (Co, Fe, Ni), Co/CNT demonstrated the highest catalytic activity for deoxygenating *Gracilaria corticata* bio-oil, lowering the oxygen content from 45.2 to 19.7% at 310 °C for 30 min, even with a high H_2 pressure of 18 MPa [80]. The addition of a second metal, such as CoMo/γ-Al_2O_3, further reduced the oxygen content from 25.8 to 8.55% in *Nannochloropsis* bio-oil [141], while a lower oxygen content of 1.9% was achieved in *Chlorella pyrenoidosa* bio-oil using mixed Ru/C and Raney-Ni catalysts [140]. Additionally, $PtSn_x/C$ ($x = 1, 2, 3$) alloys with added Sn effectively enhanced stearic acid conversion at 350 °C in hydrothermal conditions, with increased selectivity for linoleic acid as more Sn was added [89].

In HDO research on vanillin as a model oxygen-containing compound, 100% conversion of vanillin and 2-methoxy-4-methylphenol (MMP) selectivity were achieved over bimetallic Ni–Co@N-CNTs at 120 °C for 6 h under 2 MPa [147]. In comparison, vanillin conversion and MMP selectivity were only 65.3% and 36.8% over Ni@N-CNTs/CMF, while Co@N-CNTs/CMF yielded 37.4% vanillin conversion with a slightly higher MMP selectivity of 57.1% under the same conditions. XPS results indicated that electrons were transferred from Ni to Co atoms, creating a higher electron cloud density around Co atoms, which promoted selective C–O bond adsorption. DFT calculations also showed that the activation of the C–O bond was significantly enhanced at the Ni–Co interface, with the absolute adsorption energy of vanillin increasing from 0.36 eV on Ni and Co surfaces to 0.48 eV at the Ni–Co interface in Fig. 9a. Moreover, active hydrogen species (H^+) from H_2 dissociation were more easily desorbed from the Ni–Co surface, facilitating C–O bond hydrogenation (Fig. 9b). The H_2-TPD results (Fig. 9c) showed that the alloyed Ni–Co

Table 2 Recent published literatures concerning hydrothermal deoxygenation of microalgae bio-oil

Algae	Hydrothermal conditions	Catalyst	Key results	References
Scenedesmus sp.	300 °C, 2 h	MgAl-LDO3	To facilitate low-oxygen bio-oil production from microalgae through HTL, MgAl-LDOx with plentiful acidic sites is advantageous, as it promotes both DCN and DCX reactions. By employing this catalyst, the oxygen content was significantly reduced from 41.3 to 11.01%	[132]
Chlorella	300 °C, 4 h, 10 MPa H_2 pressure	SO_3H/SBA-15-NH_2	The activation energy required for the noncatalytic conversion of liquefied bio-oil to dual-stage oil decreased to 92.42 kJ/mol when using SO_3H/SBA-15-NH_2 as a catalyst. This process resulted in a reduction of the oxygen content from 34.53 to 3.06%	[133]
E. prolifera	350 °C, 20 min	Fe/HZSM-5	ZSM-5-based catalysts effectively promote the production of long-chain alkanes and benzene derivatives while significantly reducing phenols and oxygen-containing species. When using Fe/HZSM-5, the oxygen content decreased from 7.78 to 3.12%	[134]

(continued)

Table 2 (continued)

Algae	Hydrothermal conditions	Catalyst	Key results	References
Nannochloropsis sp.	350 °C, 4 h, 75 bar H_2 pressure	Ni-Re/γ-Al_2O_3	Doping Cu into Ni/γ-Al_2O_3 reduced deoxygenation activity due to the preferential adsorption of C = O on the Cu surface and the formation of a NiO-CuO alloy, which diminished the availability of active Ni metal sites. In contrast, Ni-Re/γ-Al_2O_3 effectively reduced the oxygen content in biocrude from 20.43 to 1.85%	[135]
Gracilaria corticata	300 °C, 30 min, 18 MPa H_2 pressure	GaNiFe-LDO/AC	GaNiFe-LDO/AC promotes the selectivity for heptadecane and enhances the quality of bio-oil by lowering the O/C ratio and increasing the H/C ratio. As a result, the oxygen content was significantly reduced from 36.2 to 18.1%	[136]
Spirulina	310 °C, 30 min, 18 MPa H_2 pressure	Co/CNT	Using ethanol with a Co/CNT catalyst effectively produces low-oxygen bio-oil under hydrothermal conditions, reducing the oxygen content from 45.2% in raw spirulina feedstock to 19.7% in the upgraded bio-oil phase	[80]

(continued)

Table 2 (continued)

Algae	Hydrothermal conditions	Catalyst	Key results	References
Ulva prolifera	280 °C, 15 min, 120 bar H_2 pressure	MgO–ethanol	Under hydrothermal conditions with ethanol as a solvent, phenolic compounds can be converted into ester substances, leading to a reduction in oxygen content from 35.2 to 18.6% in the MgO-ethanol mixed hydrothermal environment	[137]
Chlorella	300 °C, 30 min, 6 MPa H_2 pressure	CuO/Al-SBA-15	The hydrocarbon yield significantly increased on the catalysts, with aromatics being the main oxygen-containing compounds over Al-SBA-15, while CuO/Al-SBA-15 favored the formation of aliphatics on metal oxide-supported catalysts. The hydrocarbon yield rose from 25.1 wt% to 65.7 wt% on CuO/Al-SBA-15, and the oxygen content decreased from 34.5 to 5.71%	[138]
Chlorella	350 °C, 2 h,	Fe_2O_3/MCM-41	In the 320–350 °C subcritical water range, the conversion of palmitic acid from *Chlorella* decomposition increased by 14–29% with Fe_2O_3/MCM-41, enhancing decarboxylation selectivity to pentadecane to 62%. The oxygen content was reduced from 34.53 to 20.44%	[139]

(continued)

Table 2 (continued)

Algae	Hydrothermal conditions	Catalyst	Key results	References
C. pyrenoidosa	350 °C, 60 min, 18 MPa H_2 pressure	UGO-Ru/C:Raney-Ni(1:1)	Compared to Ru/C alone, the mixed catalysts (Ru/C + Raney-Ni) enhanced hydrogenation and facilitated both HDO and DCX reactions, reducing the oxygen content from 25.4 to 1.9%	[140]
Nannochloropsis sp.	350 °C, 60 min, 3500 kPa H_2 pressure	$CoMo/Al_2O_3$	The crude bio-oils produced over $CoMo/Al_2O_3$ had lower O/C ratios compared to the noncatalytic process, both with and without H_2 addition. $CoMo/Al_2O_3$ effectively converts fatty acids into alkanes, reducing the oxygen content to 8.5%	[141]

catalyst improved hydrogen desorption compared to monometallic Co@N-CNTs/CMF and Ni@N-CNTs/CMF.

Ni/γ-Al_2O_3 catalysts exhibited excellent deoxygenation activity, reducing the oxygen content to just 0.2% in Nannochloropsis bio-oil [135]. However, the deoxygenation activity of Ni/γ-Al_2O_3 decreased when Cu was doped into the catalyst. Transmission Electron Microscope (TEM) analysis suggested that the in situ formation of NiO–CuO during HTL covered the active Ni surface sites, reducing deoxygenation efficiency. Despite this, some metal oxides are capable of catalyzing deoxygenation, as seen in various studies. For instance, Fe_2O_3/MCM-41 was able to lower the oxygen content from 34.53 to 20.44% in biocrude produced from *Chlorella* hydrothermal reactions at 300 °C for 30 min [148]. Similarly, CuO/Al-SBA-15, a metal oxide decorated with Al, reduced the oxygen content to 5.71% in *Chlorella* biocrude at 300 °C, 30 min, and 6 MPa H_2 pressure [138].

5.2 *Acid-Based Catalysts*

The properties of water, including its K_w and density, are significantly influenced by temperature and pressure, which in turn affect the surface characteristics of solid acid catalysts. In the dehydration of 2-octanol under sub- and supercritical water conditions, TiO_2 can alter its acid site types (Brønsted/Lewis) and intensity due to changes

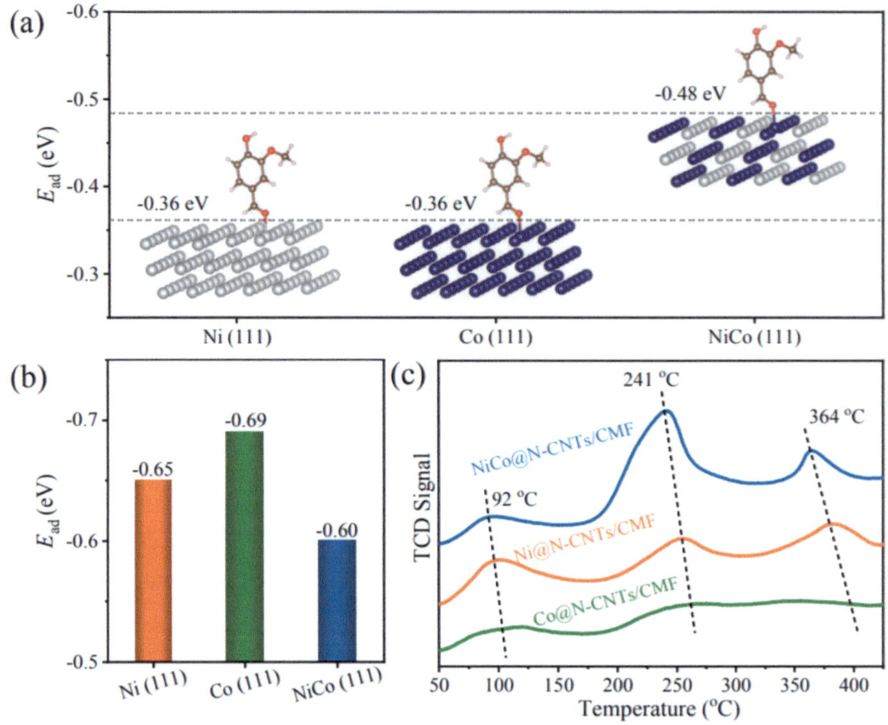

Fig. 9 **a** Optimized geometry for the configuration of vanillin adsorption on Ni (111), Co (111) and NiCo (111) surfaces. Gray, blue, brown, red, and pink spheres represent Ni, Co, C, O and H, respectively; **b** Adsorption energy of hydrogen on Ni (111), Co (111) and NiCo (111) surfaces; **c** H_2-TPD profiles of Ni@N-CNTs/CMF [147]

in the physicochemical properties of HTW [149]. As shown in Fig. 10, TiO_2 exhibits Lewis acidity in SCW with low water density, while it displays relatively stronger Brønsted acidity in subcritical water with higher density. The highest conversion of 2-octanol and selectivity for 2-octene were achieved at 390 °C. This behavior is also observed in other metal oxides, such as Nb_2O_5, Nb_2O_5/TiO_2 and WO_3/TiO_2 [150].

In addition to enhancing alcohol dehydration, solid acid catalysts are also effective for deoxygenation by promoting the esterification of organic carboxylic acids during the liquefaction reaction, thanks to their high surface area and unique crystal structure [151]. Both MgAl-LDH3 and MgAl-LDO3, which possess abundant active acid and base sites, facilitate DCX of fatty acids to produce alkanes and alkenes [132]. Furthermore, the formation of alcohols and ethers through DCN of fatty acids can also be promoted by MgAl-LDH3 and MgAl-LDO3. Notably, MgAl-LDO3, with its higher number of basic sites and activity, is more effective for DCN and DCX of fatty acids compared to MgAl-LDH3, achieving a 22.59% reduction in the O/C ratio of microalgae bio-oil—substantially more than that achieved with MgAl-LDH3. In

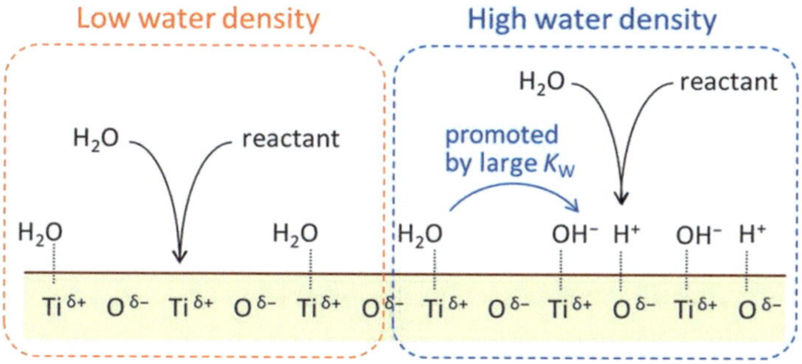

Fig. 10 Schematic image of the acidity model of TiO_2 in sub and supercritical water [157]

this context, ethers are more readily produced with the strong acidic sites of MgAl-LDH3, while alcohols tend to form with MgAl-LDO3 via DCX. Basic catalysts can also enhance deoxygenation by deprotonating alcohols through nucleophilic activation, whereas strong acid catalysts can protonate carbonyls, making them more electrophilic [152].

Besides layered double hydroxides, zeolites—conventional solid acid materials—are widely utilized for upgrading microalgae bio-oil under hydrothermal conditions due to their numerous and suitable acid sites [153–155]. A dual-functionalized SBA-15 catalyst, created by grafting sulfuric acid and basic amine ($–NH_2$) groups to the support, can promote phenol conversion. Modified SO_3H/SBA-15-NH_2 effectively reduced the oxygen content in *Chlorella* algae bio-oil to 3.06% during the upgrading process [133]. Fe/HZSM-5 was used to catalyze the HTL of *E. prolifera*, lowering the oxygen content from 7.78 to 3.12% [134]. Ni/γ-Al_2O_3-HZSM-5 was found to be more active than Ni/HZSM-5 for hydrothermal HDO of phenol, where the γ-Al_2O_3 binder introduced Lewis acidity, promoting cyclohexanol dehydration due to the close proximity of Lewis acid sites to metal Ni sites, resulting in irreversible hydrogenation of cyclohexene [103]. Analysis of the resultant bio-oil composition indicated that the zeolite catalyst altered product selectivity, significantly increasing the content of long-chain alkanes and aromatic hydrocarbons while decreasing the amount of phenolic oxygen-containing compounds. Additionally, adjusting the pore size of HZSM-5 can optimize aromatic hydrocarbon yields. A modified hierarchical HZSM-5 with larger pore sizes was able to produce approximately 44% of aromatic products (including toluene, ethylbenzene, and xylene) from the deoxygenation of stearic acid in supercritical water at 400 °C over a 60-min reaction time [156].

6 Economic Analysis for Producing Microalgae Bio-oil

To overcome the challenges facing the growth of bioliquid fuel, the supply of raw materials is the primary concern. Microalgae exhibit a significantly higher per unit area yield compared to terrestrial plants, positioning them as a promising third-generation renewable energy source. The Pacific Northwest National Laboratory (PNNL) conducted a comprehensive assessment of the feasibility of producing raw microalgae, considering factors such as land characteristics, freshwater resources, climate, and other essential resources. The findings indicate that 5.5% of the land area in the USA is suitable for cultivating microalgae species, potentially producing 20 million tons of dry microalgae by 2022 [158].

Microalgae bio-oil is rich in unsaturated fatty acids, making it ideal for hydrogenation to produce renewable diesel. The hydrogenation process constitutes less than 5% of the overall production cost. Sensitivity analyses reveal that increasing the oil content, unit yield, and lipid extraction efficiency of microalgae can significantly reduce the cost of generating microalgae diesel [159]. For instance, assuming an average microalgae yield of 13.2 g/(m^2·d) with 25% oil content, the lowest estimated price for this third-generation biodiesel is $18.63 per gallon. If the unit yield can increase to 25 g/(m^2·d), the cost could drop to $12.11 per gallon. However, to achieve competitive industrialization costs, prices must fall below $3.30 per gallon of gasoline equivalent (GGE).

Regarding HTL of microalgae, studies indicate that 57–70% of the carbon content in microalgae can be retained in the bio-oil phase, while approximately 43% appears in the water phase, alongside 2% in solid form and 4% as gas products generated during hydrothermal reactions. The low calorific value of microalgae HTL oil reaches 36 MJ/kg, with an oxygen content around 8%, which is lower than that of bio-oil produced through rapid pyrolysis. This makes the HTL oil more amenable to hydrogenation and upgrading, with 81 wt% of HTL oil convertible into hydrocarbons and only 4 wt% of chemical hydrogen required [160]. Currently, equipment investments in microalgae cultivation and harvesting account for 80% of the costs, while processing and transformation via the HTL route represent the remaining 20%. With the aforementioned assumptions about average yield and oil content, the cost of producing bio-oil through HTL is approximately half that of hydrogenation, indicating that HTL technology is a promising method for microalgae bio-oil production.

As previously mentioned, metal-based catalysts are crucial for hydrothermal HDO, particularly for phenolic oxygenates. Beyond their activity and durability, the cost of these catalysts is another critical factor for the commercial viability of HTL microalgae bio-oil. In studies examining the continuous HDO of o-cresol in supercritical water, while platinum (Pt) demonstrated superior catalytic activity compared to Raney-Ni (with turnover frequencies of 1.4 and 0.15 (mol/m^2·min) $\times 10^5$, respectively), the high cost of Pt—approximately 3000 times that of Raney-Ni—limits its industrial application [65]. Moreover, Ni-based catalysts have shown effective activity in HDN for nitrogenous heterocycles like indole [161], pyridine [50], quinoline [162], and tryptophan [78]. Notably, hydrothermal technology is also extensively employed for material synthesis. Research by Jin's group revealed that in situ synthesized, low-cost metal catalysts—including Mn [163], Zn [164], Cu [165], Ni [166], Co [167] and Al [168]—can effectively reduce CO_2 to organic compounds during hydrothermal reactions. This suggests that exploring in situ synthesis technology for catalysts may warrant further investigation.

7 Conclusions and Outlook

This part aims to summarize advancements in hydrothermal hydrodeoxygenation (HDO) of microalgae bio-oil and typical oxygen-containing model compounds. Specifically, it explores the impact of various experimental parameters, including temperature, reaction time, and solvents, on deoxygenation activity, as well as the unique role of hot, compressed HTW in the deoxygenation process. HTW conditions can enhance mass transfer, particularly in biphasic organic solvents and water systems, leading to improved conversion of oxygen-containing aromatic compounds and selectivity for deoxygenated products. Beyond its environmental benefits, water acts as a reactant in deoxygenation reactions, effectively lowering the activation energy barrier for breaking C–O bonds by donating protons.

Among the numerous experimental factors, temperature and reaction time are critical in influencing the oxygen content of algae bio-oil. Additionally, hydrogen partial pressure and the pore size of zeolite catalysts can alter product distribution in hydrothermal deoxygenation processes. For metal-based catalytic deoxygenation, the incorporation of a second metal can enhance the selective adsorption of C=O bonds, thereby improving the conversion of oxygenated compounds. However, in situ-formed metal oxides under hydrothermal conditions may cover the active sites of metals, impeding deoxygenation activity. It is also important to note that hydrogen generated in situ from HTW can compete with oxygen-containing substances for adsorption on the limited active sites of metal catalysts.

The types and intensities of acid sites in solid acid catalysts can change due to HTW conditions, with higher Lewis acidity and lower water density observed in SCW enhancing alcohol conversion and selectivity for alkenes. This article offers valuable insights for future scientific investigations and practical applications of bio-oils.

To commercialize low-oxygen-content HTL microalgae, careful selection of catalysts is essential, focusing on three key criteria: deoxygenation activity, hydrothermal stability, and cost. Given that hydrothermal technology is also applied in material synthesis, exploring in situ-formed deoxygenation catalysts during hydrothermal reactions may be a promising area for future research.

For studying deoxygenation mechanisms, in situ characterization techniques can deepen our understanding of the structure–activity relationships in catalytic deoxygenation reactions, warranting further research in this area. Theoretical investigations should involve multiscale simulations, computational fluid dynamics, micro-kinetic modeling, and DFT calculations to investigate the complex hydrothermal deoxygenation processes coupled with heat and mass transfer, diffusion, and competitive adsorption. Moreover, machine learning models can be utilized to analyze the relationships between deoxygenation activity and the intrinsic properties of HTW, such as dielectric constant (ε), water density (ρ), and ionic product (K_w). This approach could quantitatively identify key descriptors for hydrothermal deoxygenation and guide future experimental designs.

References

1. G. Kocar, N. Civas, An overview of biofuels from energy crops: current status and future prospects. Renew. Sustain. Energy Rev. **28**, 900–916 (2013)
2. Y.J. Zhou, E.J. Kerkhoven, J. Nielsen, Barriers and opportunities in bio-based production of hydrocarbons. Nat. Energy **3**, 925–935 (2018)
3. L.R. Amjith, B. Bavanish, A review on biomass and wind as renewable energy for sustainable environment. Chemosphere **293**, 133579 (2022)
4. X. Ma, Y. Mi, C. Zhao, Q. Wei, A comprehensive review on carbon source effect of microalgae lipid accumulation for biofuel production. Sci. Total. Environ. **806**, 151387 (2022)
5. H.-J. Huang, X.-Z. Yuan, Recent progress in the direct liquefaction of typical biomass. Prog. Energy Combust. Sci. **49**, 59–80 (2015)
6. A.A. Peterson, F. Vogel, R.P. Lachance, M. Froeling, M.J. Antal Jr., J.W. Tester, Thermochemical biofuel production in hydrothermal media: a review of sub- and supercritical water technologies. Energy Environ. Sci. **1**, 32–65 (2008)
7. K. Brindhadevi, S. Anto, E.R. Rene, M. Sekar, T. Mathimani, C. Nguyen Thuy Lan, A. Pugazhendhi, Effect of reaction temperature on the conversion of algal biomass to bio-oil and biochar through pyrolysis and hydrothermal liquefaction. Fuel **285**, 119106 (2021)
8. N. Sharma, K.K. Jaiswal, V. Kumar, M.S. Vlaskin, M. Nanda, I. Rautela, M.S. Tomar, W. Ahmad, Effect of catalyst and temperature on the quality and productivity of HTL bio-oil from microalgae: a review. Renew. Energy **174**, 810–822 (2021)
9. Z. Chang, P. Duan, Y. Xu, Catalytic hydropyrolysis of microalgae: influence of operating variables on the formation and composition of bio-oil. Biores. Technol. **184**, 349–354 (2015)
10. X.L. Miao, Q.Y. Wu, C.Y. Yang, Fast pyrolysis of microalgae to produce renewable fuels. J. Anal. Appl. Pyrol. **71**, 855–863 (2004)
11. M.B. Johnson, Z. Wen, Production of biodiesel fuel from the microalga *Schizochytrium limacinum* by direct transesterification of algal biomass. Energy Fuels **23**, 5179–5183 (2009)
12. R.B. Levine, T. Pinnarat, P.E. Savage, Biodiesel production from wet algal biomass through in situ lipid hydrolysis and supercritical transesterification. Energy Fuels **24**, 5235–5243 (2010)
13. R. Harun, M.K. Danquah, G.M. Forde, Microalgal biomass as a fermentation feedstock for bioethanol production. J. Chem. Technol. Biotechnol. **85**, 199–203 (2010)
14. R.P. John, G.S. Anisha, K.M. Nampoothiri, A. Pandey, Micro and macroalgal biomass: a renewable source for bioethanol. Biores. Technol. **102**, 186–193 (2011)
15. Y. Guo, T. Yeh, W. Song, D. Xu, S. Wang, A review of bio-oil production from hydrothermal liquefaction of algae. Renew. Sustain. Energy Rev. **48**, 776–790 (2015)
16. C. Yang, S. Wang, J. Yang, D. Xu, Y. Li, J. Li, Y. Zhang, Hydrothermal liquefaction and gasification of biomass and model compounds: a review. Green Chem. **22**, 8210–8232 (2020)
17. X. Gu, J.S. Martinez-Fernandez, N. Pang, X. Fu, S. Chen, Recent development of hydrothermal liquefaction for algal biorefinery. Renew. Sustain. Energy Rev. **121**, 109707 (2020)

18. M. Saber, B. Nakhshiniev, K. Yoshikawa, A review of production and upgrading of algal bio-oil. Renew. Sustain. Energy Rev. **58**, 918–930 (2016)
19. R. Ding, Y. Wu, Y. Chen, J. Liang, J. Liu, M. Yang, Effective hydrodeoxygenation of palmitic acid to diesel-like hydrocarbons over MoO_2/CNTs catalyst. Chem. Eng. Sci. **135**, 517–525 (2015)
20. P. Biller, A.B. Ross, Potential yields and properties of oil from the hydrothermal liquefaction of microalgae with different biochemical content. Biores. Technol. **102**, 215–225 (2011)
21. T.M. Brown, P. Duan, P.E. Savage, Hydrothermal liquefaction and gasification of *Nannochloropsis* sp. Energy Fuels **24**, 3639–3646 (2010)
22. S. Tang, Z. Shi, X. Tang, X. Yang, Hydrotreatment of biocrudes derived from hydrothermal liquefaction and lipid extraction of the high-lipid *Scenedesmus*. Green Chem. **21**, 3413–3423 (2019)
23. M. Snare, I. Kubickova, P. Maki-Arvela, K. Eranen, D.Y. Murzin, Heterogeneous catalytic deoxygenation of stearic acid for production of biodiesel. Ind. Eng. Chem. Res. **45**, 5708–5715 (2006)
24. J. Liang, Z. Zhang, K. Wu, Y. Shi, W. Pu, M. Yang, Y. Wu, Improved conversion of stearic acid to diesel-like hydrocarbons by carbon nanotubes-supported CuCo catalysts. Fuel Process. Technol. **188**, 153–163 (2019)
25. M.S. Vlaskin, A.V. Grigorenko, M.S. Kotelev, D.S. Kopitsyn, K.M. Mazurova, E.V. Ivanov, Composition and properties of microalgae biomass hydrothermal liquefaction products. Chem. Technol. Fuels Oils **55**, 373–377 (2019)
26. S. Boullosa-Eiras, R. Lodeng, H. Bergem, M. Stocker, L. Hannevold, E.A. Blekkana, Catalytic hydrodeoxygenation (HDO) of phenol over supported molybdenum carbide, nitride, phosphide and oxide catalysts. Catal. Today **223**, 44–53 (2014)
27. C. Zhao, J. He, A.A. Lemonidou, X. Li, J.A. Lercher, Aqueous-phase hydrodeoxygenation of bio-derived phenols to cycloalkanes. J. Catal. **280**, 8–16 (2011)
28. P.M. de Souza, R.C. Rabelo-Neto, L.E.P. Borges, G. Jacobs, B.H. Davis, T. Sooknoi, D.E. Resasco, F.B. Noronha, Role of keto intermediates in the hydrodeoxygenation of phenol over Pd on oxophilic supports. ACS Catal. **5**, 1318–1329 (2015)
29. A. Robinson, G.A. Ferguson, J.R. Gallagher, S. Cheah, G.T. Beckham, J.A. Schaidle, J.E. Hensley, J.W. Medlin, Enhanced hydrodeoxygenation of m-cresol over bimetallic Pt-Mo catalysts through an oxophilic metal-induced tautomerization pathway. ACS Catal. **6**, 4356–4368 (2016)
30. A.M. Robinson, J.E. Hensley, J.W. Medlin, Bifunctional catalysts for upgrading of biomass-derived oxygenates: a review. ACS Catal. **6**, 5026–5043 (2016)
31. S. Gea, I. Irvan, K. Wijaya, A. Nadia, A.N. Pulungan, J.L. Sihombing, R. Rahayu, Bio-oil hydrodeoxygenation over acid activated-zeolite with different Si/Al ratio. Biofuel Res. J. **9**, 1630–1639 (2022)
32. X. Li, G. Chen, C. Liu, W. Ma, B. Yan, J. Zhang, Hydrodeoxygenation of lignin-derived bio-oil using molecular sieves supported metal catalysts: a critical review. Renew. Sustain. Energy Rev. **71**, 296–308 (2017)
33. T. Cordero-Lanzac, J. Rodriguez-Mirasol, T. Cordero, J. Bilbao, Advances and challenges in the valorization of bio-oil: hydrodeoxygenation using carbon-supported catalysts. Energy Fuels **35**, 17008–17031 (2021)
34. H. Pourzolfaghar, F. Abnisa, W.M.A.W. Daud, M.K. Aroua, Atmospheric hydrodeoxygenation of bio-oil oxygenated model compounds: a review. J. Anal. Appl. Pyrol. **133**, 117–127 (2018)
35. H. Shi, Valorization of biomass-derived small oxygenates: kinetics, mechanisms and site requirements of H_2 involved hydrogenation and deoxygenation pathways over heterogeneous catalysts. ChemCatChem **11**, 1824–1877 (2019)
36. N. Akiya, P.E. Savage, Roles of water for chemical reactions in high-temperature water. Chem. Rev. **102**, 2725–2750 (2002)
37. C. Li, H. Cai, B. Zhang, W. Li, G. Pei, T. Dai, A. Wang, T. Zhang, Tailored one-pot production of furan-based fuels from fructose in an ionic liquid biphasic solvent system. Chin. J. Catal. **36**, 1638–1646 (2015)

38. A. Kruse, E. Dinjus, Hot compressed water as reaction medium and reactant—properties and synthesis reactions. J. Supercrit. Fluids **39**, 362–380 (2007)
39. M. Akizuki, T. Fujii, R. Hayashi, Y. Oshima, Effects of water on reactions for waste treatment, organic synthesis, and bio-refinery in sub- and supercritical water. J. Biosci. Bioeng. **117**, 10–18 (2014)
40. A.V. Bandura, S.N. Lvov, The ionization constant of water over wide ranges of temperature and density. J. Phys. Chem. Ref. Data **35**, 15–30 (2006)
41. S.S. Toor, L. Rosendahl, A. Rudolf, Hydrothermal liquefaction of biomass: a review of subcritical water technologies. Energy **36**, 2328–2342 (2011)
42. X. Liu, H. Zhong, C. Wang, D. He, F. Jin, CO_2 reduction into formic acid under hydrothermal conditions: a mini review. Energy Sci. Eng. **10**, 1601–1613 (2022)
43. X. Wang, Y. Yang, H. Zhong, T. Wang, J. Cheng, F. Jin, Molecular H2O promoted catalytic bicarbonate reduction with methanol into formate over $Pd_{0.5}$ $Cu_{0.5}$ C under mild hydrothermal conditions. Green Chem. **23**, 430–439 (2021)
44. V.V. Rozanov, O.V. Krylov, Hydrogen spillover in heterogeneous catalysis. Usp. Khim. **66**, 117–130 (1997)
45. C. Zhang, C. Jia, Y. Cao, Y. Yao, S. Xie, S. Zhang, H. Lin, Water-assisted selective hydrodeoxygenation of phenol to benzene over the Ru composite catalyst in the biphasic process. Green Chem. **21**, 1668–1679 (2019)
46. X. Yang, Y. Liang, X. Zhao, Y. Song, L. Hu, X. Wang, Z. Wang, J. Qiu, Au/CNTs catalyst for highly selective hydrodeoxygenation of vanillin at the water/oil interface. RSC Adv. **4**, 31932–31936 (2014)
47. Z. Li, Y. Wu, G. Tang, D. Zhang, J. Lu, Comparison between heat transfer to supercritical water in a smooth tube and in an internally ribbed tube. Int. J. Heat Mass Transf. **84**, 529–541 (2015)
48. Z. Zhu, H. Tan, J. Wang, S. Yu, K. Zhou, Hydrodeoxygenation of vanillin as a bio-oil model over carbonaceous microspheres-supported Pd catalysts in the aqueous phase and Pickering emulsions. Green Chem. **16**, 2636–2643 (2014)
49. J. Fu, X. Lu, P.E. Savage, Catalytic hydrothermal deoxygenation of palmitic acid. Energy Environ. Sci. **3**, 311–317 (2010)
50. Y. Guo, X. Liu, P. Duan, D. Xu, R. Luque, Catalytic hydrodenitrogenation of pyridine under hydrothermal conditions: a comprehensive study. ACS Sustain. Chem. Eng. **9**, 362–374 (2021)
51. W. Song, Y. He, S. Lai, W. Lai, X. Yi, W. Yang, X. Jiang, Selective hydrodeoxygenation of lignin phenols to alcohols in the aqueous phase over a hierarchical Nb_2O_5 supported Ni catalyst. Green Chem. **22**, 1662–1670 (2020)
52. R.C. Nelson, B. Baek, P. Ruiz, B. Goundie, A. Brooks, M.C. Wheeler, B.G. Frederick, L.C. Grabow, R.N. Austin, Experimental and theoretical insights into the hydrogen-efficient direct hydrodeoxygenation mechanism of phenol over Ru/TiO_2. ACS Catal. **5**, 6509–6523 (2015)
53. Y. Yoon, R. Rousseau, R.S. Weber, D. Mei, J.A. Lercher, First-principles study of phenol hydrogenation on Pt and Ni catalysts in aqueous phase. J. Am. Chem. Soc. **136**, 10287–10298 (2014)
54. Y. Das, H. Liu, F. Tian, S. Kota, R. Zhang, L.F. Chalak, Study of conversion of bio-oil model compounds in supercritical water using density functional theory. Sci. Rep. **10**, 9247 (2020)
55. J. Engelhardt, P. Lyu, P. Nachtigall, F. Schueth, A.M. Garcia, The influence of water on the performance of molybdenum carbide catalysts in hydrodeoxygenation reactions: a combined theoretical and experimental study. ChemCatChem **9**, 1985–1991 (2017)
56. A.L. Jongerius, R.W. Gosselink, J. Dijkstra, J.H. Bitter, P.C.A. Bruijnincx, B.M. Weckhuysen, Carbon nanofiber supported transition-metal carbide catalysts for the hydrodeoxygenation of guaiacol. ChemCatChem **5**, 2964–2972 (2013)
57. F. Liu, C. Okolie, R.M. Ravenelle, J.C. Crittenden, C. Sievers, P.C.A. Bruijnincx, B.M. Weckhuysen, Silica deposition as an approach for improving the hydrothermal stability of an alumina support during glycerol aqueous phase reforming. Appl. Catal. A General **551**, 13–22 (2018)

58. G. Zhu, W. Wang, K. Wu, S. Tan, L. Tan, Y. Yang, Hydrodeoxygenation of cresol on MoS_2/amorphous carbon composites synthesized by a one-step hydrothermal method: the effect of water on their activity and structure. Ind. Eng. Chem. Res. **55**, 12173–12182 (2016)
59. W. Wang, S. Tan, K. Wu, G. Zhu, Y. Liu, L. Tan, Y. Huang, Y. Yang, Hydrodeoxygenation of cresol as a model compound for bio-oil on MoS_2: effects of water and benzothiophene on the activity and structure of catalyst. Fuel **214**, 480–488 (2018)
60. P.D. Coan, M.B. Griffin, P.N. Ciesielski, J.W. Medlin, Phosphonic acid modifiers for enhancing selective hydrodeoxygenation over Pt catalysts: the role of the catalyst support. J. Catal. **372**, 311–320 (2019)
61. S. Chen, G. Zhou, H. Xie, Z. Jiao, X. Zhang, Hydrodeoxygenation of methyl laurate over the sulfur-free Ni/γ-Al_2O_3 catalysts. Applied Catalysis a-General **569**, 35–44 (2019)
62. K.-K. Sun, G.-P. Lu, J.-W. Zhang, C. Cai, The selective hydrogenolysis of C–O bonds in lignin model compounds by Pd–Ni bimetallic nanoparticles in ionic liquids. Dalton Trans. **46**, 11884–11889 (2017)
63. S. Li, L. Guo, Stability and activity of a co-precipitated Mg promoted Ni/Al_2O_3 catalyst for supercritical water gasification of biomass. Int. J. Hydrogen Energy **44**, 15842–15852 (2019)
64. S. Li, P.E. Savage, L. Guo, Stability and activity maintenance of Al_2O_3 and carbon nanotube-supported Ni catalysts during continuous gasification of glycerol in supercritical water. J. Supercrit. Fluids **135**, 188–197 (2018)
65. J.G. Dickinson, P.E. Savage, Stability and activity of Pt and Ni catalysts for hydrodeoxygenation in supercritical water. J. Mol. Catal. A Chem. **388**, 56–65 (2014)
66. X. Liu, M. Yang, Z. Deng, A. Dasgupta, Y. Guo, Hydrothermal hydrodeoxygenation of palmitic acid over Pt/C catalyst: mechanism and kinetic modeling. Chem. Eng. J. **407**, 126332 (2021)
67. S. Chung, Q. Liu, U.A. Joshi, J.R. Regalbuto, A.A. Boateng, M.A. Smith, C.G. Coe, Using polyfurfuryl alcohol to improve the hydrothermal stability of mesoporous oxides for reactions in the aqueous phase. J. Porous Mater. **25**, 407–414 (2018)
68. C. Tian, B. Li, Z. Liu, Y. Zhang, H. Lu, Hydrothermal liquefaction for algal biorefinery: a critical review. Renew. Sustain. Energy Rev. **38**, 933–950 (2014)
69. C. Gai, Y. Zhang, W.-T. Chen, P. Zhang, Y. Dong, An investigation of reaction pathways of hydrothermal liquefaction using *Chlorella pyrenoidosa* and *Spirulina platensis*. Energy Convers. Manage. **96**, 330–339 (2015)
70. D.L. Barreiro, W. Prins, F. Ronsse, W. Brilman, Hydrothermal liquefaction (HTL) of microalgae for biofuel production: state of the art review and future prospects. Biomass Bioenergy **53**, 113–127 (2013)
71. K. Anastasakis, A.B. Ross, Hydrothermal liquefaction of the brown macro-alga *Laminaria saccharina*: effect of reaction conditions on product distribution and composition. Biores. Technol. **102**, 4876–4883 (2011)
72. J.D. Sheehan, P.E. Savage, Modeling the effects of microalga biochemical content on the kinetics and biocrude yields from hydrothermal liquefaction. Biores. Technol. **239**, 144–150 (2017)
73. P.-G. Duan, S.-K. Yang, Y.-P. Xu, F. Wang, D. Zhao, Y.-J. Weng, X.-L. Shi, Integration of hydrothermal liquefaction and supercritical water gasification for improvement of energy recovery from algal biomass. Energy **155**, 734–745 (2018)
74. I. Nava-Bravo, S.B. Velasquez-Orta, I. Monje-Ramirez, L.P. Guereca, A.P. Harvey, R. Cuevas-Garcia, I. Yanez-Noguez, M.T. Orta-Ledesma, Catalytic hydrothermal liquefaction of microalgae cultivated in wastewater: influence of ozone-air flotation on products, energy balance and carbon footprint. Energy Convers. Manage. **249**, 114806 (2021)
75. D. Xu, P.E. Savage, Effect of reaction time and algae loading on water-soluble and insoluble biocrude fractions from hydrothermal liquefaction of algae. Algal Res. Biomass Biofuels Bioprod. **12**, 60–67 (2015)
76. S. Koley, M.S. Khadase, T. Mathimani, H. Raheman, N. Mallick, Catalytic and non-catalytic hydrothermal processing of *Scenedesmus obliquus* biomass for bio-crude production—a sustainable energy perspective. Energy Convers. Manage. **163**, 111–121 (2018)

77. W. Zhang, J. Li, T. Liu, S. Leng, L. Yang, H. Peng, S. Jiang, W. Zhou, L. Leng, H. Li, Machine learning prediction and optimization of bio-oil production from hydrothermal liquefaction of algae. Biores. Technol. **342**, 126011 (2021)
78. Y. Guo, Z. Chen, R.M. Rioux, P.E. Savage, Hydrothermal reaction of tryptophan over Ni-based bimetallic catalysts. J. Supercrit. Fluids **143**, 336–345 (2019)
79. S.B. Velasquez-Orta, R. Garcia-Estrada, I. Monje-Ramirez, A. Harvey, M.T. Orta Ledesma, Microalgae harvesting using ozoflotation: effect on lipid and FAME recoveries. Biomass Bioenergy **70**, 356–363 (2014)
80. B. Liu, Z. Wang, L. Feng, Effects of reaction parameter on catalytic hydrothermal liquefaction of microalgae into hydrocarbon rich bio-oil. J. Energy Inst. **94**, 22–28 (2021)
81. J. Ni, W. Leng, J. Mao, J. Wang, J. Lin, D. Jiang, X. Li, Tuning electron density of metal nickel by support defects in Ni/ZrO_2 for selective hydrogenation of fatty acids to alkanes and alcohols. Appl. Catal. B Environ. **253**, 170–178 (2019)
82. Y. Shi, E. Xing, Y. Cao, M. Liu, K. Wu, M. Yang, Y. Wu, Tailoring product distribution during upgrading of palmitic acid over bi-functional metal/zeolite catalysts. Chem. Eng. Sci. **166**, 262–273 (2017)
83. J. Fu, X. Lu, P.E. Savage, Hydrothermal decarboxylation and hydrogenation of fatty acids over Pt/C. Chemsuschem **4**, 481–486 (2011)
84. S.A.W. Hollak, M.A. Ariens, K.P. de Jong, D.S. van Es, Hydrothermal deoxygenation of triglycerides over Pd/C aided by in situ hydrogen production from glycerol reforming. Chemsuschem **7**, 1057–1062 (2014)
85. C. Miao, O. Marin-Flores, S.D. Davidson, T. Li, T. Dong, D. Gao, Y. Wang, M. Garcia-Perez, S. Chen, Hydrothermal catalytic deoxygenation of palmitic acid over nickel catalyst. Fuel **166**, 302–308 (2016)
86. M. Watanabe, T. Iida, H. Inomata, Decomposition of a long chain saturated fatty acid with some additives in hot compressed water. Energy Convers. Manage. **47**, 3344–3350 (2006)
87. D.R. Vardon, B.K. Sharma, H. Jaramillo, D. Kim, J.K. Choe, P.N. Ciesielski, T.J. Strathmann, Hydrothermal catalytic processing of saturated and unsaturated fatty acids to hydrocarbons with glycerol for in situ hydrogen production. Green Chem. **16**, 1507–1520 (2014)
88. Z. Zhang, Q. Yang, H. Chen, K. Chen, X. Lu, P. Ouyang, J. Fu, J.G. Chen, In situ hydrogenation and decarboxylation of oleic acid into heptadecane over a Cu–Ni alloy catalyst using methanol as a hydrogen carrier. Green Chem. **20**, 197–205 (2018)
89. T.M. Yeh, R.L. Hockstad, S. Linic, P.E. Savage, Hydrothermal decarboxylation of unsaturated fatty acids over $PtSn_x$ C catalysts. Fuel **156**, 219–224 (2015)
90. J.G. Dickinson, J.T. Poberezny, P.E. Savage, Deoxygenation of benzofuran in supercritical water over a platinum catalyst. Appl. Catal. B Environ. **123**, 357–366 (2012)
91. M. Yang, J. Zhang, Y. Guo, Supercritical water gasification of phenol over Ni–Ru bimetallic catalyst: intermediates and kinetics. J. Supercrit. Fluids **160**, 12–20 (2020)
92. Q. Lin, K. Wu, Y. Chen, Y. Tang, Y. Wu, M. Yang, Catalytic hydrothermal liquefaction of microalgae for bio-oil production over silylated SBA-15 with high hydrothermal stability. Ind. Eng. Chem. Res. **56**, 14454–14462 (2017)
93. S. Wang, S. Zhao, X. Cheng, L. Qian, B. Barati, X. Gong, B. Cao, C. Yuan, Study on two-step hydrothermal liquefaction of macroalgae for improving bio-oil. Biores. Technol. **319**, 124176 (2021)
94. A.Y. Bunch, X. Wang, U.S. Ozkan, Hydrodeoxygenation of benzofuran over sulfided and reduced $Ni\text{-}Mo/\gamma\text{-}Al_2O_3$ catalysts: effect of H_2S. J. Mol. Catal. A Chem. **270**, 264–272 (2007)
95. Y. Romero, F. Richard, Y. Reneme, S. Brunet, Hydrodeoxygenation of benzofuran and its oxygenated derivatives (2,3-dihydrobenzofuran and 2-ethylphenol) over $NiMoP/Al_2O_3$ catalyst. Appl. Catal. A General **353**, 46–53 (2009)
96. A. Banerjee, S.H. Mushrif, Reaction pathways for the deoxygenation of biomass-pyrolysis-derived bio-oil on Ru: a DFT study using furfural as a model compound. ChemCatChem **9**, 2828–2838 (2017)
97. A.Y. Bunch, U.S. Ozkan, Investigation of the reaction network of benzofuran hydrodeoxygenation over sulfided and reduced $Ni\text{-}Mo/Al_2O_3<$ catalysts. J. Catal. **206**, 177–187 (2002)

98. M.C. Edelman, M.K. Maholland, R.M. Baldwin, S.W. Cowley, Vapor-phase catalytic hydrodeoxygenation of benzofuran. J. Catal. **111**, 243–253 (1988)
99. Y.-J. Ding, C.-X. Zhao, Z.-C. Liu, Catalytic hydrothermal liquefaction of rice straw for production of monomers phenol over metal supported mesoporous catalyst. Biores. Technol. **294**, 122097 (2019)
100. Y. Guo, W. Song, J. Lu, Q. Ma, D. Xu, S. Wang, Hydrothermal liquefaction of *Cyanophyta*: evaluation of potential bio-crude oil production and component analysis. Algal Res. Biomass Biofuels Bioprod. **11**, 242–247 (2015)
101. R. Gao, Y. Liu, B. Liu, Z. Xu, Novel utilization of pyrolysis products produced from waste printed circuit boards: catalytic cracking and synthesis of graphite carbon. J. Clean. Prod. **236**, 117662 (2019)
102. M. Saidi, A. Jahangiri, Refinery approach of bio-oils derived from fast pyrolysis of lignin to jet fuel range hydrocarbons: reaction network development for catalytic conversion of cyclohexanone. Chem. Eng. Res. Des. **121**, 393–406 (2017)
103. C. Zhao, S. Kasakov, J. He, J.A. Lercher, Comparison of kinetics, activity and stability of Ni/HZSM-5 and Ni/Al_2O_3 HZSM-5 for phenol hydrodeoxygenation. J. Catal. **296**, 12–23 (2012)
104. R.R. Barton, M. Carrier, C. Segura, J.L.G. Fierro, S. Park, H.H. Lamb, N. Escalona, S.W. Peretti, Ni/HZSM-5 catalyst preparation by deposition-precipitation. Part 2. Catalytic hydrodeoxygenation reactions of lignin model compounds in organic and aqueous systems. Appl. Catal. A General **562**, 294–309 (2018)
105. E.W. Becker, Micro-algae as a source of protein. Biotechnol. Adv. **25**, 207–210 (2007)
106. J.R. Vallentyne, Biogeochemistry of organic matter. 2. Thermal reaction kinetics and transformation products of amino compounds. Geochim. Cosmochim. Acta **28**, 157–188 (1964)
107. N. Sato, A.T. Quitain, K. Kang, H. Daimon, K. Fujie, Reaction kinetics of amino acid decomposition in high-temperature and high-pressure water. Ind. Eng. Chem. Res. **43**, 3217–3222 (2004)
108. D. Klingler, J. Berg, H. Vogel, Hydrothermal reactions of alanine and glycine in sub- and supercritical water. J. Supercrit. Fluids **43**, 112–119 (2007)
109. D.K. Alargov, S. Deguchi, K. Tsujii, K. Horikoshi, Reaction behaviors of glycine under super- and subcritical water conditions. Orig. Life Evol. Biosph. **32**, 1–12 (2002)
110. Y. Ogata, E.I. Imai, H. Honda, K. Hatori, K. Matsuno, Hydrothermal circulation of seawater through hot vents and contribution of interface chemistry to prebiotic synthesis. Orig. Life Evol. Biosph. **30**, 527–537 (2000)
111. J.S. Cox, T.M. Seward, The reaction kinetics of alanine and glycine under hydrothermal conditions. Geochim. Cosmochim. Acta **71**, 2264–2284 (2007)
112. J. Li, X.G. Wang, M.T. Klein, T.B. Brill, Spectroscopy of hydrothermal reactions, 19: pH and salt dependence of decarboxylation of α-alanine at 280–330°C in an FT-IR spectroscopy flow reactor. Int. J. Chem. Kinet. **34**, 271–277 (2002)
113. W. Abdelmoez, T. Nakahasi, H. Yoshida, Amino acid transformation and decomposition in saturated subcritical water conditions. Ind. Eng. Chem. Res. **46**, 5286–5294 (2007)
114. S. Changi, M. Zhu, P.E. Savage, Hydrothermal reaction kinetics and pathways of phenylalanine alone and in binary mixtures. Chemsuschem **5**, 1743–1757 (2012)
115. H. Li, Z. Liu, Y. Zhang, B. Li, H. Lu, N. Duan, M. Liu, Z. Zhu, B. Si, Conversion efficiency and oil quality of low-lipid high-protein and high-lipid low-protein microalgae via hydrothermal liquefaction. Biores. Technol. **154**, 322–329 (2014)
116. H.K. Reddy, T. Muppaneni, S. Ponnusamy, N. Sudasinghe, A. Pegallapati, T. Selvaratnam, M. Seger, B. Dungan, N. Nirmalakhandan, T. Schaub, F.O. Holguin, P. Lammers, W. Voorhies, S. Deng, Temperature effect on hydrothermal liquefaction of *Nannochloropsis gaditana* and *Chlorella* sp. Appl. Energy **165**, 943–951 (2016)
117. J. Zhang, A. Dasgupta, Z. Chen, D. Xu, P.E. Savage, Y. Guo, Supercritical water gasification of phenol over Ni–Ru bimetallic catalysts. Water Res. **152**, 12–20 (2019)
118. L. Yang, Y. Li, P.E. Savage, Near- and supercritical ethanol treatment of biocrude from hydrothermal liquefaction of microalgae. Biores. Technol. **211**, 779–782 (2016)

119. S. Brand, F. Hardi, J. Kim, D.J. Suh, Effect of heating rate on biomass liquefaction: differences between subcritical water and supercritical ethanol. Energy **68**, 420–427 (2014)
120. H. Huang, X. Yuan, G. Zeng, J. Wang, H. Li, C. Zhou, X. Pei, Q. You, L. Chen, Thermochemical liquefaction characteristics of microalgae in sub- and supercritical ethanol. Fuel Process. Technol. **92**, 147–153 (2011)
121. X. Yuan, J. Wang, G. Zeng, H. Huang, X. Pei, Z. Liu, M. Cong, Comparative studies of thermochemical liquefaction characteristics of microalgae using different organic solvents. Energy **36**, 6406–6412 (2011)
122. Y. Liu, X.-Z. Yuan, H.-J. Huang, X.-L. Wang, H. Wang, G.-M. Zeng, Thermochemical liquefaction of rice husk for bio-oil production in mixed solvent (ethanol-water). Fuel Process. Technol. **112**, 93–99 (2013)
123. Y. Chen, Y. Wu, P. Zhang, D. Hua, M. Yang, C. Li, Z. Chen, J. Liu, Direct liquefaction of *Dunaliella tertiolecta* for bio-oil in sub/supercritical ethanol–water. Biores. Technol. **124**, 190–198 (2012)
124. X. Peng, X. Ma, Y. Lin, J. Wang, X. Chen, Investigation on solvolysis liquefaction mechanism of *Chlorella pyrenoidosa* in ethanol-water solvent based on major model compounds: protein and starch. J. Anal. Appl. Pyrol. **141**, 104621 (2019)
125. X. Peng, X. Ma, Y. Lin, Investigation on characteristics of liquefied products from solvolysis liquefaction of *Chlorella pyrenoidosa* in ethanol–water systems. Energy Fuels **30**, 6475–6485 (2016)
126. T.W. Walker, A.K. Chew, H. Li, B. Demir, Z.C. Zhang, G.W. Huber, R.C. Van Lehn, J.A. Dumesic, Universal kinetic solvent effects in acid-catalyzed reactions of biomass-derived oxygenates. Energy Environ. Sci. **11**, 617–628 (2018)
127. B. Jin, P. Duan, Y. Xu, F. Wang, Y. Fan, Co-liquefaction of micro- and macroalgae in subcritical water. Biores. Technol. **149**, 103–110 (2013)
128. C. Gai, Y. Li, N. Peng, A. Fan, Z. Liu, Co-liquefaction of microalgae and lignocellulosic biomass in subcritical water. Biores. Technol. **185**, 240–245 (2015)
129. Z. He, D. Xu, L. Liu, Y. Wang, S. Wang, Y. Guo, Z. Jing, Product characterization of multi-temperature steps of hydrothermal liquefaction of *Chlorella* microalgae. Algal Res. Biomass Biofuels Bioprod. **33**, 8–15 (2018)
130. P. Duan, P.E. Savage, Catalytic hydrotreatment of crude algal bio-oil in supercritical water. Appl. Catal. B Environ. **104**, 136–143 (2011)
131. J. Arun, K.P. Gopinath, R. Sivaramakrishnan, N.V. Madhav, K. Abhishek, V.G.K. Ramanan, A. Pugazhendhi, Bioenergy perspectives of cattails biomass cultivated from municipal wastewater via hydrothermal liquefaction and hydro-deoxygenation. Fuel **284**, 118963 (2021)
132. Z. Shen, T. Ma, L. Fei, A. Li, J. Liu, Z. Xu, X. Hu, Y. Sun, MgAl-LDH/LDO-catalyzed hydrothermal deoxygenation of microalgae for low-oxygen biofuel production. Acs ES&T Eng. **1**, 989–999 (2021)
133. C. Yu, Z. Qi, J. Bian, R. Song, W. Wang, C. Li, Insight into acid-base bifunctional catalysts for microalgae liquefaction and bio-oil pyrolysis: product characteristics, energy recovery and kinetics. J. Anal. Appl. Pyrol. **155**, 105086 (2021)
134. J. Li, G. Wang, Z. Wang, L. Zhang, C. Wang, Z. Yang, Conversion of *Enteromorpha prolifera* to high-quality liquid oil via deoxy-liquefaction. J. Anal. Appl. Pyrol. **104**, 494–501 (2013)
135. K. Pongsiriyakul, W. Kiatkittipong, S. Adhikari, J.W. Lim, S.S. Lam, K. Kiatkittipong, A. Dankeaw, P. Reubroycharoen, N. Laosiripojana, K. Faungnawakij, S. Assabumrungrat, Effective Cu/Re promoted Ni-supported γ-A_2O_3 catalyst for upgrading algae bio-crude oil produced by hydrothermal liquefaction. Fuel Process. Technol. **216**, 106670 (2021)
136. Y. Li, C. Zhu, J. Jiang, Z. Yang, W. Feng, L. Li, Y. Guo, J. Hu, Catalytic hydrothermal liquefaction of *Gracilaria corticata* macroalgae: effects of process parameter on bio-oil up-gradation. Biores. Technol. **319**, 124163 (2021)
137. J. Xu, X. Dong, Y. Wang, Hydrothermal liquefaction of macroalgae over various solids, basic or acidic oxides and metal salt catalyst: products distribution and characterization. Ind. Crops Prod. **151**, 112458 (2020)

138. J. Li, X. Fang, J. Bian, Y. Guo, C. Li, Microalgae hydrothermal liquefaction and derived biocrude upgrading with modified SBA-15 catalysts. Biores. Technol. **266**, 541–547 (2018)
139. J. Bian, Q. Zhang, P. Zhang, L. Feng, C. Li, Supported Fe_2O_3 nanoparticles for catalytic upgrading of microalgae hydrothermal liquefaction derived bio-oil. Catal. Today **293–294**, 159–166 (2017)
140. Y. Xu, P. Duan, B. Wang, Catalytic upgrading of pretreated algal oil with a two-component catalyst mixture in supercritical water. Algal Res. **9**, 186–193 (2015)
141. P. Duan, P.E. Savage, Hydrothermal liquefaction of a microalga with heterogeneous catalysts. Ind. Eng. Chem. Res. **50**, 52–61 (2011)
142. A. Veses, B. Puertolas, J. Manuel Lopez, M. Soledad Callen, B. Solsona, T. Garcia, Promoting deoxygenation of bio-oil by metal-loaded hierarchical ZSM-5 zeolites. ACS Sustain. Chem. Eng. **4**, 1653–1660 (2016)
143. J. Zhang, K. Wang, M.W. Nolte, Y.S. Choi, R.C. Brown, B.H. Shanks, Catalytic deoxygenation of bio-oil model compounds over acid-base bifunctional catalysts. ACS Catal. **6**, 2608–2621 (2016)
144. A.N.K. Lup, F. Abnisa, W.M.A.W. Daud, M.K. Aroua, A review on reaction mechanisms of metal-catalyzed deoxygenation process in bio-oil model compounds. Appl. Catal. A General **541**, 87–106 (2017)
145. Y. Shi, E. Xing, K. Wu, J. Wang, M. Yang, Y. Wu, Recent progress on upgrading of bio-oil to hydrocarbons over metal/zeolite bifunctional catalysts. Catal. Sci. Technol. **7**, 2385–2415 (2017)
146. B. Patel, P. Arcelus-Arrillaga, A. Izadpanah, K. Hellgardt, Catalytic hydrotreatment of algal biocrude from fast hydrothermal liquefaction. Renew. Energy **101**, 1094–1101 (2017)
147. D. Wang, W. Gong, J. Zhang, M. Han, C. Chen, Y. Zhang, G. Wang, H. Zhang, H. Zhao, Encapsulated Ni–Co alloy nanoparticles as efficient catalyst for hydrodeoxygenation of biomass derivatives in water. Chin. J. Catal. **42**, 2027–2037 (2021)
148. J. Bian, Q. Zhang, P. Zhang, L. Feng, C. Li, Supported Fe_2O_3 nanoparticles for catalytic upgrading of microalgae hydrothermal liquefaction derived bio-oil. Catal. Today **293**, 159–166 (2017)
149. M. Akizuki, Y. Oshima, Solid acid/base catalysis in sub- and supercritical water. Ind. Eng. Chem. Res. **57**, 5495–5506 (2018)
150. M. Akizuki, Y. Oshima, Acid catalytic properties of TiO_2, Nb_2O_5, and NbO_x/TiO_2 in supercritical water. J. Supercrit. Fluids **141**, 173–181 (2018)
151. G. Tian, Y. Xu, Z. Xu, Z. Tian, L. Lin, Effect of aluminum on the mechanical stress stability of WO_x/ZrO_2 superacid. Chin. J. Catal. **29**, 415–417 (2008)
152. Y. Jeon, W.S. Chi, J. Hwang, D.H. Kim, J.H. Kim, Y.-G. Shul, Core-shell nanostructured heteropoly acid-functionalized metal-organic frameworks: bifunctional heterogeneous catalyst for efficient biodiesel production. Appl. Catal. B Environ. **242**, 51–59 (2019)
153. Y. Xu, X. Zheng, H. Yu, X. Hu, Hydrothermal liquefaction of *Chlorella pyrenoidosa* for bio-oil production over Ce/HZSM-5. Biores. Technol. **156**, 1–5 (2014)
154. M. Zheng, L.C. Schideman, G. Tommaso, W.-T. Chen, Y. Zhou, K. Nair, W. Qian, Y. Zhang, K. Wang, Anaerobic digestion of wastewater generated from the hydrothermal liquefaction of *Spirulina*: toxicity assessment and minimization. Energy Convers. Manage. **141**, 420–428 (2017)
155. I. Hita, T. Cordero-Lanzac, F.J. García-Mateos, M.J. Azkoiti, J. Rodríguez-Mirasol, T. Cordero, J. Bilbao, Enhanced production of phenolics and aromatics from raw bio-oil using HZSM-5 zeolite additives for PtPd/C and NiW/C catalysts. Appl. Catal. B **259**, 118112 (2019)
156. Q. Lu, Y. Gong, J. Lu, J. Liu, Y. Wu, Hydrothermal catalytic upgrading of model compounds of algae-based bio-oil to monocyclic aromatic hydrocarbons over hierarchical HZSM-5. Ind. Eng. Chem. Res. **59**, 20551–20560 (2020)
157. M. Akizuki, Y. Oshima, Effect of water properties on selectivity for 1-octene and 2-octanol reaction systems in sub- and supercritical water using a TiO_2 catalyst. J. Supercrit. Fluids **123**, 76–81 (2017)

158. M.S. Wigmosta, A.M. Coleman, R.J. Skaggs, M.H. Huesemann, L.J. Lane, National microalgae biofuel production potential and resource demand. Water Resour. Res. **47**, 1–13 (2011)
159. R. Davis, A. Aden, P.T. Pienkos, Techno-economic analysis of autotrophic microalgae for fuel production. Appl. Energy **88**, 3524–3531 (2011)
160. R.E. Davis, D.B. Fishman, E.D. Frank, M.C. Johnson, S.B. Jones, C.M. Kinchin, R.L. Skaggs, E.R. Venteris, M.S. Wigmosta, Integrated evaluation of cost, emissions, and resource potential for algal biofuels at the national scale. Environ. Sci. Technol. **48**, 6035–6042 (2014)
161. X. Liu, Y. Guo, H. He, L. Zheng, L. Kong, A comprehensive study of indole catalytic hydrodenitrogenation under hydrothermal conditions. AIChE J. **68**, e17531 (2022)
162. D. Xie, X. Liu, H. Lv, Y. Guo, Products, pathways, and kinetics for catalytic hydrodenitrogenation of quinoline in hydrothermal condition. J. Supercrit. Fluids **182**, 105509 (2022)
163. L. Lyu, X. Zeng, J. Yun, F. Wei, F. Jin, No catalyst addition and highly efficient dissociation of H_2O for the reduction of CO_2 to formic acid with Mn. Environ. Sci. Technol. **48**, 6003–6009 (2014)
164. F. Jin, X. Zeng, J. Liu, Y. Jin, L. Wang, H. Zhong, G. Yao, Z. Huo, Highly efficient and autocatalytic H_2O dissociation for CO_2 reduction into formic acid with zinc. Sci. Rep. **4**, 4503 (2014)
165. H. Zhong, F. Jin, B. Wu, H. Chen, G. Yao, Hydrothermal conversion of CO_2 into formic acid on the catalysis of Cu effect of water properties on selectivity for, in 2nd International Symposium on Aqua Science, Water Resource and Low Carbon EnergySanya, People's Republic of China, 2009, pp. 213–216.
166. G. Yao, F. Chen, Z. Huo, F. Jin, Hydrazine as a facile and highly efficient hydrogen source for reduction of $NaHCO_3$ into formic acid over Ni and ZnO catalysts. Int. J. Hydrogen Energy **41**, 9135–9139 (2016)
167. L. Lyu, F. Jin, H. Zhong, H. Chen, G. Yao, A novel approach to reduction of CO_2 into methanol by water splitting with aluminum over a copper catalyst. RSC Adv. **5**, 31450–31453 (2015)
168. G. Yao, X. Zeng, Y. Jin, H. Zhong, J. Duo, F. Jin, Hydrogen production by water splitting with Al and in-situ reduction of CO_2 into formic acid. Int. J. Hydrogen Energy **40**, 14284–14289 (2015)

Review of Heterogeneously Catalyzed Denitrogenation and Demetallization Reactions for Renewable Process in Algal Bio-oil Refinery

In recent years, there has been a growing concern over energy deficiency, ecological and environmental deterioration, including climate change and global warming [1–4]. This concern has prompted the need for developing renewable and sustainable energy resources, as they are closely related to global security and the economy [5]. With the goal of achieving carbon peak and carbon neutralization, the efficient utilization of biomass energy has become crucial. Biomass energy is currently the only renewable carbon source available, and its utilization, in combination with carbon capture and storage (CCS) technology, can help achieve carbon neutralization and negative carbon emissions. Additionally, it can reduce reliance on traditional fossil fuels [6].

Biomass energy, as the fourth-largest energy source globally, has the potential to alleviate the energy crisis caused by the limited reserves of fossil fuels. It is also considered a promising future source for large-scale production of liquid fuel [7]. Among the various alternative liquid energy resources, biofuel or biocrude oil, derived from biomass, has gained significant attention and research interest. It is a potentially carbon–neutral feedstock, originating from renewable biomass resources, and has the potential to reduce the global dependence on petroleum-based fuels [7].

Microalgae have emerged as a promising and innovative biomass feedstock for biofuel production, surpassing traditional food crops and lignocellulosic biomass, for several compelling reasons. Firstly, certain strains of microalgae have the potential to yield biofuels at least 60 times higher than that of soybeans, approximately 15 times more productive than terrestrial plants, and nearly 5 times that of palm oil per acre of land annually. This remarkable productivity makes microalgae a highly efficient and viable option for biofuel production.

Secondly, microalgae have an incredibly short growth cycle, with up to 2–5 generations occurring within a span of just 24 h. This rapid growth rate is significantly faster than that of traditional terrestrial biomass, allowing for frequent and efficient cultivation of microalgae for biofuel production.

Y. Guo et al., *Hydrothermal Processing of Biomass for Hydrogen and Bio-oil Production*, https://doi.org/10.1007/978-981-96-6336-1_6

Furthermore, microalgae can be cultivated in diverse environments such as wastewater or saline-alkali zones. This ability eliminates the need for valuable agricultural land, ensuring that microalgae cultivation does not compete with food production. This characteristic is particularly advantageous, as it allows for the utilization of land that is otherwise unsuitable for traditional agriculture.

Lastly, microalgae play a vital role in mitigating the greenhouse effect. Through the process of photosynthesis, microalgae are capable of capturing carbon dioxide (CO_2) from the atmosphere, thus helping to reduce the levels of this greenhouse gas. This unique ability makes microalgae a potential solution for alleviating the negative impacts of greenhouse gas emissions [8–10].

The conversion of microalgal biocrude oil into biodiesel is widely recognized as a promising green alternative to fossil-oriented diesel. This conversion can be achieved through various technologies, including fermentation [11, 12], transesterification [13, 14], pyrolysis [15, 16], gasification [17, 18], liquefaction [19, 20], and carbonization [21, 22]. Among these methods, thermochemical conversion, particularly HTL, is considered to be an encouraging way to process feedstock with high conversion efficiency [23].

Panahi et al. [24] have pointed out that producing commercial microalgae-derived biocrude oil still faces challenges related to economic feasibility given the current knowledge and technology availability. They emphasize the need to address these economic challenges.

Additionally, the integration of exergoenvironmental methods and life cycle assessment (LCA) analyses has demonstrated that incorporating heat integration and recovering waste streams can substantially enhance the energy efficiency performance of bioenergy systems [25–27]. This further indicates that HTL technology is an economically viable method for converting microalgal materials into biocrude oil. Notably, HTL technology does not require energy consumption for drying wet raw materials, which typically contain over 90% water, as the conversion takes place under hydrothermal conditions (200–400 °C, 10–25 MPa) [28–30]. This characteristic of HTL technology contributes to its economic appeal for biocrude oil production.

Microalgal feedstock, obtained from the aquatic food chain, accumulates nutrients during its growth, resulting in the production of significant amounts of protein and lipid. However, during thermochemical conversion processes like hydrothermal liquefaction, these proteins and lipids can be converted into nitrogenous and oxygenous compounds. These compounds can have adverse effects on the storage and utilization of biofuels [31, 32].

The presence of nitrogen-containing compounds in biofuels can lead to the generation of pollutants such as NOx during combustion. Additionally, these compounds can cause catalyst poisoning during subsequent biocrude upgrading processes [33]. To ensure the quality of transportation fuels, regulatory bodies such as the American Society for Testing and Materials (ASTM) and European standards (EN) have set limits on the maximum nitrogen content. The U.S. Department of Energy has set a goal for the nitrogen content in upgraded algal fuels to be below 0.05% (500 ppm).

To achieve this goal, further development of denitrogenation technologies is necessary. These technologies aim to remove nitrogen-containing compounds from the biofuel to obtain high-quality commercial biofuels that meet regulatory standards.

In addition to nitrogenous compounds, the presence of significant amounts of Al, Ca, Fe, K, Mg, and Na (ranging from 4.3–160.8 mg/L), along with trace amounts of Cu, Mn, Ni, and Zn (ranging from 0.4–2.1 mg/L) in microalgal biocrude oil, cannot be overlooked. These metals and their salts can directly or indirectly lead to scaling and corrosion of equipment [34], catalyst deactivation [35], and degradation of product quality [36], thereby impacting the safety and economic feasibility of biorefinery operations [37]. It is essential for the metal content to be below 1000 ppm to meet utilization standards. Therefore, demetallization is crucial and necessary for the successful operation of algal biorefineries. In addition to nitrogenous compounds, the presence of metallic content in microalgal biocrude oil is a significant consideration. This includes a high content of elements such as Al, Ca, Fe, K, Mg, and Na (ranging from 4.3 to 160.8 mg/L), as well as trace amounts of Cu, Mn, Ni, and Zn (0.4–2.1 mg/L). These metals and their salts in microalgal biocrude oil can directly or indirectly lead to scaling and corrosion of related equipment, catalyst deactivation, and degradation of product quality, ultimately affecting the safety and economic viability of biorefinery operations [37].

Meeting utilization standards requires that the metal content in biocrude oil can be less than 1000 ppm. Therefore, demetallization is crucial for the successful and efficient operation of algal biorefineries. This process aims to remove or reduce the metal content in the biocrude oil to meet the required utilization standards and ensure the overall quality and safety of the biofuel production process.

It is true that catalytic HDO strategies for biocrude oil have been extensively reviewed [38, 39]. Lup et al. [40] have provided a comprehensive discussion on the mechanisms and kinetics of heterogeneous metal-catalyzed deoxygenation of oxygenated model compounds with various functional groups found in biocrude oil. However, there are a limited number of review papers specifically focusing on the denitrogenation and demetallization of microalgae biocrude.

Existing reviews mainly concentrate on the effect of operating conditions on denitrogenation and demetallization separately, lacking in-depth mechanistic insights [41]. To bridge this gap, our aim is to systematically summarize the nitrogen-containing substances present in biocrude and the role of metal catalysts in the denitrogenation reactions. We also aim to elucidate the catalytic mechanisms involved in the denitrogenation of real algal biomass and model compounds under hydrothermal conditions. Additionally, we will review the parametric influences of important process conditions on demetallization during the hydrothermal conversion of microalgae biocrude and the associated catalytic reactions. This article intends to provide a comprehensive understanding of catalyst development strategies and contribute to the fundamental knowledge of microalgae biocrude upgrading. It aims to serve as a valuable resource for further scientific research in this field.

1 Denitrogenation of Microalgal Biocrude Oil

This section separately discusses the mechanisms of breaking C–N bonds, the catalytic activity of metal-based materials for denitrogenation, and the nitrogen removal of actual microalgal biocrude oil under hydrothermal conditions.

1.1 Nitrogen-Containing Substances

Microalgae, unlike terrestrial biomass, acquire nutrients from the aquatic food chain during their growth process. This means that, in addition to lipids and saccharides, they contain significant amounts of nitrogen-containing substances, such as proteins. In the HTL reaction procedure, proteins can undergo conversion into nitrogenous heterocyclic compounds found in biocrude, namely indole, pyridine, and quinoline [42, 43]. These nitrogen-containing heterocyclic compounds are primarily formed through the Maillard reaction between saccharides and amino acids [44].

Moreover, due to the peptide bonds present in all amino acids, which are C–N bonds connecting the carboxyl and amine groups, hydrothermal conditions facilitate rapid hydrolysis of these bonds. Consequently, decarboxylation and deamination reactions occur, leading to the formation of ammonia, organic amines, and aliphatic amines [45–48]. In the hydrothermal conversion of *Nannochloropsis* sp., approximately 15% of the biocrude oil produced at 500 °C, as identified through GC–MS analysis, consisted of nitrogenous polycyclic aromatic hydrocarbons (PaHs) compounds, including indoles, quinolines, and carbazoles (as depicted in Fig. 1) [46]. These compounds exhibit high refractoriness even under severe SCW conditions, posing a significant challenge for the upgrading of the derived biocrude oil.

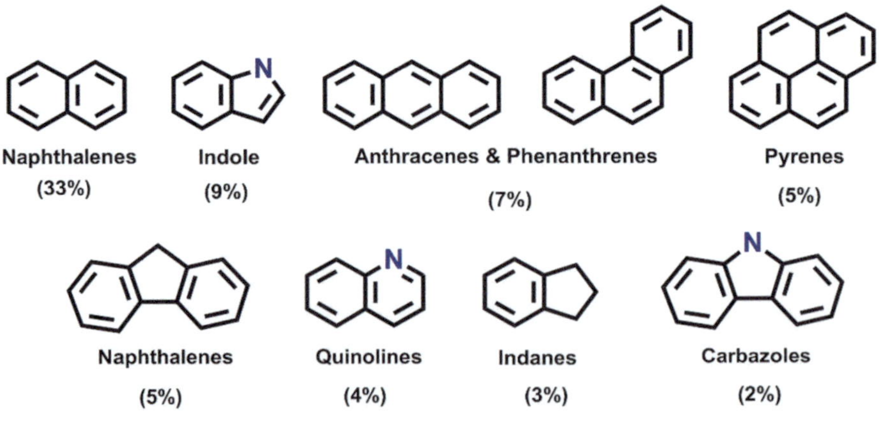

Fig. 1 PaHs in the biocrude oil produced at 500 °C [46]

1.1.1 C–N Bond in Nitrogen-Containing Substances

Among the various types of nitrogen-containing compounds, there are two distinct types of C–N bonds: C(sp^3)–N for aliphatic amines and C(sp^2)–N for aniline derivatives. The presence of an amino group in aliphatic amines (C(sp^3)–N) makes it relatively easy to remove as a leaving group. However, in aromatic anilines, there is a conjugation effect between the lone pair electron on the amino group and the aromatic ring. Consequently, the bond energy of the conjugated C(sp^2)–N bond in aniline is higher compared to the C(sp^3)–N bond in aliphatic amines, resulting in greater stability for aniline [49]. Therefore, in order to break the C(sp^2)–N bond, the aromatic ring needs to undergo hydrogenation to saturate it first. This requirement has been further supported by DFT calculation results [50].

Mechanism of Breaking C(sp^3)–N Bond

In the case of aliphatic amines, the Hoffman degradation mechanism is widely accepted for deamination reactions. Studies on the HDN reaction of 2-methylpiperidine using Ni_2P and NiMoS catalysts have revealed that the primary mechanism involved is SN_2 nucleophilic substitution. However, a minor fraction of the reaction also involves C(sp^3)–N bond cleavage through an E_2 elimination pathway. The kinetics of C(sp^3)–N bond cleavage are influenced by the structure of the nitrogen-containing substrate. Nitride molecules with low steric hindrance exhibit S_{N2} nucleophilic substitution as the dominant mechanism in the HDN reaction, while nitride molecules with significant steric hindrance primarily undergo denitrogenation through the E_2 elimination mechanism [51].

Furthermore, research by Rota [52] demonstrated that in the HDN reaction of aliphatic amines with β-H, approximately 70%–80% of the denitrogenated products are obtained through an elimination mechanism, while the remaining 20%–30% are formed through nucleophilic substitution. In certain conditions, unimolecular HDN mechanisms such as E_1 elimination and S_{N1} nucleophilic substitution can also occur. As shown in Fig. 2, if the amine group is connected to a tertiary carbon only, C(sp^3)–N bond cleavage can proceed through either E_1 elimination or S_{N1} nucleophilic substitution. The occurrence of E_1 elimination and S_{N1} nucleophilic substitution reactions is dependent on the stability of the α-carbocation linked to the nitrogen-containing molecule.

Some researchers have proposed an explanation for the cleavage of the C(sp^3)–N bond from the perspective of organometallic chemistry. According to this perspective, the cleavage of the C(sp^3)–N bond is attributed to the metal atom or ion rather than the acid site of the catalyst [53]. In this mechanism, the nitrogen-containing compound first adsorbs onto the surface of the metal catalyst, forming a metal alkyl intermediate compound. Subsequently, the fatty amine is formed through hydrogenation and demetallization reactions. In the HDN reaction of piperidine, the cleavage of the C–N bond can occur via two methods:

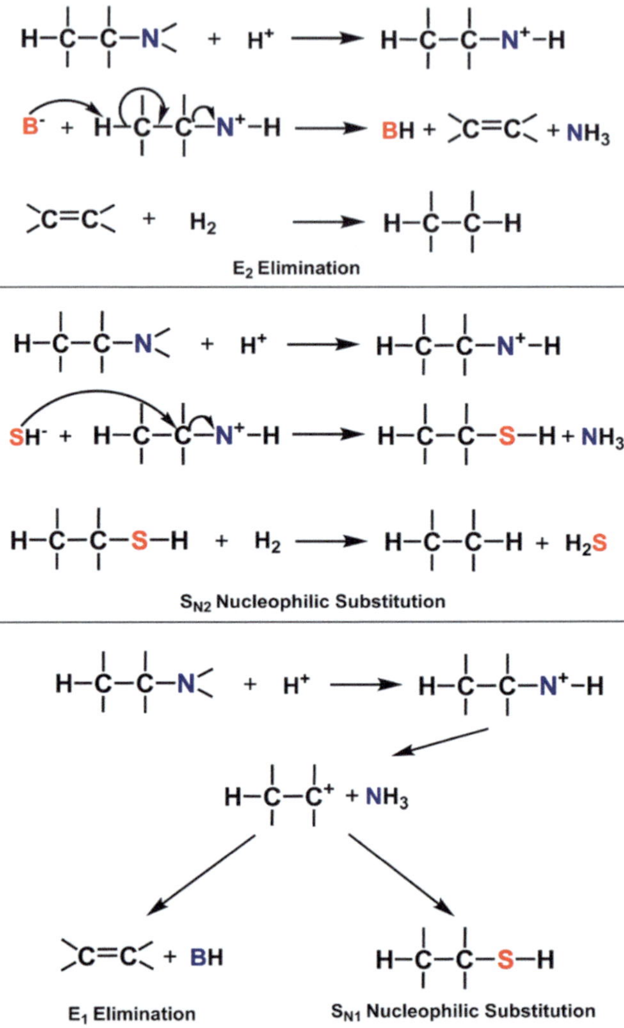

Fig. 2 Nucleophilic and substitution mechanisms for breaking C–N bond [51]

(a) Formation of a cyclic alkyl metal species from a metal aziridine intermediate, which is then hydrogenated to open the ring.
(b) Formation of a metal carbene species from an imine complex intermediate, which is then hydrogenated to open the ring (see Fig. 3).

Mechanism of Breaking C(Sp^2)–N Bond

In the HDN reactions of heterocyclic nitrogen-containing compounds, aniline serves as an intermediate. Generally, there are two pathways for the HDN reaction of aniline.

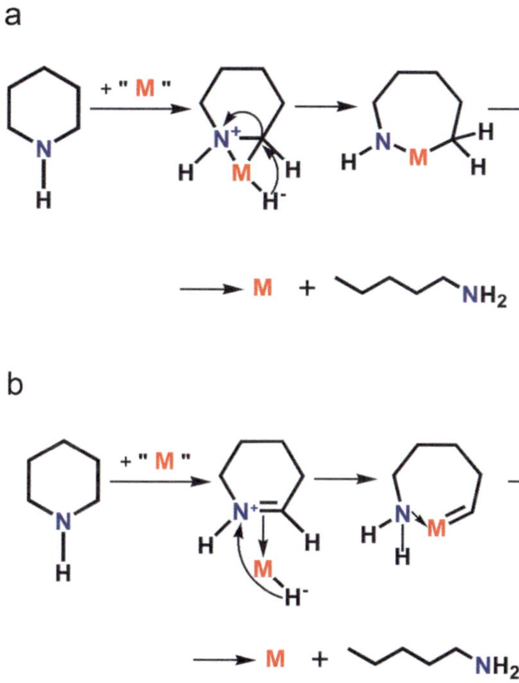

Fig. 3 Metal-mediated C–N bond scission mechanism [53]. (**a**) Cyclic alkyl metal species from metal aziridine intermediate formed first, then hydrogenated to open ring. (**b**) Metal carbene species from imine complex intermediate formed first, then hydrogenated to open ring

As shown in Fig. 4a, one pathway involves the hydrogenation of aniline to cyclohexylamine, followed by the cleavage of the C(sp^2)–N bond to form cyclohexene. Finally, cyclohexene is further hydrogenated to cyclohexane. The other pathway involves the direct cleavage of the C(sp^2)–N bond in aniline to form benzene and ammonia. The lone pair electrons on the nitrogen atom in aniline have a conjugation effect with the benzene ring, resulting in a significant energy barrier for the direct denitrogenation of aniline [54, 55]. Regarding the mechanism of the direct hydrogenolysis of aniline, different theories have been proposed. The non-planar structure of tetrahydroaniline may lead to denitrogenation through a trans elimination pathway. Zhao's research found that aniline is hydrogenated to tetrahydroaniline, which then undergoes NH3 elimination to form cyclohexadiene. Finally, cyclohexane and benzene are rapidly formed [56]. Another explanation suggests that aniline is hydrogenated to form dihydroaniline, which then undergoes deamination to form benzene [57], as shown in Fig. 4b.

The cleavage of the C(sp^2)–N bond in aniline on a MoP (001) catalyst surface has been studied, and DFT calculations have shown that the energy barrier for direct C(sp^2)–N bond breaking is 2.07 eV, which is difficult to overcome [58]. However, in the presence of co-adsorbed surface hydrogen, the energy barrier for C(sp^2)–N bond breaking decreases to 1.73 eV, making this reaction more energetically favorable.

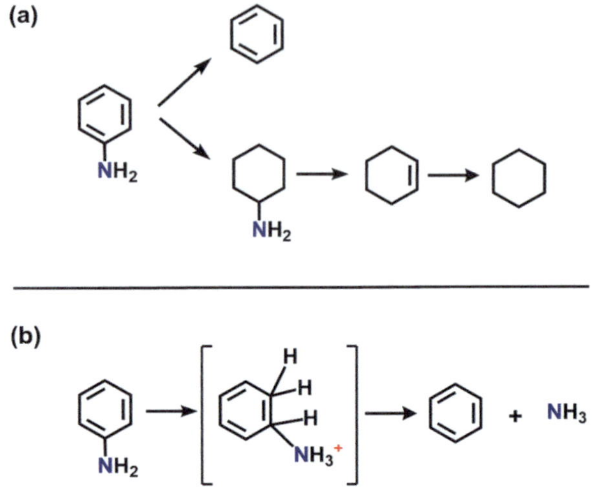

Fig. 4 HDN reaction paths of aniline [55, 57], (**a**) Hydrogenation pathway of aniline HDN reaction; (**b**) Direct hydrogenolysis pathway of aniline HDN reaction

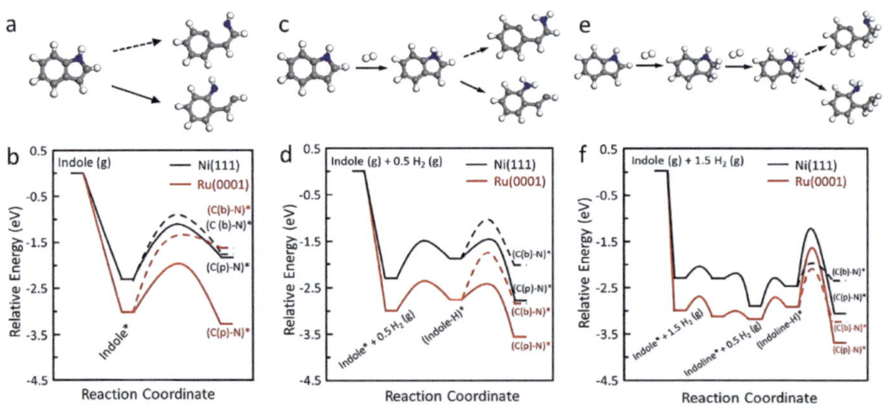

Fig. 5 Indole reaction path and corresponding energetics. (**a–b**) Indole direct ring-opening reaction; (**c–d**) hydrogen attacking nitrogen atom followed by bond breaking; (**e–f**) hydrogenation of C=C double bond on the pyrrole ring followed by hydrogen attacking nitrogen atom and bond breaking [59]

Furthermore, recent research has found that the behavior of indole ring opening depends on the presence or absence of hydrogen. If no hydrogen is provided, the cleavage of the C(p)–N bond to the aromatic ring is more favorable. However, if hydrogen is provided, the pyrrole ring in indole will be hydrogenated first, forming indoline. Subsequently, the C(b)–N bond to the aromatic ring is more likely to be broken, as shown in Fig. 5 [59].

1.1.2 HDN Reaction of Nitrogenous Model Compounds in Biocrude Oil

In the following section, we will provide a detailed explanation of the denitrogenation mechanisms in the HDN reactions of heterocyclic nitrogen-containing model compounds in the organic phase, based on previously published research.

Heterocyclic nitrogen-containing compounds can be classified into two categories: five-membered ring compounds and six-membered ring compounds. In five-membered ring compounds, the lone pair electrons on the nitrogen atom participate in the formation of the conjugated π bond of the heterocyclic ring. This results in relatively weak basicity. On the other hand, in six-membered ring compounds, the lone pair electrons on the nitrogen atom do not participate in the conjugated π bond, resulting in relatively strong basicity. Consequently, the nitrogen atom in six-membered ring compounds is more readily accessible to the hydrogen proton on the catalyst surface, facilitating HDN reactions.

Considering that the energy of the C=N bond in heterocyclic rings is nearly twice that of the C–N bond in alkylamines (615 kJ/mol versus 305 kJ/mol) [60], it is necessary to hydrogenate the C=N bond first in order to saturate the nitrogen-containing heterocyclic ring. After saturation, the C–N bond can be broken, leading to the denitrogenation of the compound.

Pyridine and Quinoline

In six-membered heterocyclic nitrogen-containing compounds such as pyridine and quinoline, a nitrogen atom replaces carbon in the benzene ring. In pyridine, one of the three valence p electrons of nitrogen participates in the formation of a conjugated aromatic ring with carbon atoms. The lone pair electrons readily react with a proton or Lewis acid, indicating the basic nature of these six-member heterocyclic compounds. The presence of conjugation in the benzene ring results in a decrease in the electron cloud density and alkalinity of the N atom in quinoline and its derivatives compared to pyridine.

Figure 6 illustrates a representative reaction pathway of pyridine in organic solvents [61]. The C–N bond in pyridine, due to its aromaticity, is resistant to breaking. Consequently, the first step in the hydrogenation reaction of pyridine is the conversion to piperidine, followed by ring opening to form fatty amine. Eventually, the C–N bond in fatty amine can be cleaved through hydrogenolysis. It is important to note that the hydrogenation and ring-opening steps are considerably slower than the other steps of pyridine HDN when using a NiMo/γ-Al_2O_3 catalyst at 543 K and 3.1 MPa hydrogen pressure [62]. Furthermore, hydrogenolysis is an irreversible process, while the hydrogenation from pyridine to piperidine is reversible. Experimental observations have also identified the presence of an olefin intermediate (C_5H_{10}) in this reaction network. Under high partial pressures of H_2S, thiocycloalkanes are detected, suggesting the occurrence of partial substitution reactions during HDN [63]. It is worth mentioning that in the HDN reaction of pyridine and piperidine, a small amount of N-amyl piperidine byproduct could be formed

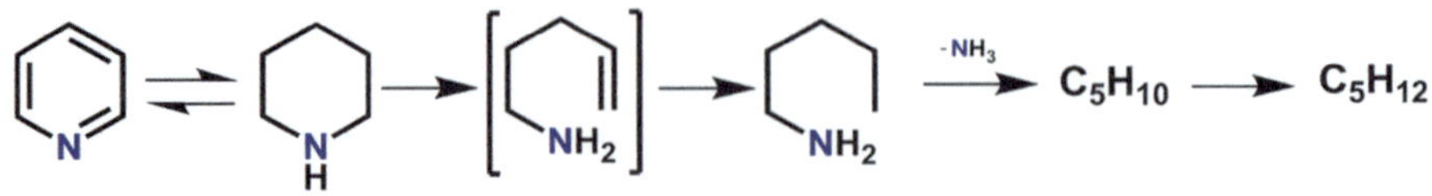

Fig. 6 HDN reaction pathways of pyridine [61]

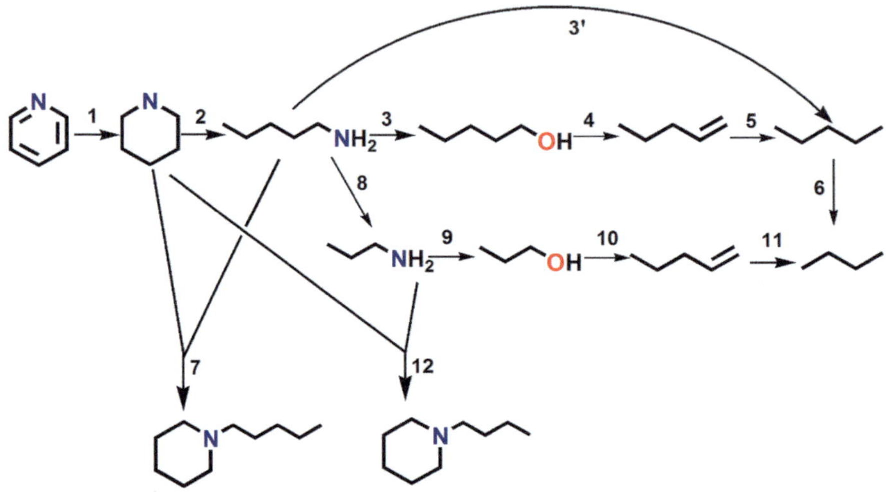

Fig. 7 Reaction pathways of pyridine HDN in hydrothermal condition [65]

through a disproportionation reaction between piperidine and amyl amine over Ni/γ-Al_2O_3, particularly at low reaction temperatures (250 °C) and high concentrations of piperidine or amyl amine [64].

Duan and Savage [65] conducted a study on the hydrogenation of pyridine under SCW conditions at 420 °C and 4 MPa hydrogen pressure. The hydrogenation reaction network of pyridine proposed by them is illustrated in Fig. 7. In contrast to the hydrogenation of pyridine in the organic phase, under hydrothermal conditions, water acts as a reactant in the reaction. Water molecules directly attack the amino group linked to the α-carbon, replacing it with a hydroxyl group to generate an oxygenated compound. Furthermore, Luo's research also indicated the introduction of oxygen atoms into the products during the hydrogenation process under hydrothermal conditions, possibly due to the weakening of the hydrogen bond between water molecules as the temperature increases [66].

Quinoline (Q), which contains both a pyridine ring and a benzene ring, is suggested to have a distinct hydrogenation and HDN mechanism compared to pyridine, as depicted in Fig. 8 [67]. Two primary pathways are involved: in the first pathway, the pyridine ring in Q is hydrogenated to form 1,2,3,4-tetrahydroquinoline (THQ-1), followed by the generation of o-propenyl aniline (OPA) through the ring-opening reaction of THQ-1. Finally, the denitrogenated product propyl benzene (PB) is

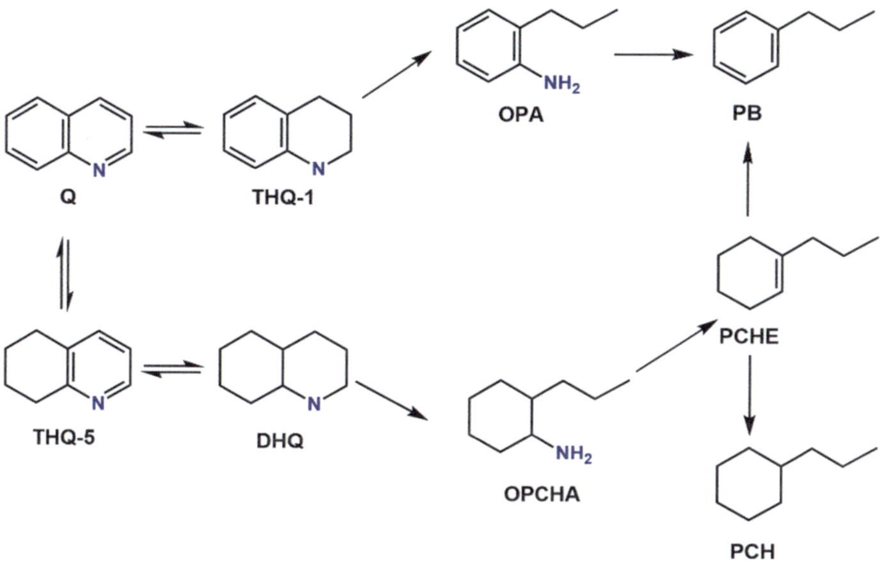

Fig. 8 HDN pathways of quinolone [67]

formed through the deamination of OPA. In the second HDN pathway, Q is first hydrogenated to produce 5,6,7,8-tetrahydroquinoline (THQ-5). Due to the lower aromaticity of the pyridine ring compared with the benzene ring in Q, the formation of THQ-1 predominates in the hydrogenation reactions of Q [64]. Furthermore, research has indicated that 40% of Q was converted through the first pathway at 370 °C and 6.5 kPa of H_2S over a NiMo/γ-Al_2O_3 catalyst [54]. Following hydrogenation, o-propylcyclohexylamine (OPCHA) is generated from ring-opening reactions of the saturated intermediate decahydroquinoline (DHQ), and eventually propyl cyclohexene (PCH) is formed through the deamination of OPCHA.

Indole and Carbazole

Nitrogen heterocyclic compounds, specifically five-membered heterocycles such as indoles and carbazole, are compounds in which the β carbon atom of cyclopentadiene is replaced by a nitrogen atom. Indole, like the pyridine ring, exhibits directional characteristics, with its nitrogen atom being sp^2 hybridized. To satisfy Hückel's $4n + 2$ rule, the lone pair of electrons in the p orbital of nitrogen in pyrrole participates in conjugation with the 4π electrons in the pyrrole ring, resulting in increased stability of the entire system. In the structure of indole, when a hydrogen atom bonded to the nitrogen is removed as a proton, the P-π conjugation effect between nitrogen and the aromatic ring enhances the stability of the resulting nitrogen anion.

The representative HDN reaction network of indole is illustrated in Fig. 9 [Tayakout-Fayolle, C. Geantet, [68]]. The denitrogenation of indole can occur

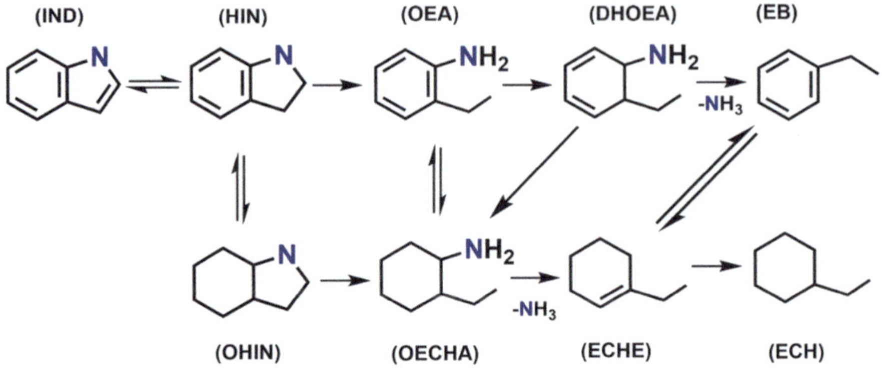

Fig. 9 HDN reaction pathways of indole [68]

through two different pathways: one pathway proceeds via o-ethylaniline (OEA), while the other pathway proceeds via octahydroindole (OHIN). The OEA pathway has been identified as dominant [69, 70]. Initially, indole undergoes rapid hydrogenation to form 2,3-dihydroindole (HIN), which then undergoes ring opening to produce OEA. OEA subsequently undergoes hydrogenation to yield dihydro-o-ethylaniline (DHOEA). The active intermediate DHOEA further yields the final product ethylbenzene through a deamination reaction, or it can form the intermediate o-ethylcyclohexylamine (OECHA) through a hydrogenation reaction. In the second HDN pathway, HIN continues to undergo hydrogenation to form OHIN, which then undergoes ring opening to produce OECHA. OECHA results in the production of ethylcyclohexene (ECHE) through a deamination reaction, and ECHE is further hydrogenated to yield ethylcyclohexane (ECH). It has been observed that under typical hydrogenation conditions, the likelihood of dehydrogenation of ECHE to ethylbenzene (EB) is quite low, so EB is primarily obtained through the hydrogenolysis of OEA. Indole and indoline exhibit stronger adsorption on NiMo/γ-Al_2O_3 catalyst than aniline, potentially leading to severe inhibition of aniline conversion in the presence of indole and indoline. This observation suggests that the hydrogenation-saturated OHIN pathway is as significant as the pathway from HIN to OHIN [71]. Additionally, the HDN of hexahydroquinone, as reported by Kim and Massoth, has demonstrated that the HDN products primarily result from OEA instead of OHIN at 310 °C and 31 atm hydrogen pressure when using CoMo/γ-Al_2O_3 as the catalyst [72].

During the hydrothermal HDN process of indole over Pt/γ-Al_2O_3 catalyst, two parallel pathways occur after the formation of nitrogenous intermediates, as depicted in Fig. 10: (1) direct removal of the nitrogen atom from indole, and (2) introduction of an oxygen atom via hydrogenolysis reaction, followed by its removal to indirectly form hydrocarbons. This mechanism is similar to the HDN mechanism of pyridine under hydrothermal conditions [65, 66]. It is noteworthy that the aforementioned HDN investigations of indole and pyridine utilized high-pressure H_2 as the hydrogen source for hydrogenation and hydrogenolysis. Our research group recently studied

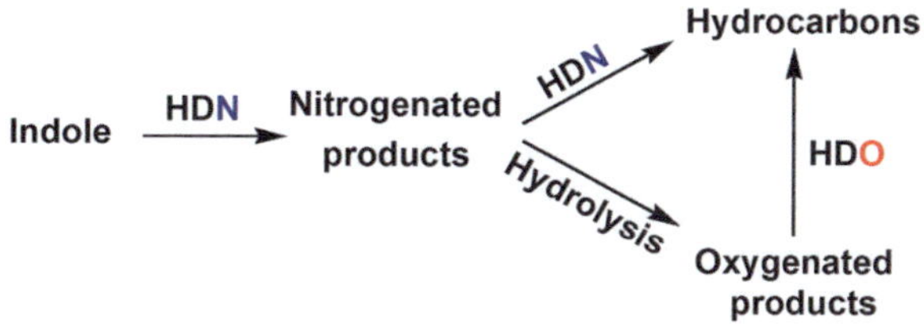

Fig. 10 HDN reaction of indole under hydrothermal condition [66]

the hydrothermal HDN of indole and found that Ru-containing catalysts (Ni–Ru) facilitated the production of hydrocarbons when used with added formic acid as a hydrogen source. The primary pathway involves the hydrogenation of indole to indoline, and the ring opening of indoline to form alkyl anilines occurs faster than the formation of alkyl benzenes. DFT calculations confirmed that Ru is more active than Ni for the ring opening of indole and the deamination of o-toluidine. Additionally, 2-ethylcyclohexanone, an oxygen-containing compound, was detected when formic acid was used as the hydrogen donor, indicating that water can participate in the HDN reaction as a reactant under hydrothermal conditions [71].

In another publication by our research group [73], tryptophan was chosen as a model amino acid and studied under hydrothermal conditions using Ni-based bimetallic nano catalysts (NiRu, NiCu, NiPd, NiPt) to investigate denitrogenation and deoxygenation simultaneously. The results showed that the removal of oxygen atoms is much easier than the removal of nitrogen atoms under supercritical water conditions. Ni and NiCu catalysts exhibited the lowest selectivity for the formation of denitrogenated compounds, while NiRu showed the highest selectivity. Catalysts containing noble metals also produced monosubstituted alkylaromatics due to the deamination of aromatic amines, with the NiPt catalyst providing the highest molar yield (~7%). The catalytic and non-catalytic hydrothermal reaction paths for tryptophan, based on the observed product distributions, are shown in Fig. 11.

The possible reaction pathway for the HDN reaction of carbazole (CARB) on Mo/γ-Al_2O_3 catalyst is depicted in Fig. 12 [74]. Due to the influence of the space baffle effect, the benzene ring in CARB is more susceptible to hydrogenation, leading to the formation of 1,2,3,4-tetrahydrocarbazole (THCARB), which can then be further hydrogenated to 1,2,3,4,4a-hexahydrocarbazole (HH-CARB). HH-CARB can undergo two different paths of conversion: one pathway involves ring opening to form O-cyclohexylaniline (OCHAN), while the other pathway entails the formation of perhydrocarbazole through hydrogenation. Additionally, O-clohexylamine can be formed primarily through the decomposition of OCHAN, and this compound can remove ammonia to produce bicyclohexyl (BCH). Due to its high reactivity, BCH can undergo further decomposition to yield α-thylbicyclo [4,4,0] decane and hexylcyclohexane.

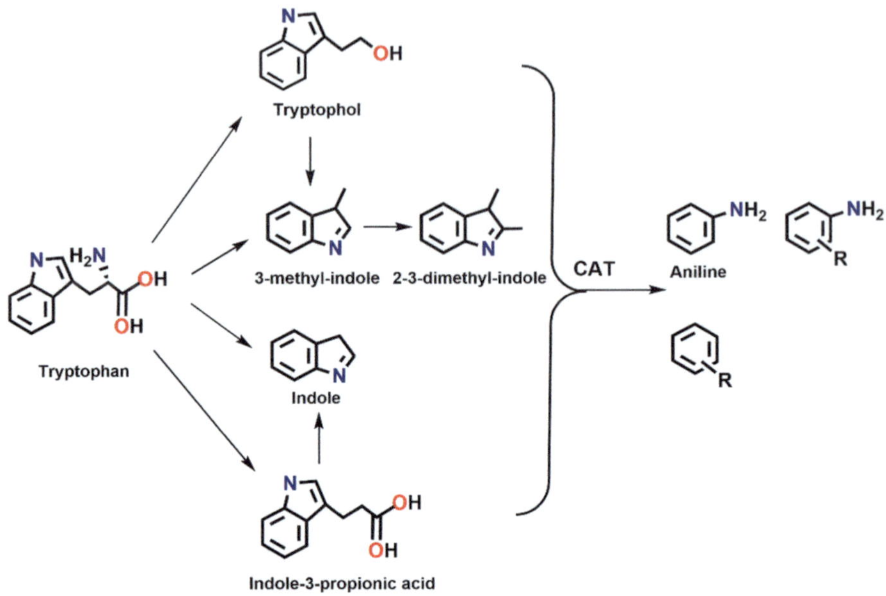

Fig. 11 Reaction network of tryptophan under hydrothermal condition with Ni-based catalyst [73]

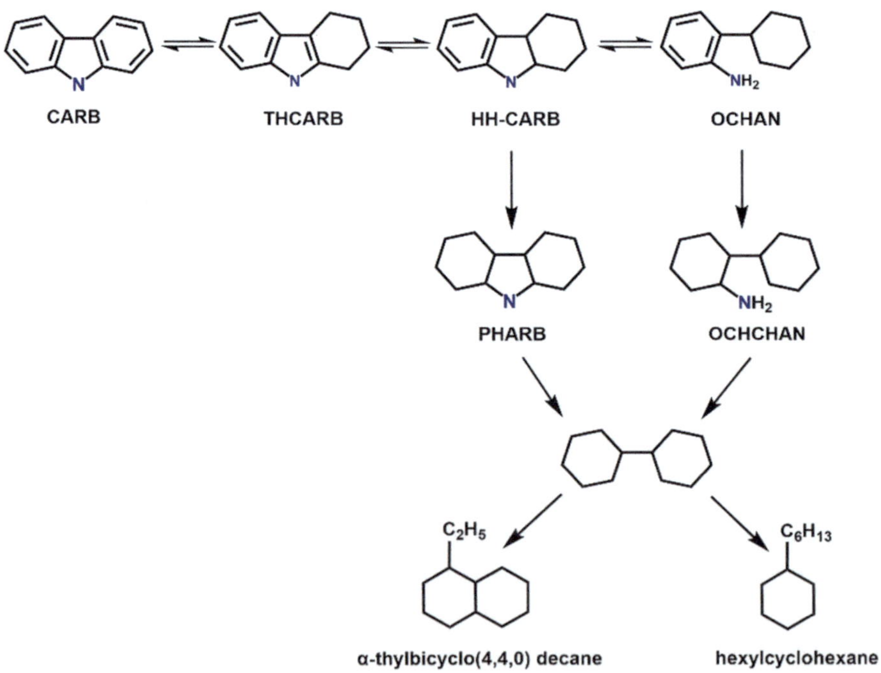

Fig. 12 Carbazole HDN reaction network [74]

1.1.3 Denitrogenation of Real Microalgae and Its Biocrude Under Hydrothermal Condition

In the published literature, the catalysts for biocrude oil upgrading mainly consist of conventional industrial upgrading catalysts such as CoMo, NiMo, HZSM-5, and Ni, as well as some noble metal catalysts like Pt, Pd, and Ru [75]. Among these catalysts, Raney Ni has shown the highest catalytic activity for the hydrothermal upgrading of Chlorella biocrude oil compared to Pt/C, Pd/C, Ru/C, Mo_2C, MoS_2, Co–Mo/γ-Al_2O_3, HZSM-5, and Ni/SiO_2-γ-Al_2O_3. The nitrogen content in the resulting biocrude decreased from 8.0% to 4.1% at 350 °C after 240 min of hydrothermal treatment, and even lower values of 2.0% were achieved at 400 °C with a mixture of Raney Ni and Ru/C catalysts [75].

In another study comparing the denitrogenation activity of Co/Mo, Pt, and Ni catalysts, Ni exhibited the highest activity for the hydrothermal upgrading of *Nannochloropsis* sp. biocrude oil at 400 °C and 6 MPa hydrogen pressure. The nitrogen content in the upgraded biocrude oil sharply decreased from 10.5% to 1.6% [76]. However, the addition of Pt/C in hydrothermal liquefaction of *Nannochloropsis* sp. did not significantly reduce the nitrogen content in the resulting biocrude oil, even at more severe conditions [77].

Further investigations compared the catalytic activity of different noble metal catalysts (Pt/C, Pd/C, Ru/C, Rh/C, sulfide Pt/C, and Pt/γ-Al_2O_3) and conventional upgrading catalysts (sulfide CoMo/γ-Al_2O_3), metal carbides (Mo_2C), metal sulfide (Mo_2S) catalysts, and noble metal oxide (PtO_2) for the hydrodenitrogenation (HDN) of pyridine under hydrothermal conditions. Pt/γ-Al_2O_3 exhibited the highest activity, achieving 100% conversion at 400 °C, 60 min reaction time, 0.025 g/cm^3 water density, and 5.5 MPa H_2. The major products of this reaction were butane and pentane, with smaller amounts of methane, ethane, and propane [65].

In the case of quinoline, a nitrogen-containing compound, catalytic HDN experiments using sulfide-NiMo/γ-Al_2O_3 under partial oxidation conditions (0.55 O_2/H_2O ratio, 350–450 °C, 25–35 MPa) in hydrothermal conditions resulted in an 85% nitrogen removal rate [78].

Table 1 summarizes recent research on the catalytic HDN of microalgae and its biocrude under hydrothermal conditions.

1.2 *Catalysts for HDN Reaction*

In the realm of reactions involving nitrogen-containing compounds, the cleavage of the C–N bond necessitates the critical steps of hydrogenation and hydrogenolysis. However, achieving these processes without a catalyst is quite challenging. Therefore, the primary focus of this section is to delve into the catalytic mechanisms and performance of various catalysts in the process of HDN.

Table 1 Recent published literatures concerning HDN of microalgae and its biocrude

Feedstock	Hydrothermal conditions	Catalysts	Key results	References
Crude algal bio-oil produced from HTL of *Nannochloropsis* sp.	400 °C; 3.4 MPa H_2; 4 h; 80% catalyst loading; stainless-steel batch mini reactor	5% Pd/C	Compared to the crude bio-oil, the higher heating value (HHV) increased from 37.07 to 43.79 MJ/kg, and the nitrogen content decreased from 4.46 to 1.99 wt.%	[79]
Chlorella pyrenoidosa HTL oil with hydrothermal pretreatment	0.3 g of catalyst (0.15 g Ru/C + 0.15 g Pt/Al_2O_3); 6 MPa H_2; 400 °C for 4 h; 58-mL Autoclave reactor	Ru/C + Pt/Al_2O_3	Among the tested catalysts, Ru/C and Pt/$A_{l2}O_3$ showed the best performance in terms of denitrogenation. With the use of these catalysts, the nitrogen content decreased from 4.0 to 1.8 wt.%	[80]
Crude algal bio-oil obtained from HTL of *Nannochloropsis* sp.	0.15 g/cm^3 water density; 25 wt.% catalyst loading; 3.4 MPa H_2; 400 °C; 4 h; stainless-steel bomb reactors	Pt/C	The product oil, with a higher heating value of approximately 42 MJ/kg, exhibited lower nitrogen content. In comparison to the crude bio-oil, the nitrogen content decreased from 4.89 to 2.24 wt.%	[77]
Chlorella p. HTL oil with hydrothermal pretreatment	3.0 g of pretreated oil; 0.3 g of catalyst; water density of 0.025 g/cm^3 at 400 °C; 6 MPa H_2; 4 h; 58 mL autoclave reactor	Raney-Ni	Among the catalysts tested, Raney-Ni demonstrated the highest efficiency in denitrogenation. When used with the pretreated oil, the nitrogen content decreased from 4.1 to 1.6 wt.%	[75]

(continued)

Table 1 (continued)

Feedstock	Hydrothermal conditions	Catalysts	Key results	References
Algal bio-oil produced from HTL of *Nannochloropsis* sp.	0.50 g of crude bio-oil; 0.10 g/ cm^3 water density; 3500 kPa H_2; 430 °C; 6 h; 20% wt.% catalyst loading; stainless-steel batch mini reactor	HZSM-5	The use of HZSM-5 catalyst resulted in an increase in the higher heating value (HHV) of the bio-oil. Additionally, the nitrogen content of the bio-oil decreased by approximately half	[81]
Crude *Chlorella p.* oil	6 MPa H_2; 0.025 g/cm^3 water density; 400 °C; 1 h; 40 wt.% catalyst loading; mini stainless-steel autoclave reactor	Pt/ γ-Al_2O_3	The researchers were able to obtain an upgraded oil with a nitrogen content of 4.7 wt.%, which is significantly lower compared to the nitrogen content of crude algal oil, which was 7.3 wt.%	[79]
Crude *C. pyrenoidosa* bio-oils from HTL, alcoholysis, pyrolysis, and hydropyrolysis	400 °C; 4 h; 10wt% catalyst loading; ρ_{H2O} = 0.025 g/cm^3; 6 MPa H_2; 58 mL custom-made stainless-steel autoclave reactor	Ru/C + Mo_2/C	For all crude bio-oils, Ru/C + Mo_2/C can reduce N content significantly. As for HTL oil, N content reduced from 7.05 to 2.87%	[82]
Chlorella p. HTL oil with hydrothermal pretreatment	400 °C; 240 min; 6 MPa H_2; 10 wt.% catalyst loading; 0.025 g/cm^3 water density; 58 mL batch reactor	HY (5% Na_2O)	HY (5% Na_2O) showed the highest activity toward denitrogenation. Compared with pretreated bio-oil, N content reduced from 4.58 to 1.67 wt%	[83]

(continued)

Table 1 (continued)

Feedstock	Hydrothermal conditions	Catalysts	Key results	References
Duckweed (*Lemna minor*) HTL bio-oil with hydrothermal pretreatment	400 °C; 1 h; 6 MPa H_2; 10 wt%AC; 1.5 mL freshly deionized water; 37 mL autoclave reactor	Activated carbons	Bamboo stem AC produced an upgraded bio-oil with the highest yield, the highest fraction of material boiling below 350 °C, and the highest energy density. Meanwhile, N content reduced from 1.81 to 1.32 wt.%	[84]
Duckweed biocrude produced from the HTL of *Lemna minor*	1.5 g of biocrude; 0.15 g of catalyst; 3.5 mL of freshly deionized water; 0.042 mol H_2; 350 °C; 4 h; 17 mL autoclave reactor	Ru/C	Ru/C showed activity for denitrogenation and N content decreased from 4.7 to 4.5 wt.%. The oil produced with Ru/C also showed the best property	[85]
Water-insoluble biocrudes from HTL of *Nannochloropsis* microalgae	1.86 ml of water; 0.0148 g catalyst; 400 °C; 60 min; 4.1 ml mini-batch reactors	Pt/C	Pt/C had the best effect on upgrading the water-insoluble biocrude. It provided the lowest N (2.47 wt%) content, while the N content of biocrude was 4.85	[86]

1.2.1 Metal Sulfide

Sulfur vacancy sites or under-coordinated lattice points are commonly recognized as the active catalytic sites of metal sulfides. Aromatic substances can adhere to the catalyst surface at these active sites through π or δ bonds. The hydrorefining capabilities of metal sulfides have been the subject of extensive research, leading to the proposal of various catalyst models. Notably, the rim-edge [78], Co–Mo–S [87], corner-edge [88], and remote control theory [88] models have gained widespread acceptance among the diverse range of proposed models.

(1) **Rim-Edge Model**

The radius-edge model, first proposed by Chianelli et al. [89], provides an explanation for the selectivity observed in the hydrogenation and hydrogenolysis processes of MoS catalysts, as depicted in Fig. 13. According to this model, there are two distinct types of active sites: the rim and the edge. The rim site is situated at the edge positions of the top and bottom layers of MoS_2 platelets, and it predominantly facilitates hydrogenation reactions. On the other hand, the edge site is located at the edge position of the middle layer of the catalyst, serving as the active site for hydrogenolysis. The sulfur atoms on the top and bottom planes are more resistant to removal compared to the sulfur atoms on the rim and edge, while the molybdenum atoms on the edges are more likely to be exposed. Consequently, the vacancies present on the edge of the MoS_2 structure are considered to be the actual active sites responsible for hydrogenolysis reactions.

(2) **Co–Mo–S Model**

The Co–Mo–S structural model, as depicted in Fig. 14, was initially proposed by Dagge and has gained widespread acceptance. This model is based on the observation that Co atoms can be found at the edge positions of MoS2 nanostructures and the recognition that Co–Mo–S structures play a crucial role in enhancing reactivity. Within the catalyst, Co exists in three distinct phases, as shown in Fig. 14. The most catalytically significant phase, known as Co–Mo–S, possesses a texture similar to that of MoS_2, with incorporated Co atoms. The Co–Mo–S phase is non-stoichiometric with regard to the Co/Mo ratio, and it cannot be defined in the conventional crystallographic sense. Previous studies have established that, for unpromoted MoS_2, only the edge sites, rather than the basal plane sites, are active [90, 91]. Therefore, it has been proposed that Co atoms located at the edge positions create new and more active sites by facilitating the formation of sulfur vacancies at or adjacent to the Co sites [92, 93].

(3) **Corner-Edge Model**

The Co–Mo–S structural model, as illustrated in Fig. 15, categorizes the entire catalyst into three different positions: corner, edge, and basal [94]. The corner sites are

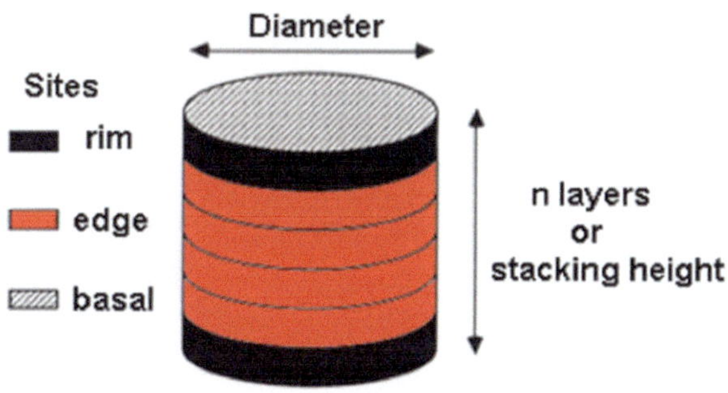

Fig. 13 Rim-edge model of MoS_2 [89]

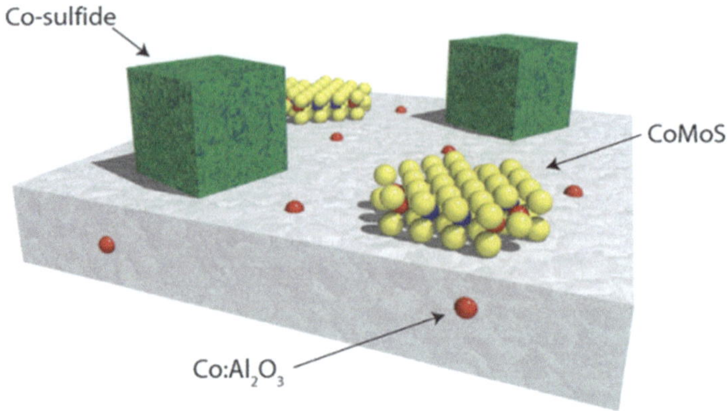

Fig. 14 Illustration of the phases in a sulfided Co–Mo catalyst [88]

situated at the vertices of the polygon, the edge sites are located along the sides of the polygon, and the basal sites are positioned within the polygon. The sulfur atoms at the basal sites are fully coordinated and therefore inactive. In the context of the hydrogenation of heterocycles or aromatic rings in the HDN reaction, the corner sites are identified as the active sites. Conversely, the edge sites are recognized as the active sites for C–N bond cleavage [95].

(4) **Remote Control Model**

The remote control model [96], shown in Fig. 16, proposes that Mo exists in the form of MoS_2, and the catalyst's activity and selectivity are correlated with the Mo and MoS_2 phases, which can create a synergistic effect through their interaction during the reaction. In this model, the activated hydrogen from CoS_x is transferred to MoS_2 and reduces it, thereby enhancing the intrinsic activity of MoS_2. The reduced MoS_2

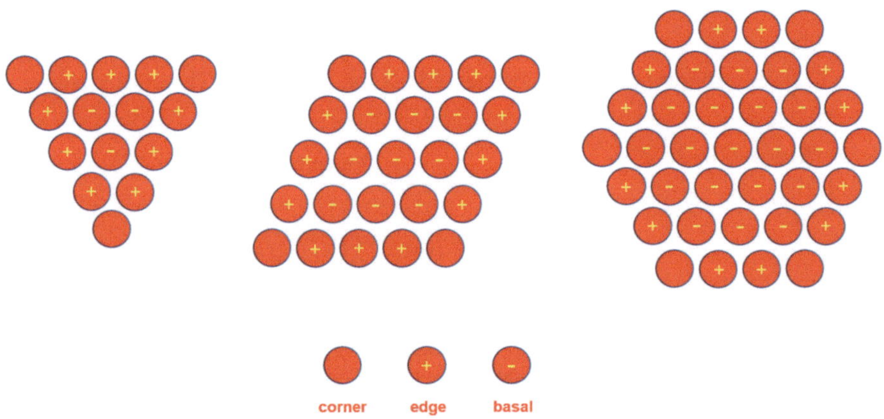

Fig. 15 Corner-edge model [94]

possesses two active centers. One type of active center is formed by the removal of a sulfur atom, resulting in an active center with an unsaturated coordination number of 3, which is responsible for hydrogenation reactions. Under more severe reduction conditions, MoSH groups are generated near coordinatively unsaturated sites (CUS), and these groups serve as the active centers for HDN reactions.

(5) Interconversion Between Hydrogenation and Hydrogenolysis Sites

The catalyst acid site theory suggests that nitrogen-containing compounds react with acid sites on the catalyst surface through two mechanisms: accepting a proton (B acid) from the catalyst surface and providing lone pair electrons to an electron-deficient (L acid) catalyst surface. The difficulty of the reaction is inversely related to the basicity of the nitrogen-containing compounds. During HDN reactions, the HDN catalyst can provide two types of active centers simultaneously [97]. One type is an active center with a sulfur-deficient structure on the Mo catalyst, which promotes hydrogenation reactions. The other type is B acid sites that enhance hydrogenolysis reactions [97]. In summary, the hydrogenation and hydrogenolysis of C–N bonds occur on different active sites of the catalyst. Moreover, these two different active sites can be transformed into each other. Since hydrogenation reactions occur at anionic vacancies, H_2S can facilitate the transformation of a hydrogenation site (Site I) into a hydrogenolysis site (Site II) (refer to Fig. 17).

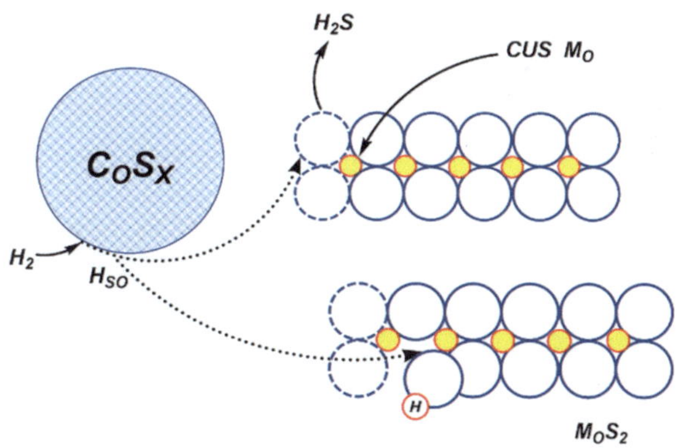

Fig. 16 Remote control model [96]

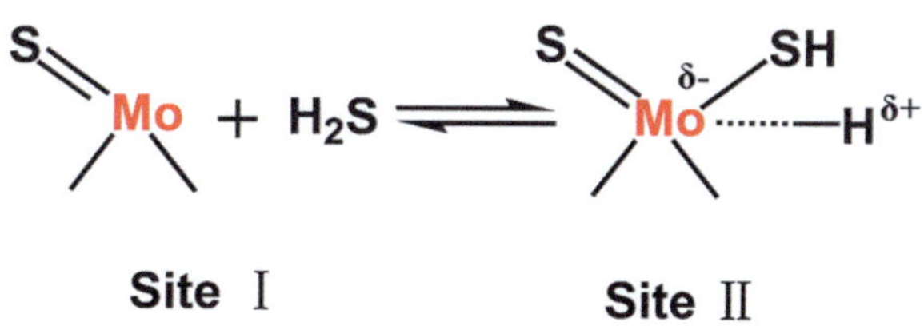

Fig. 17 Interconversion of two sites on metal sulfide [97]

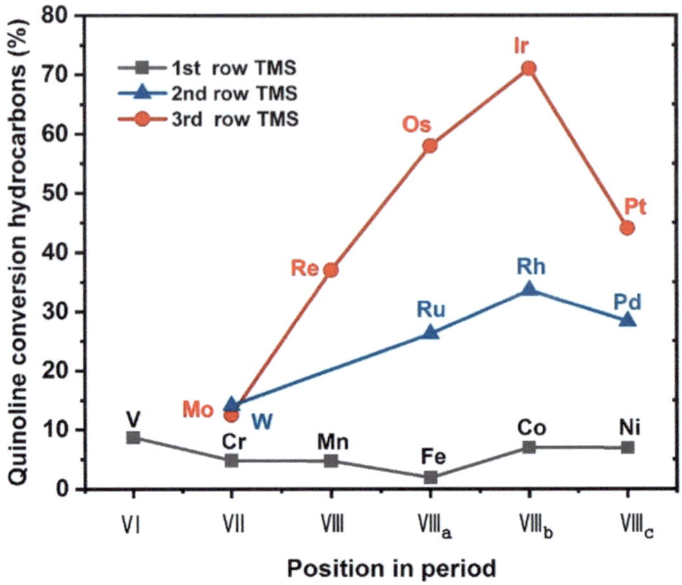

Fig. 18 Relationship between the quinoline HDN conversion rate and the position of different metals in the periodic table [98]

Catalytic Performance of Metal Sulfide

Eijsbouts et al. conducted a systematic study on the catalytic HDN of quinoline using different transition metal sulfide catalysts supported on activated carbon [98]. They investigated the relationship between quinoline HDN conversion and the position of different metals in the periodic table, and the results are shown in Fig. 18. The study revealed that second- and third-row transition metal sulfides (TMS) formed volcano curves with maximum conversions for quinoline to C8, C9 hydrocarbons at Rh/C and Ir/C catalysts. On the other hand, first-row transition metal sulfides (Mo/C and W/C) with low quinoline hydrogenation activity also exhibited low quinoline conversion to hydrocarbons. In summary, noble metals with good hydrogenation activity also demonstrated relatively high HDN activity. Specifically, Ru and Re showed the best selectivity for propylbenzene. The HDN rate of quinoline over Ru sulfide supported on Y-zeolite reached 70% at 3.45 MPa and 350 °C, which was higher than the catalytic activity of conventional Ni–Mo/γ-Al_2O_3 catalysts [99]. Additionally, a 2 wt% Pt/SiO_2-Al_2O_3 catalyst exhibited higher HDN activity compared to conventional CoMo and NiMo/γ-Al_2O_3 catalysts [100].

1.2.2 Transition Metal Carbides and Nitrides

Transition metal nitrides and carbides are interstitial compounds in which nitrogen and carbon atoms enter the lattice of transition metals. These compounds possess a combination of properties observed in covalent solids, ionic crystals, and transition metals [101]. In recent years, both transition metal carbides and nitrides have gained significant attention as promising catalysts for biocrude oil upgrading. They are considered inexpensive catalysts that exhibit superior activity compared to transition metals such as Co, Mo, Ni, and others. These catalysts show similar performance to conventional metal sulfide and noble metal catalysts [101].

(1) **Structure of Transition Metal Carbides and Nitrides**

The structure of metal carbides and nitrides belongs to the category of interstitial compounds, which is similar to that of the metal and can have a face-centered cubic (fcc), hexagonal close-packed (hcp), or simple hexagonal (hex) crystal structure, depending on the specific metal involved. In these compounds, nitrogen or carbon atoms enter the interstitial spaces between metal atoms, and their structural properties are determined by both geometric and electronic factors [102], as illustrated in Fig. 19.

Theoretical calculations of transition metal carbides and nitrides suggest that they possess metal bonds, covalent bonds, and ionic bonds simultaneously [103]. The metal bonding characteristics are related to the rearrangement of the metal–metal bond in these compounds, while the covalent bonding characteristics are formed due to interactions between the metal and the non-metal atoms. Ionic bonding characteristics arise from electron transfer between the metal and the non-metal atoms. Experimental techniques such as X-ray photoelectron spectroscopy (XPS), near-edge X-ray absorption fine structure (NEXAFS), and others generally support the idea that the charge transfer direction in these compounds is from metal atoms to non-metal atoms, and the ionicity is determined by the number of charge transfers, which increases from group VI B to group VII B metals. This trend is ultimately related to the decrease in electronegativity of the corresponding metal.

The formation of nitrides and carbides modifies the d-band nature of the corresponding metal, leading to catalytic properties similar to those of group VIII noble metals due to similar electronic structure [104].

(2) **Catalytic Performance of Transition Metal Carbides and Nitrides**

Transition metal carbide and nitride catalysts have demonstrated excellent hydrodenitrogenation (HDN) activity and high selectivity for the formation of aromatic products [105]. A study on the catalytic HDN activity of transition metal carbides and nitrides, including molybdenum, tungsten, vanadium, niobium, titanium, and commercial sulfided Ni–Mo/Al_2O_3, was conducted using quinoline as the model compound at 3.1 MPa H_2 and 370 °C [106]. The results, shown in Fig. 20, revealed that the carbides and nitrides were active for the HDN of quinoline, with activity in the order of group 6 > group 5 > group 4. Furthermore, a linear relationship between HDN activity and the crystallite and particle size of molybdenum carbide was observed, indicating that the HDN reaction over Mo carbide is structure-sensitive.

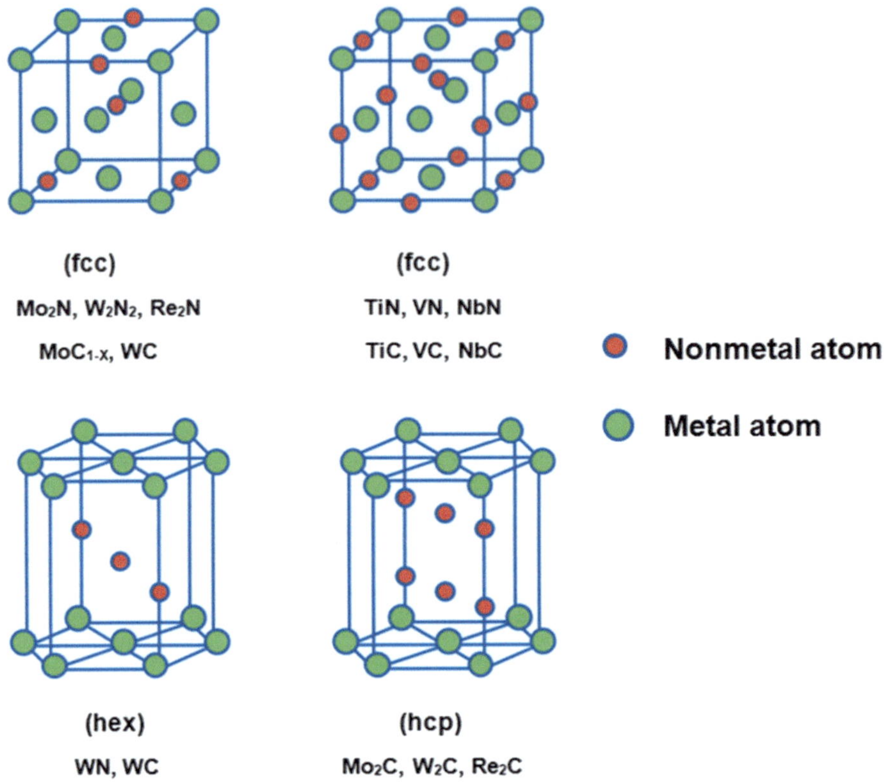

Fig. 19 Typical crystal structure of transition metal nitride and carbide [102]

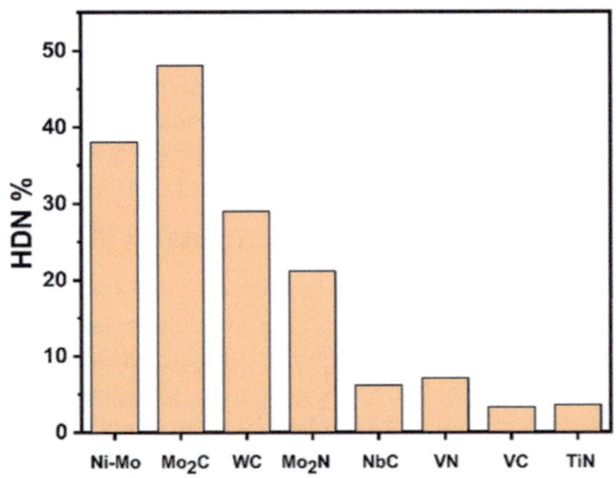

Fig. 20 Comparison of the steady-state HDN activities of different metal carbides and nitrides [106]

Another study investigated the HDN catalytic performance of bimetallic (Co–Mo) and trimetallic (Co/Ni–W) carbide catalysts in a continuous flow reactor using pyridine as a model compound [107]. As depicted in Fig. 21, the study found that the incorporation of Co into Mo carbide improved HDN activity compared to industrial sulfide catalysts. However, a higher molar ratio of Co/Mo resulted in segregation and poor catalytic stability of both the Co–Mo and Co(Ni)–W carbide catalysts.

The work of Green and colleagues focused on the HDN activity of bimetallic Co and Mo carbides using pyridine as a model compound [108, 109]. The Co/Mo ratio of the bimetallic carbides varied between 0.2 and 0.65, and the highest activity and stability were observed for the CoMo carbide with a Co/Mo ratio of 0.4. Additionally, the study found that Mo nitrides (Mo_2N) exhibited 5–10 times higher activities for pyridine HDN compared to sulfided metal catalysts [110–112]. It was suggested that two types of HDN sites exist on the Mo_2N surfaces: modest activity sites on the particles and high activity sites at grain boundaries. Furthermore, the HDN activity of Mo2N appeared to be structure-sensitive.

Colling and Thompson [111] proposed that the catalytic properties of supported Mo nitrides depended on the size and composition of the Mo nitride domains. They suggested that the most active sites are located at the perimeters of two-dimensional, raft-like domains, with regions near the perimeter appearing to be nitrogen deficient. Lower activity sites were associated with the γ–Mo_2N crystallite surfaces. DFT calculations suggested that the activation energy of the hydrogenation step of carbazole was higher than that of the C–N bond cleavage and denitrogenation step on the Mo_2N (110) slab. The successive hydrogenation proceeded on the Mo atom, and one C–N bond of the hydrogenated carbazoles on the same Mo atom, while

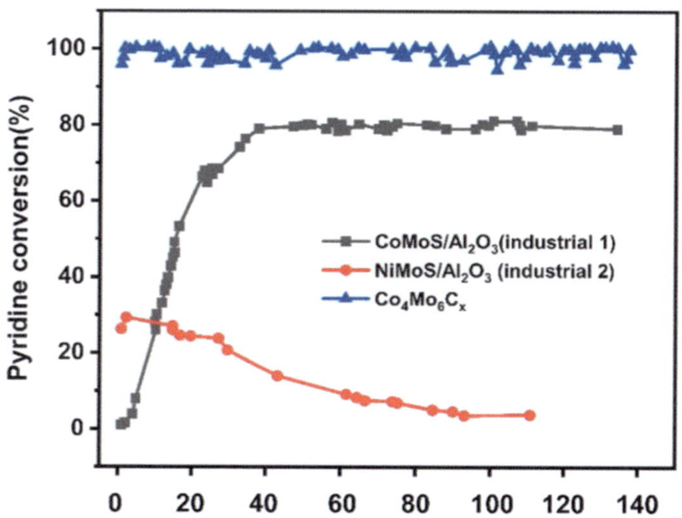

Fig. 21 Comparison of catalytic activity of $Co_4Mo_6C_x$ and industrial catalysts [107]

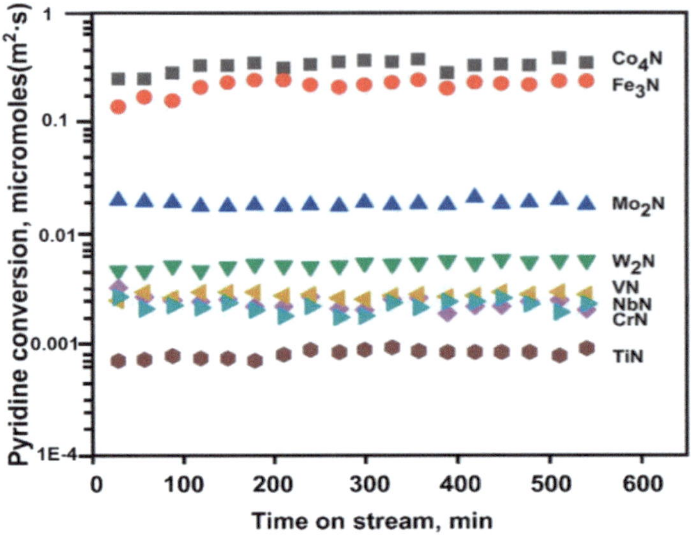

Fig. 22 Activity of metal nitrides for pyridine HDN [109]

the nitrogen removal of the hydrogenated carbazoles and the dissociative hydrogen adsorption took place on the bridge position of the Mo atoms [50].

The catalytic activity of Mo_2N during indole HDN was initially and significantly increased by adding phosphorus to the catalyst [113]. When comparing the activities of various group IV–VIII bulk metal nitride catalysts for pyridine HDN, it was observed that Co_4N and Fe_3N exhibited higher specific activity for pyridine HDN at atmospheric pressure compared to Mo_2N [114], as shown in Fig. 22.

The comparison between γ-Al_2O_3 supported Mo_2N and NiMoN for the HDN of 3-methyl pyridine involved testing Mo_2N and a set of Mo2N-based bimetallic catalysts with high surface area (~140 m^2/g) after 4 h on stream for the HDN of pyridine at 300 °C with an H_2 pressure of 3.0 MPa [115]. As depicted in Fig. 23, the Co–Mo catalyst exhibited the highest catalytic activity with 89% conversion and 72.5% denitrogenation, which is significantly higher than that shown by the commercial sulfided CoMo/γ-Al_2O_3 catalyst, which displayed 40% conversion and 16.5% denitrogenation [116].

1.2.3 Transition Metal Phosphide

Phosphorus can form a metal bond or a covalent bond with transition metals; thus, the transition metal phosphide can be ionic, covalent, or metal-type compounds.

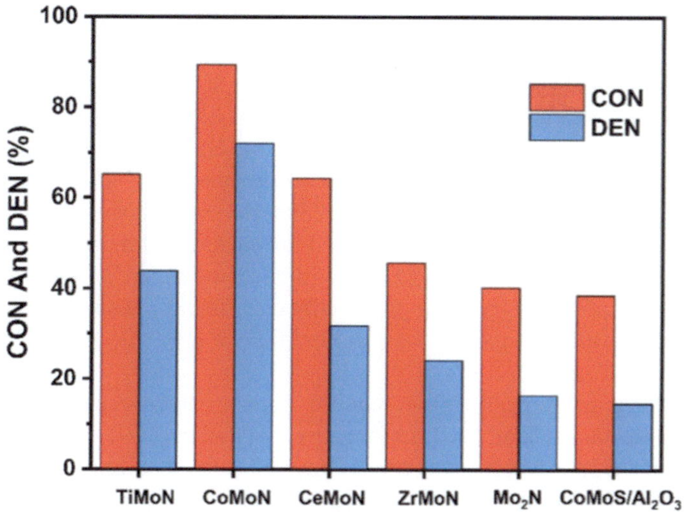

Fig. 23 Conversion (CON) and denitrogenation (DEN) of pyridine HDN over various catalysts [116]

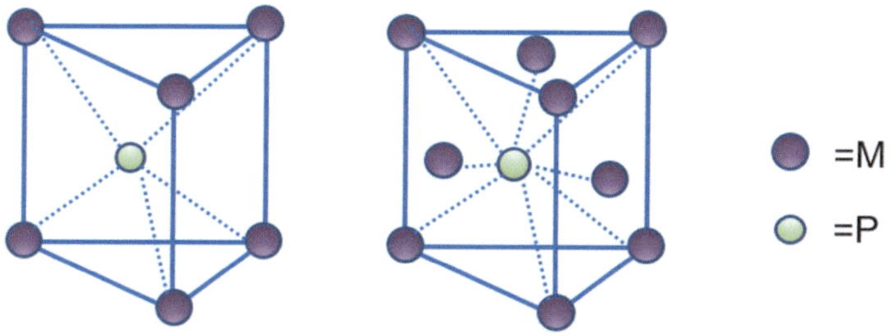

Fig. 24 Schematic diagram of phosphide prism structure [118]

Structure of Transition Metal Phosphide

Although physical properties of transition metal phosphides are similar to nitrides and carbides, their crystal structure is totally different. For metal nitrides and carbides, carbon and nitrogen atoms form a unit cell between metal atomic gaps. However, for metal phosphides, since the atomic radius of phosphorus is greater than nitrogen and carbon ($RP = 0.109$ nm > $RC = 0.071$ nm > $RN = 0.065$ nm) [117], the phosphorus atom cannot enter the interstitial sites in the close-packed metal crystal structure. Instead, the metal atoms form a triangular prism surrounding the phosphorus atom with the nearest phosphorus and the metal atoms coordinated tetrahedrally as shown in Fig. 24.

1.2.4 Catalytic Performance of Transition Metal Phosphide

Stinner's research [119] found that in the HDN reaction of amphetamine, the activity of MoP was six times higher than that of $MoS_2/\gamma\text{-}Al_2O_3$. In another study by Clark et al. [120], different catalysts were tested for the HDN reaction of quinoline at 3.1 MPa and 370 °C in a three-phase (quinoline, dibenzothiophene, and benzofuran) upflow fixed-bed reactor. The results showed that tungsten phosphide (WP) demonstrated better HDN performance than tungsten carbide (WC), tungsten nitride (W_2N), and tungsten disulfide (WS_2) and even better than the commercial Ni–Mo–S–Al_2O_3 catalyst. When comparing the catalytic activity of different phosphates for the HDN of quinolone, the following order was observed: Fe_2P < CoP < MoP < WP < Ni_2P [120].

A DFT study on the competitive reaction between hydrogenation and denitrogenation of pyridine on MoP (010) surface revealed that pyridine can adsorb on MoP (010) in both side-on and end-on modes. The most stable configuration, η_4 (N_1, C_2, C_3, C_5), in the side-on mode facilitates the hydrogenation of pyridine, while denitrogenation plays a dominant role in the competition between denitrogenation and hydrogenation of pyridine [121].

The research on silica-supported nickel phosphide (Ni_2P) catalyst for the HDN of carbazole revealed that the metal phosphide catalyst favored the formation of ring-opened (hexylcyclohexane) and ring-contracted (2-methylcyclopentyl cyclohexane) products, indicating the presence of both metal and Brønsted acid sites on the catalyst surfaces. The Ni_2P/SiO_2 catalyst demonstrated excellent stability during the carbazole HDN reaction compared to Fe_2P/SiO_2, CoP/SiO_2, MoP, WP, and Ni–Mo–S/$\gamma\text{-}Al_2O_3$, as evidenced by the phase purity and average crystallite size of the supported metal phosphide particles (Fig. 25) [122]. Furthermore, in the HDN research of aliphatic nitrile propionitrile, the SiO_2-supported Ni_2P catalyst was found to initiate the hydrogenolysis of both C–N bonds and C–C bonds, leading to the production of a significant amount of methane and ethane products [123].

The use of zeolite Y supported nickel phosphides has been investigated as an alternative catalyst for the HDN of nitrogen-containing heterocycles. This catalyst has demonstrated comparable quinoline denitrogenation activities to noble metal and transition metal catalysts, as reported in the study by Torrente-Murciano, C. J. Chuck, V. P. Ting [Zeolite Y supported nickel phosphide catalysts for the hydrodenitrogenation of quinoline as a proxy for crude bio-oils from hydrothermal liquefaction of microalgae]. Furthermore, TiO_2 has shown promise as a catalyst promoter in the HDN of quinoline and decahydroquinoline, enhancing both the hydrogenation and C–N bond cleavage activities of nickel phosphide [promoting effect of TiO_2 on the hydrodenitrogenation performance of nickel phosphide]. Additionally, the addition of TiO_2 in MoP/MCM-41 has been found to enhance the C–N bond cleavage in quinoline while suppressing its dehydrogenation activity [effect of TiO_2 on hydrodenitrogenation performances of MCM-41 supported molybdenum phosphides]. These findings suggest the potential of TiO_2 as a catalyst promoter in the HDN of nitrogen-containing heterocycles.

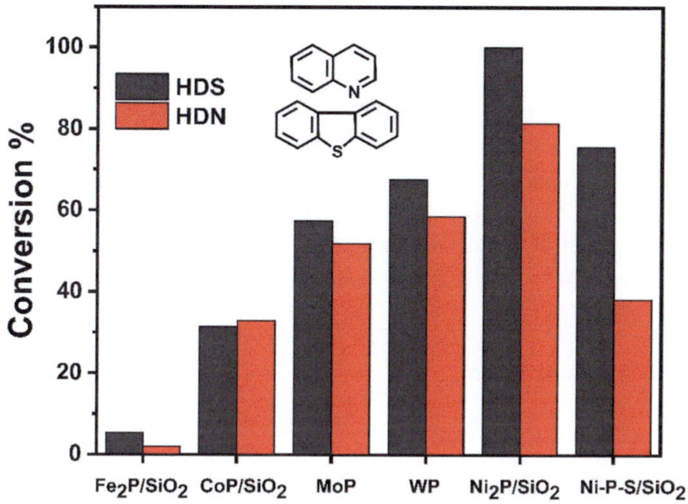

Fig. 25 Comparison of different phosphides and a commercial sulfide-NiMo/Al_2O_3 catalysts in the hydroprocessing of quinoline [123]

1.2.5 Transition and Noble Metal Catalyst

The d-orbitals in Group VIII metals, such as Co, Ni, Pd, Pt, and Ru, have an underfilled electronic configuration and exhibit a cubic crystal structure, making them suitable for HDN reactions [124]. However, the catalytic performance of a single metal active component is often not sufficient, and higher catalytic activity can be achieved by incorporating multiple metal active components.

For example, doping Mo_2C with platinum (Mo_2CePt) has been shown to enhance HDN efficiency compared to Mo_2C alone [125]. In a comparison of the HDN activity of monometallic Pt and bimetallic PtMo catalysts supported on NaY zeolite, it was found that the quinoline conversion over the bimetallic PtMo catalyst was four times higher than that of the monometallic Pt catalyst under similar conditions [126]. The selectivity for HDN products, such as propylbenzene, propyl cyclohexane, and propyl cyclohexene, was also significantly increased over the bimetallic catalyst.

In the case of Ni_2P catalysts, the addition of Ce has been shown to improve both quinoline hydrogenation and C–N bond cleavage activities, particularly in the denitrogenation of quinoline to propyl cyclohexane [127]. XRD and XPS measurements indicated the coexistence of Ce^{4+} and Ce^{3+} species on the surface of the Ce–Ni_2P catalyst, and the addition of Ce led to a decrease in the particle size of Ni_2P, resulting in increased specific surface areas. This can be attributed to the strong interaction between the Ce species and Ni_2P [59, 128].

The cooperation of Ir or Pt with RuS_2 in bimetallic catalysts has also been found to improve catalytic behavior in the HDN reaction of quinoline. Ir and Pt serve as active hydrogen suppliers, enhancing selectivity toward target products [129].

Similarly, the addition of Ru, Pd, and Pt to pure Ni catalysts has been shown to enhance the hydrocarbons selectivity from tryptophan HDN and improve benzene selectivity in the HDN of indole [73, 130].

In terms of stability, a Co–Ni/C bimetallic catalyst exhibited high catalytic activity for converting phenol, with 95% activity retained after the third use under SCW conditions (500 °C and 30 min) [131]. However, the conversion of phenol decreased from 100 to 67% with $Ni_{20}Zn_{15}/C$ at 450 °C and 30 min, likely due to the agglomeration of the nanoparticle under SCW conditions [132]. These findings suggest that the choice of the second metal in a bimetallic catalyst can greatly impact its stability.

1.2.6 Catalyst Deactivation

During the production of biocrude, the presence of N- and S-containing substances in microalgae biomass can lead to the poisoning of catalytic active sites, ultimately deactivating the catalyst [133]. Additionally, the hydrothermal conditions can induce phase transitions in the physical structure of the catalyst support and the metal itself [134, 135]. In this section, we will explore the significant challenge of catalyst deactivation in the catalytic upgrading of biocrude, specifically focusing on the impact of N and S-containing species as well as water.

(1) **Nitrogen Effect**

Catalyst poisoning caused by the adsorption of nitrogen-containing substances can be attributed to several factors, including the coverage of active sites, blockage of pore channels, and reduction of the catalyst's surface area [133]. During the HTL process of microalgae, nitrogen is present in various forms such as aliphatic amines, nitrogen heterocyclic ring compounds, or polycyclic aromatic hydrocarbons. These nitrogen-containing compounds can adsorb onto the L and B acid centers of the catalyst surface, leading to a decrease in conversion rate and product selectivity [136]. For instance, aliphatic amines can combine with Ni and Co atoms to form metal nitrides, which can deposit on the catalyst's active sites and cause catalyst deactivation [137]. Hegedus and Mathe also observed the poisoning of noble metal catalysts (Pd/C, Ru/C, and Rh/C) by nitrogen elements in pyrrole during pyrrole hydrogenation [138].

Nitrogen-containing substances like quinoline and acridine have a tendency to diffuse into the micropores of HY zeolite and preferentially adsorb onto the acid sites. This leads to severe deactivation and poisoning of the catalysts due to increased coke deposition [139]. Furthermore, it has been suggested in recent literature that nitrides with stronger basicity are more easily adsorbed on the acid sites of catalysts, thus reducing the catalyst's service life [140]. The pKa values of different types of nitrides are shown in Table 2. Vogelaar et al. demonstrated that both unreacted nitrogen-containing compounds and nitrogenous polymers can adsorb on the catalyst surface, blocking the catalyst's pore channels and reducing its surface area and pore volume [141].

Table 2 Different types of nitrogen compounds

Compound	Structure	pKa
Ethylamine		10.66
Pyridine	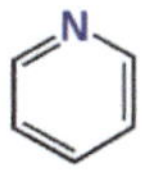	5.2
Piperdine	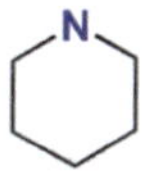	11.1
Quinoline	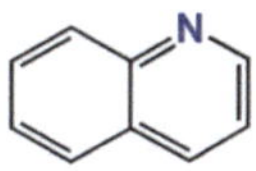	4.9
Tetrehydroquinoline	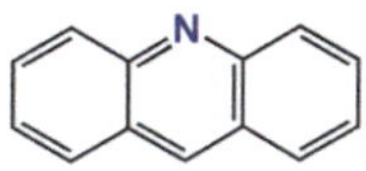	5.0
Acridine	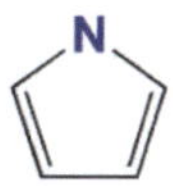	0.4
Indole	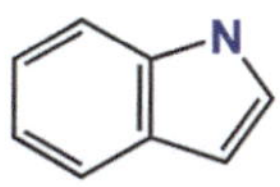	-3.6
Indoline	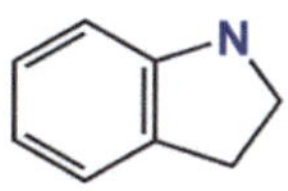	5.0
Carbazole	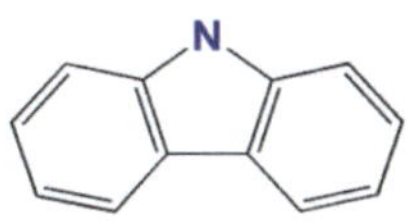	-6.0
Aniline	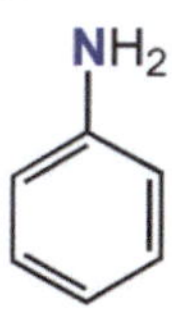	5.0

The strong metal-support interactions between the active metal and support can also impact the hydrodenitrogenation activity [142]. In one study, Pt/TiO_2 catalysts used in SCW conditions (420 °C and 30 MPa) showed less than 2% Pt loss and no collapse of TiO_2, with propylamine conversion remaining at 100% after 5 h on stream. Additionally, the analysis of oxygen fugacity (FO_2)-PH revealed that the thermodynamic stability of catalysts can be influenced by the PH value of the solution, which can alter the oxidation state or dissolution of the metal and metal metallic oxide during hydrodenitrogenation under hydrothermal conditions. Mapiour's research also indicated that increasing the partial pressure could slow down the catalyst's deactivation in hydrodenitrogenation reactions [143].

(2) **Sulfur Effect**

Sulfur, an essential element for microalgae growth, has been recognized as a known catalyst poison in both laboratory and field studies of catalytic biocrude upgrading [144]. However, the removal of sulfur from microalgae biocrude oil presents challenges. Roussis et al. [145] observed a decrease in sulfur content in raw *Nannochloropsis* sp. biocrude oil from 0.6% to 0.4% after hydrothermal treatment at 300 °C for 30 min, with the sulfur content remaining at 0.4% even when the temperature was increased to 400 °C. Another study showed that the sulfur element in *Chlorella* sp. biocrude oil remained at 0.97% after the HTL process (300 °C and 90 min) [146].

Osada et al. [147] characterized sulfur-poisoned Ru catalysts after hydrothermal reaction. According to the X-ray photoelectron spectroscopy (XPS) analysis, sulfur in the Ru metal was present as RuS_2, $Ru(SO_3)_2$, and $Ru(SO_4)_2$. The experimental results indicated that the sulfur content on the catalyst's surface decreased with increasing water density, suggesting that high water density may inhibit the adsorption of sulfur at the active metal sites, effectively alleviating catalyst sulfur poisoning. In another study on the catalytic hydrothermal reaction of ethanol, Dreher et al. [148] introduced sulfoxide as a sulfur-containing compound into the reaction mixture to investigate the influence of sulfur on Ru/C catalyst. The results showed that under supercritical conditions, sulfur blocked 40% of the active sites due to strong adsorption. A subsequent study [149] demonstrated that the poisoning of the Ru catalyst by sulfur is due to the irreversible adsorption of S^{2-} on the catalyst's active sites. Osada et al. [149] immersed a 2% Ru/TiO_2 catalyst in sulfuric acid solution to form a sulfur-poisoned catalyst marked as S–Ru/TiO_2. After treating S–Ru/TiO_2 with 573K subcritical water for 10 h in a continuous system, the catalytic activity of the treated catalyst was higher compared to the untreated S–Ru/TiO_2 catalyst, and gas production increased from 21% to 57.1%. The XPS characterization results showed that 3/4 of the sulfur on the catalyst surface was removed after treatment by subcritical water, indicating a promising regeneration method for sulfur-poisoned catalysts.

Although the aforementioned studies are not specifically focused on denitrogenation of biocrude, they highlight the impact of sulfur on catalysts used for algal biocrude upgrading and the denitrogenation of nitrogenous model compounds. Therefore, the inevitable effect of trace amounts of sulfur-containing species in algae cells on the catalysts should be addressed in future studies.

Since sulfur adsorption requires specific adsorption sites, sulfur resistance can be improved by controlling the number of adsorption sites on the catalyst's surface. According to Wang's research, adding tungsten (W) into the monometallic catalyst Ni/γ-Al_2O_3 for sulfur adsorption can significantly enhance catalytic stability in a sulfur resistance test [150]. Additionally, research utilizing DFT calculations confirmed that the interaction between the metal and sulfur can be effectively weakened by doping a second metal into the monometallic catalyst, leading to a substantial decrease in the adsorption energy (-Eads) for sulfur from 5.62 eV on Ni surface to 4.63 eV on NiAu surface [151].

(3) **Water Effect**

The deactivation of metal catalysts can be attributed to the presence of water. For example, water has been found to decrease the catalytic activity of NiMo/γ-Al_2O_3 to one-third of the activity of the fresh catalyst. Additionally, water causes a small loss in the specific surface area of the γ-Al_2O_3 support under hydrothermal conditions. This can lead to a significant decrease in the surface area and pore volume of Pt/γ-Al_2O_3 catalyst, with a reduction of 39% and 27%, respectively, at 510 °C under SCW conditions. γ-Al_2O_3 can also be transformed to AlOOH to cover metal active sites at 330 °C under hydrothermal conditions [152]. Furthermore, increasing the amount of water in a reaction can lead to a reduction in conversion and selectivity, and catalyst characterization results indicate that the main reasons for water deactivation of the Cu/Zn catalyst were water occlusion of active sites due to the low solubility of water in the substrate, promotion of crystal growth, and catalyst agglomeration in the presence of water [153].

In addition to the support γ-Al_2O_3 and metals such as Ni, Mo, Cu, and Zn, the deactivation of molecular sieve ZSM-5 has been reported. The aluminum in the ZSM-5 can react with water molecules to form aluminum hydroxide, resulting in dealumination and a decrease in the crystallinity of the ZSM-5 framework [154]. Furthermore, in a kinetic study of the deactivation of a Pd/C catalyst, researchers found that the catalyst was deactivated 1.5 times faster in pure methanol than in a 1:1 water–methanol mixture [155].

Introducing a promoter can improve hydrothermal stability. For example, the hydration of γ-Al_2O_3 can be significantly inhibited by adding SiO_2, which can cover the γ-Al_2O_3 surface to prevent the aggregation of the active metal [156]. Surface modification can also prevent catalyst deactivation. For instance, an introduced inert silica layer can act as a protective layer, preventing the detachment of Cu^{2+} ions from ZSM-5 ion exchange positions and the dealumination of the zeolite during the hydrothermal aging process [157]. Additionally, the hydrothermal stability of copper-ion-exchanged SSZ-13 (Cu-SSZ-13) can be remarkably improved by introducing a very small amount of Ce, which tends to locate at ion exchange sites of Cu-SSZ-13 to fill the defect sites and stabilize the framework, resulting in enhanced hydrothermal stability [158].

2 Demetallization of Microalgal Biocrude Oil

Despite being recognized as a preferred approach for converting microalgae into biocrude oil, HTL still encounters a number of obstacles. Among these challenges, the notable issue lies in the crude bio-oil containing elevated levels of metals.

2.1 *Metal-Containing Substances*

Utilizing microalgal bio-oil with high metal content can result in various issues, such as scaling and corrosion of associated equipment, catalyst deactivation, and diminished product quality. Additionally, the viscosity of biocrude oil may increase during storage due to the catalyzation of polymerization reactions by alkali metals present in the biocrude oil [159]. Therefore, significant consideration should be given to the demetallization of biocrude oil.

2.1.1 Metal-Containing Substances in Algae-Derived Biocrude Oil

The metal content in biocrude oil can reach up to 1000 ppm, with the main metal elements including Fe, Ca, Cu, Zn, Al, Mn, Mg, Si, and Ni. Extensive reports have detailed the metal content of microalgal biocrude under specific hydrothermal conditions, as depicted in Table 3. For example, in the HTL process of a low-lipid high-protein microalgae (*Nannochloropsis* sp.) and a high-lipid low-protein microalgae (*Chlorella* sp.), ICP-OES analysis detected Fe, Cu, Na, Ca, Mg, K, Al, and Zn in the biocrude oil of *Nannochloropsis* sp., while only Fe, Na, Ca, and K were detected in the oil of *Chlorella* sp. [160]. Furthermore, the distribution of K, Na, Mg, and Ca in the product streams for biocrude obtained from non-catalytic hydrothermal liquefaction of the brown macroalgae Laminaria Saccharina at 350 °C with a 15 min residence time has been studied [161]. It was observed that, in most cases, K and Na were predominantly found in the aqueous phase, while Ca and Mg mainly existed in the solid residue.

Table 3 Concentrations (ppm) of metals in algae HTL biocrude oil

Algae	Experimental conditions	Fe	Al	Ni	Cu	Zn	Mn	Si	Mg	Ca	References
Nannochloropsis	a	399	< 10	29	< 10	26	< 10	< 1	–	–	[162]
Nannochloropsis	b	688	5	65	27	133	13	12	47	119	[163]
Nannochloropsis	c	850	10	21	58	87	9	–	29	25	[164]
Spirulina	d	1260	< 50	80	< 20	30	–	–	–	–	[165]
Reed	c	2276	5	43	41	59	10	52	94	240	[166]
Sapphire	c	1673	1.5	65	27	133	4	12	6	48	[166]

2.1.2 Metalloporphyrin

Metals in algal biocrude oil are commonly linked with naphthenic acids in the form of organometallic compounds to create metal porphyrins. Porphyrin refers to a complex conjugated ring structure in which four pyrrole rings are symmetrically connected to each other with four methylene groups. The fundamental framework of the porphyrin is referred to as the porphyrin or porphyrin core, as illustrated in Fig. 26a. A cavity is present in the center of porphyrin, and the pyrrole nitrogen atom inside forms a complex with the metal to create a metal porphyrin. Its typical structural formula is presented in Fig. 26b. The combination of different metals and porphyrins results in structural variations in metalloporphyrins, leading to distinct reaction behaviors during thermal transformation.

In addition to metal porphyrins, metals also exist in the form of metal non-porphyrins, which are similar to metal porphyrins except that the nitrogen atoms interacting with the central metal atom in the pyrrole ring are replaced by other heteroatoms (such as oxygen or sulfur) to form a four-coordinated complex, as portrayed in Fig. 27 [167, 168]. During the hydrodemetallization process, metal non-porphyrins exhibit diverse reaction behaviors depending on the metal, resulting in varying deposition positions of metal impurities. For instance, sulfuring CoMo/

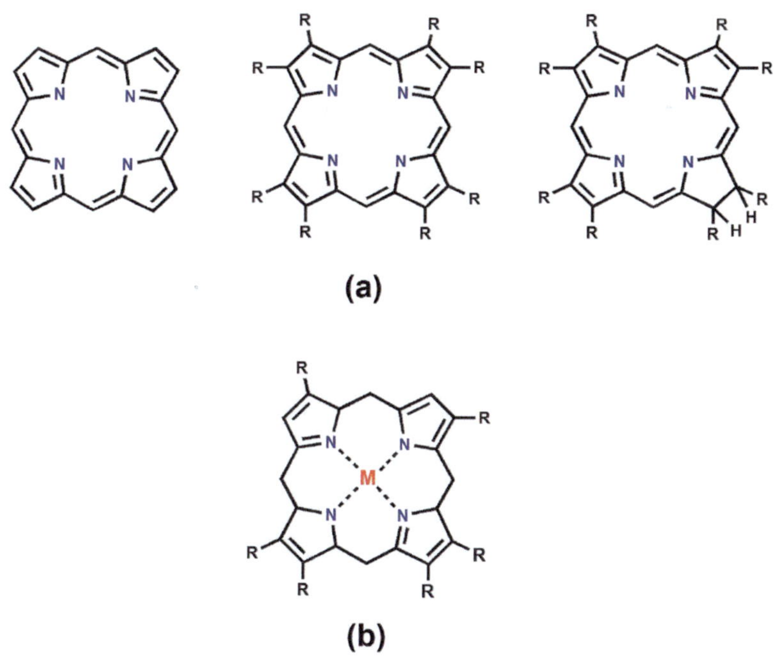

Fig. 26 **a** Structure of porphin, porphyrins, and chlorin **b** structure of metalloporphyrin [168]

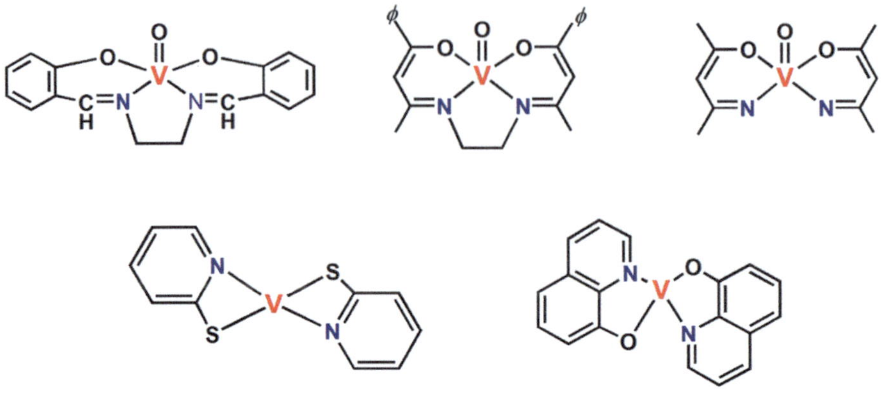

Fig. 27 Structures of metal non-porphyrins [167]

g-Al_2O_3 catalysts for demetallization of Ni-tetra(3-methylphenyl) porphyrin (Ni-T3MPP) revealed that the metal was deposited on the surface and pores of the catalyst.

2.1.3 Iron in Microalgae Biocrude Oil

In a study conducted by Jarvis et al. [169] on metals in microalgae HTL biocrude, it was found that the concentrations of several metals (such as Ca, Fe, K, Mg, and Na) in *Tetraselmis* sp. microalgae feed were all higher than 1000 ppm. However, only Fe remained at a high concentration in the HTL biocrude. A total of 138 unique iron porphyrin compounds were identified in the two biocrudes derived from the HTL of *Tetraselmis* sp. and cyanobacteria.

Among the trace elements, iron has received more attention in terms of its influence on microalgae physiology. Iron plays a significant role in algal growth, metabolism, and function. In microalgae, Fe is mainly present in the form of heme, iron-containing porphyrin, or Fe–S clusters in various metalloproteins involved in metabolic pathways. Metalloporphyrin, which involves the coordination of a single metallic ion into the central 4-N cavity, provides higher stability by replacing two hydrogen atoms linked to the two nitrogen atoms [170, 171]. Additionally, flocculating agents used in the flocculated sedimentation process of biomass seriflux are a major source of Fe. Furthermore, some microalgae store Fe in ferritin located in the chromatoplast.

Existing literature suggests that removing Fe is challenging. For example, enrichment of Fe on activated carbon supported metal catalyst surfaces was observed after hydro-upgrading of microalgae HTL biocrude [172]. The Fe concentration in microalgal oil from hemin HTL initially increases and then decreases with increasing reaction temperature and holding time. Hemin can decompose into DCM-soluble compounds only when the hydrothermal environment temperature exceeds 250 °C [173]. This behavior suggests that hemin decomposes to form smaller Fe porphyrinic fragments that are soluble in DCM. These Fe-containing compounds further degrade at more severe conditions and ultimately deposit onto the solid phase.

2.2 *Demetallization Chemistry*

To gain a deeper understanding of demetallization reactions, this section separately discusses the interaction between metal-containing compounds and catalyst surfaces, experimental conditions, and the kinetic analysis of the demetallization process. This approach provides an overview of demetallization chemistry.

2.2.1 Mechanism of Catalytic Demetallization

Metalloporphyrins primarily interact with the active center of the catalyst surface through the π electron system of the ring plane. Hayes et al. [174] proposed that the diffusion behavior of metalloporphyrins before contacting the catalyst surface is mainly influenced by molecular size, steric hindrance, and temperature. At higher temperatures, porphyrin molecules can diffuse to the inner surface of the catalyst and distribute uniformly in the radial direction. Additionally, there are two other types of interactions between vanadium porphyrin and the adsorption surface as shown in Fig. 28. Firstly, V^{4+} in the $V = O$ group of vanadium porphyrin (VO-TPP) can interact with Lewis base sites on the catalyst surface, resulting in excess negative charge on the porphyrin ring. This negative charge can be neutralized by π bond interaction with electron acceptor sites. Secondly, the $V = O$ group can interact with an acceptor through its oxygen atom, reducing the electron density on the porphyrin ring and promoting electron delocalization of vanadium. The mechanism of HDM of metalloporphyrins is widely explained through three steps: hydrogenation of metalloporphyrins to produce intermediate products, reversible dehydrogenation of the intermediate product to metalloporphyrin, and removal of metals by continuous and irreversible hydrogenation of intermediate products [175]. The presence of hydrogen sulfide can promote the HDM reaction. There are two main explanations for this observation. One is that hydrogen sulfide may coordinate with the metal or metal group at the center of the porphyrin ring, weakening the M–N bond, allowing for hydrogenation and ring opening of the porphyrin macrocycle. The other explanation suggests that sulfur vacancies associated with hydrocracking during the HDM process can be reversibly converted into saturation sites, promoting hydrogenation of metalloporphyrins [176]. According to the second explanation, the conversion of the active site on the catalyst surface and the mechanism of HDM can be depicted as shown in Fig. 29: (1) metalloporphyrins adsorb on the catalyst saturation site through porphyrin molecules, and protons from the –SH group on the sulfurized catalyst are added to the metalloporphyrin ring; (2) hydrogenation of adsorbed metalloporphyrins to form intermediate chlorin; (3) hydrogenation of metal-nitrogen (M–N) bonds on sulfur vacancies on catalysts to remove metals. It is evident that the presence of

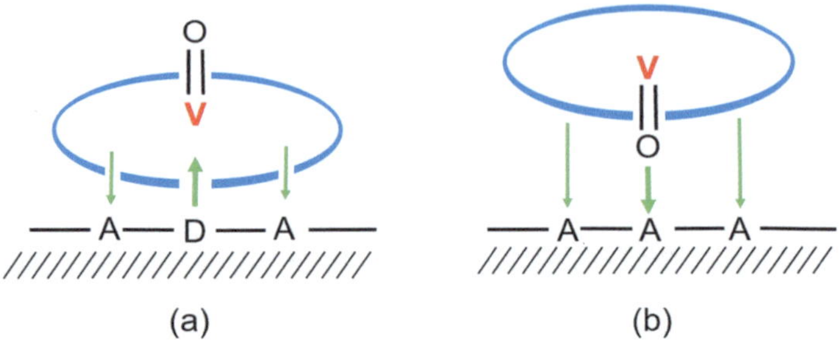

Fig. 28 Diagram of the interaction between VO-TPP and surface of catalyst [175], (**a**) Interaction between vanadium porphyrin and adsorption surface (Mechanism 1); (**b**) Interaction between vanadium porphyrin and adsorption surface (Mechanism 2)

hydrogen sulfide promotes the first two steps of the hydrogenation reaction. Hydrogen sulfide may also directly interact with the central metal in metalloporphyrins, weakening metal-nitrogen bonds, accelerating hydrogenation, and ring-opening reactions [177].

2.2.2 Demetallization in Hydrothermal Condition

Recent studies have highlighted the influence of solvents on the demetallization of biocrude [178]. The role of solvents primarily manifests in two aspects: the dilution effect and the hydrogen transfer effect [179, 180]. The rate of demetallization reaction is closely related to the solvent's ability to supply hydrogen [181]. Solvents rich in hydrogen with strong hydrogen transfer abilities can enhance the hydrogenation of metalloporphyrin compounds. SCW, a hydrogen-rich donor, is used in the demetallization of the oil phase. Some researchers propose that SCW can directly participate in the demetallization process as a hydrogen donor [182, 183]. Mandal et al. [178] achieved a 95.02% conversion of nickel porphyrin under SCW conditions without a catalyst. During the process, the central metal may convert into metal hydroxides, metal oxides, and metal ions. In another study, they investigated the removal of vanadium porphyrins under SCW conditions. At 490 °C and a reaction time of 180 min, the conversion reached 90.5 wt% and increased with longer reaction times and higher temperatures [180]. Dailey et al. [184] studied the influence of different solvents on the metal content in biocrude obtained from the HTL of microalgae. They found that when methyl tert-butyl ether (MTBE) was used as a solvent, the Fe and Na contents could be further reduced to 500 ppm and 50 ppm, respectively, whereas dichloromethane (DCM) only showed a 50% reduction in Fe and a 69% reduction in Na in microalgal crude oil. They also examined the effects of different HTL process variables (temperature, holding time, and water loading) on metal elements in biocrude and the aqueous phase co-product. The results showed that the concentrations of Zn, Cu, and Ni did not vary significantly with water density at 350 °C and 60 min, but these metal concentrations decreased by 50% when the temperature was raised to 450 °C.

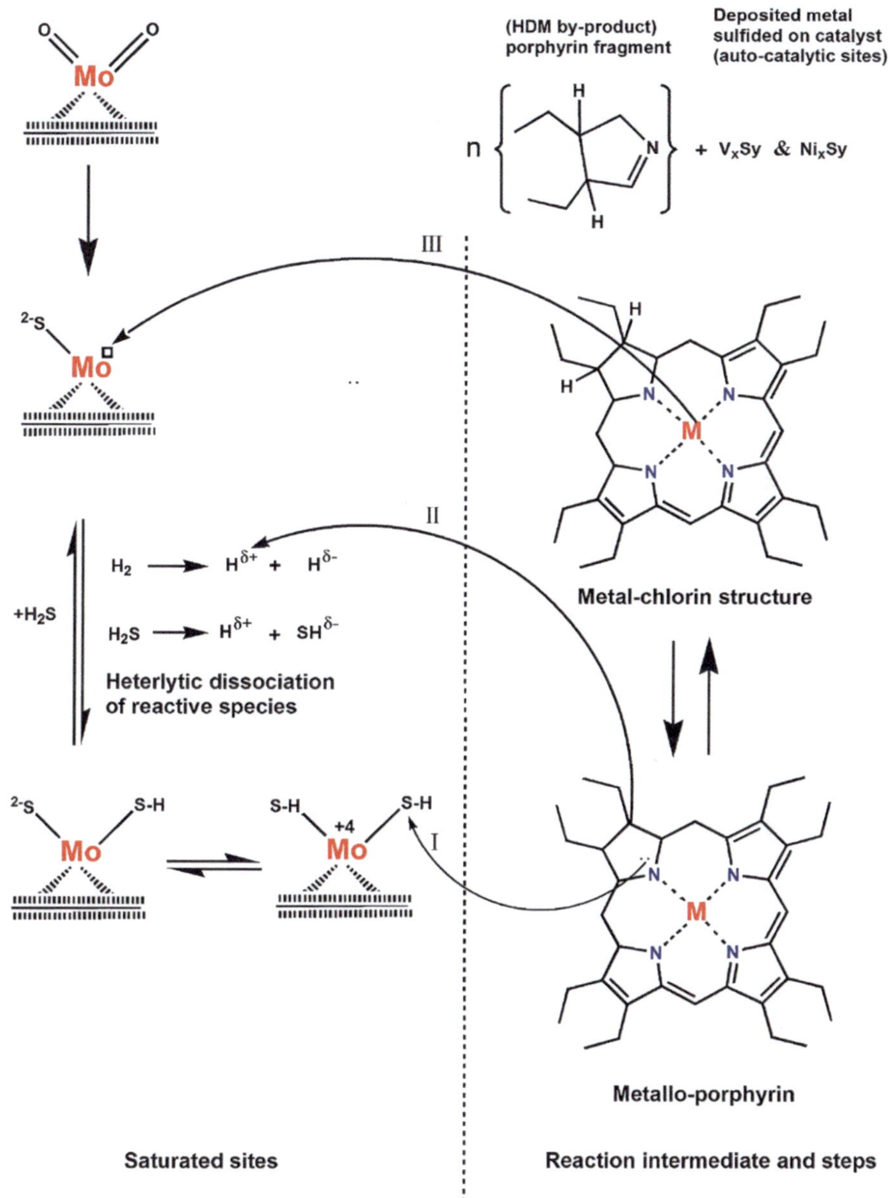

Fig. 29 Proposed reaction mechanism for HDM in presence of high partial pressure of H_2S: (I) adsorption of metal-porphyrin ring to saturated sites (delocalization of electron density); (II) hydrogenation of porphyrin structure; and (III) demetallation of the metalloporphyrin complex molecule [176]

2.2.3 Factors Affecting Demetallization

Indeed, several factors directly impact the hydrogenation and demetallization processes, including the catalyst used, the structure of metalloporphyrins, reaction conditions, and the amount of hydrogen loading [185]. These factors play crucial roles in determining the efficiency and selectivity of the demetallization reactions. By optimizing these factors, researchers can enhance the overall performance of the demetallization process and improve the quality of the biocrude or other products obtained.

(1) **Catalyst Effect**

Catalysts utilized for HDM typically consist of a transition metal serving as the active component and a mildly acidic support, providing active sites for hydrogenation and hydrogenolysis [186]. Research on HDM catalysts has been focused on optimizing the support structure to enhance the HDM rate and prolong the catalyst's working life [177, 187]. The pore size of the support material, such as $\gamma\text{-}Al_2O_3$, has been shown to impact the HDM rate, with studies indicating that the HDM rate of feed oil increased from 70.6% to 76.2% with an increase in support pore size from 14 to 40 nm [188]. Additionally, extensive research suggests that the reaction mechanism of metalloporphyrins is not correlated with the chemical state of the catalysts (oxidation, sulfide, and reduction states), indicating that these states do not alter the HDM reaction process [189, 190]. This knowledge is essential for the ongoing development and optimization of HDM catalysts for efficient demetallization processes.

(2) **Structure of Metalloporphyrins**

The molecular structure of metalloporphyrins, including the central metal atom and porphyrin, directly influences the rate of catalytic hydrogenation [191]. For instance, Agrawal et al. studied the HDM reactions of nickel etioporphyrin (Ni-ETIO) and vanadyl etioporphyrin (VO-ETIO) and observed that VO-ETIO exhibits higher catalytic hydrogenation activity at 315 °C and 9.69 MPa hydrogen pressure. They also indicated that for both nickel and vanadium HDM, the mechanism comprises three major kinetic steps: hydrogenation of metal porphyrin to the reaction intermediate metal chlorin, reversible dehydrogenation of the intermediate, and demetallization of the reaction intermediate [191]. Their study highlighted that the molecular structure of porphyrins has chemical or steric effects on the reaction process [192]. Furthermore, the presence of methyl and ethyl groups in metalloporphyrins can hinder the interaction between metal atoms and the active sites of catalysts, thereby impeding demetallization. In comparison to VO-ETIO metal porphyrin, T3MPP metalloporphyrins have a lower reduction potential and are more prone to hydrogenation reactions [192]. These findings underscore the significance of considering the molecular structure of metalloporphyrins in understanding and optimizing catalytic hydrogenation and demetallization processes.

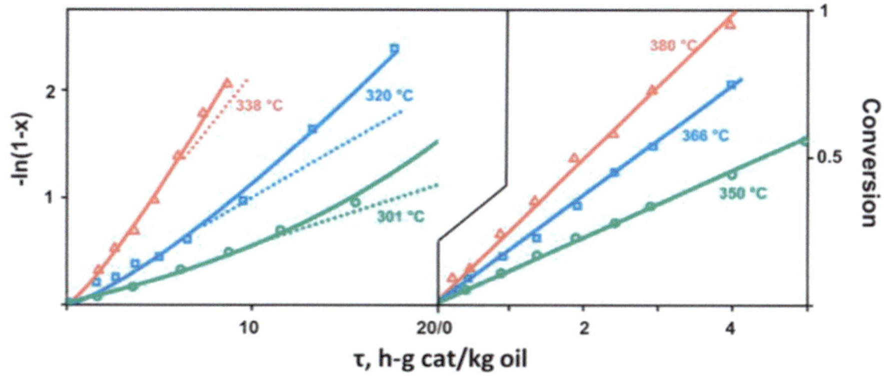

Fig. 30 First-order plots of Ni-porphyrin disappearance as function of temperature (33 ppm of Ni in feed, CoMo/γ-Al_2O_3, 6.99 MPa) [193]

(3) **Reaction Temperature and Hydrogen Loading**

Indeed, it has been observed that the HDM of metalloporphyrins can occur even without a catalyst [177, 187]. In non-catalytic HDM reactions, temperature and hydrogen partial pressure are the primary factors influencing the reactions. Chen et al. considered the effects of temperature, H_2 pressure, and the concentration of V and Ni porphyrins on the overall kinetics [193]. As shown in Fig. 30, they found that at temperatures above 350 °C, the kinetics follow a first-order reaction, while below 350 °C, a curvature in the plot suggests a different kinetic behavior. In the intrinsic kinetic HDM research of nickel and vanadium porphyrins, nearly complete demetallization (98%) of microalgae Spirulina biocrude oil was achieved over Mo/γ-Al2O3 as a catalyst at 300 °C under hydrothermal conditions [165]. However, at temperatures ranging from 400 to 410 °C, the HDM catalyst could rapidly deactivate due to the deposition of coke [194].

Furthermore, for VO-ETIO, the hydrogenation step is first order with respect to hydrogen pressure, the dehydrogenation rate is independent of hydrogen pressure, and the hydrogenolysis step is second order with respect to hydrogen pressure [191]. These insights into the non-catalytic HDM reactions and the specific kinetics of metalloporphyrins provide valuable information for understanding and optimizing the demetallization processes.

3 Conclusions and Future Directions

The high concentration of nitrogen and metal-containing compounds in algae and other protein-rich biomass presents a significant challenge for the production of clean liquid fuels. Unlike deoxygenation and desulfurization, less attention has been paid to the denitrogenation and demetallization processes of biocrude. To address this gap, this article systematically summarizes the specific reaction mechanisms and catalytic activities of various metal-based catalysts for denitrogenation and demetallization under hydrothermal conditions.

In the denitrogenation process, several mechanisms, including E_1, E_2, SN_1, SN_2, and direct hydrogenolysis, are involved in breaking the carbon–nitrogen bonds. Catalysts need to provide active sites for both hydrogenation and hydrogenolysis. For HDN processes, it has been found that the hydrogenolysis of the carbon–nitrogen bond with hydrogen gas is the rate-determining step. In the hydrothermal HDN process, water serves as both a hydrogen donor and a reactant. However, some catalysts can be deactivated by nitrogen and sulfur compounds and can also be negatively affected by water. Competitive adsorption of nitrogen-containing heterocyclic substances on Lewis acid sites of sulfided base metal hydrotreating catalysts and on Brønsted acid sites on zeolitic catalysts has been observed.

Regarding demetallization, the effect of water partial pressure is moderate, while the effect of temperature is considerably higher under supercritical water conditions. High temperatures promote the diffusion of metalloporphyrins in the catalyst pores, enhancing the interaction between metal-containing compounds and the catalyst surface and facilitating the demetallization reaction. Batch holding time, metalloporphyrin structure, and solvents also have a significant impact on the demetallization process.

Developing non-noble metal catalysts is an economic and practical approach for hydrodenitrogenation and demetallization. However, the stability and renewability of these catalysts remain major challenges for industrial applications. Additionally, a fundamental understanding of the mechanisms of these catalytic reactions is still lacking. Suggestions for future research include developing more detailed reaction kinetic models, exploring the structure-catalysis relationship through in situ characterization methods, modifying the properties of catalyst supports, and utilizing advanced catalysts such as metal–organic frameworks and single-atom catalysts. DFT calculations can be employed to reveal the atomic-level catalytic mechanisms and predict important catalyst properties. Combining bimetallic synergistic effects with other strategies can also lead to the development of more active and stable denitrogenation and demetallization catalysts.

Furthermore, the environmental and economic impacts of producing biodiesel after deep denitrogenation and demetallization should be comprehensively assessed using methods such as exergoenvironmental analysis and life cycle assessment to detail environmental pollution, resource consumption, and economic costs.

References

1. B.H. Ba, C. Prins, C. Prodhon, Models for optimization and performance evaluation of biomass supply chains: an operations research perspective. Renew. Energy **87**, 977–989 (2016)
2. F. Ecer et al., Assessment of renewable energy resources using new interval rough number extension of the level based weight assessment and combinative distance-based assessment. Renew. Energy **170**, 1156–1177 (2021)
3. H.K.S. Panahi et al., A review on green liquid fuels for the transportation sector: a prospect of microbial solutions to climate change. Biofuel Res. J.-BRJ **6**(3), 995–1024 (2019)
4. W. Stafford et al., Forestry biorefineries. Renew. Energy **154**, 461–475 (2020)
5. J. Zakzeski et al., The catalytic valorization of lignin for the production of renewable chemicals. Chem. Rev. **110**(6), 3552–3599 (2010)
6. D. Niu et al., Evaluation of provincial carbon neutrality capacity of China based on combined weight and improved TOPSIS model. Sustainability. **13**(5), 2777 (2021)
7. A. Tursi, A review on biomass: importance, chemistry, classification, and conversion. Biofuel Res. J.-BRJ **6**(2), 962–979 (2019)
8. Z. Du et al., Microwave-assisted pyrolysis of microalgae for biofuel production. Biores. Technol. **102**(7), 4890–4896 (2011)
9. P.M. Schenk et al., Second generation biofuels: high-efficiency microalgae for biodiesel production. Bioenergy Res. **1**(1), 20–43 (2008)
10. Y. Yang et al., Synergetic conversion of microalgae and CO_2 into value-added chemicals under hydrothermal conditions. Green Chem. **21**(6), 1247–1252 (2019)
11. J.-H. Ju et al., Regulation of lipid accumulation using nitrogen for microalgae lipid production in *Schizochytrium sp.* ABC101. Renew. Energy **153**, 580–587 (2020)
12. D. Yu et al., Pyrolysis of oleaginous yeast biomass from wastewater treatment: kinetics analysis and biocrude characterization. Renew. Energy **150**, 831–839 (2020)
13. M.B. Johnson, Z. Wen, Production of biodiesel fuel from the microalga Schizochytrium limacinum by direct transesterification of algal biomass. Energy Fuels **23**(10), 5179–5183 (2009)
14. R.B. Levine, T. Pinnarat, P.E. Savage, Biodiesel production from wet algal biomass through in situ lipid hydrolysis and supercritical transesterification. Energy Fuels **24**(9), 5235–5243 (2010)
15. L.A. Andrade et al., Characterization and product formation during the catalytic and non-catalytic pyrolysis of the green microalgae Chlamydomonas reinhardtii. Renew. Energy **119**, 731–740 (2018)
16. Z. Gong et al., Catalytic pyrolysis of chemical extraction residue from microalgae biomass. Renew. Energy **148**, 712–719 (2020)
17. M.A. Adnan et al., Gasification of wet microalgae to produce H_2-rich syngas and electricity: a thermodynamic study considering exergy analysis. Renew. Energy **147**, 2195–2205 (2020)
18. D. Macri et al., Supercritical water gasification of biomass and agro-food residues: energy assessment from modelling approach. Renew. Energy **150**, 624–636 (2020)
19. P. SundarRajan et al., An insight into carbon balance of product streams from hydrothermal liquefaction of Scenedesmus abundans biomass. Renew. Energy **151**, 79–87 (2020)
20. Y. Zhou et al., Production of high-quality biofuel via ethanol liquefaction of pretreated natural microalgae. Renew. Energy **147**, 293–301 (2020)
21. C.I. Aragon-Briceno et al., Hydrothermal carbonization of wet biomass from nitrogen and phosphorus approach: a review. Renew. Energy **171**, 401–415 (2021)
22. J. Pages-Diaz et al., Anaerobic bio-methane potential of the liquors from hydrothermal carbonization of different lignocellulose biomasses. Renew. Energy **157**, 182–189 (2020)
23. P.K. Kanaujia et al., Review of analytical strategies in the production and upgrading of bio-oils derived from lignocellulosic biomass. J. Anal. Appl. Pyrol. **105**, 55–74 (2014)
24. H.K.S. Panahi et al., Recent updates on the production and upgrading of bio-crude oil from microalgae. Bioresour. Technol. Rep. **7**, 100216 (2019)

25. M. Aghbashlo et al., Exergoenvironmental analysis of bioenergy systems: a comprehensive review. Renew. Sustain. Energy Rev. **149**, 111399 (2021)
26. S. Soltanian et al., A critical review of the effects of pretreatment methods on the exergetic aspects of lignocellulosic biofuels. Energy Convers. Manage. **212**, 112792 (2020)
27. M.A. Rosen, Environmental sustainability tools in the biofuel industry (2018)
28. J.-X. Wang et al., Production of biodiesel through transesterification of soybean oil using lithium orthosilicate solid catalyst. Fuel Process. Technol. **104**, 167–173 (2012)
29. I.V. Kozhevnikov, A.L. Nuzhdin, O.N. Martyanov, Transformation of petroleum asphaltenes in supercritical water. J. Supercrit. Fluids **55**(1), 217–222 (2010)
30. Z. Fang et al., Catalytic hydrothermal gasification of cellulose and glucose. Int. J. Hydrogen Energy **33**(3), 981–990 (2008)
31. S.K. Kirkok, et al., Dioxin and dibenzofuran like molecular analogues from the pyrolysis of biomass materials-the emerging challenge in bio-oil production. BMC Chem. **15**(1) (2021)
32. F. Li, S.C. Srivatsa, S. Bhattacharya, A review on catalytic pyrolysis of microalgae to high-quality bio-oil with low oxygeneous and nitrogenous compounds. Renew. Sustain. Energy Rev. **108**, 481–497 (2019)
33. D. Dong, S. Jeong, F.E. Massoth, Effect of nitrogen compounds on deactivation of hydrotreating catalysts by coke. Catal. Today **37**(3), 267–275 (1997)
34. N. Gao et al., Modified nickel-based catalysts for improved steam reforming of biomass tar: a critical review. Renew. Sustain. Energy Rev. **145**, 111023 (2021)
35. A. Ochoa et al., Coke formation and deactivation during catalytic reforming of biomass and waste pyrolysis products: a review. Renew. Sustain. Energy Rev. **119**, 109600 (2020)
36. M.T.H. Siddiqui et al., Dual-application of novel magnetic carbon nanocomposites as catalytic liquefaction for bio-oil synthesis and multi-heavy metal adsorption. Renew. Energy **172**, 1103–1119 (2021)
37. K. Kohli et al., Deactivation of hydrotreating catalyst by metals in resin and asphaltene parts of heavy oil and residues. Fuel **175**, 264–273 (2016)
38. T. Li, H. Li, C. Li, Progress in effects of microenvironment of carbon-based catalysts on hydrodeoxygenation of biomass. ChemCatChem **13**(4), 1074–1088 (2021)
39. X. Wang et al., Hydrodeoxygenation of lignin-derived phenolics—a review on the active sites of supported metal catalysts. Green Chem. **22**(23), 8140–8168 (2020)
40. A.N.K. Lup et al., A review on reaction mechanisms of metal-catalyzed deoxygenation process in bio-oil model compounds. Appl. Catal. A **541**, 87–106 (2017)
41. S. Jinsheng, Y. Jie, S. Ming, Review of denitrogenation of algae biocrude produced by hydrothermal liquefaction. Trans. Tianjin Univ. **23**(4), 301–314 (2017)
42. M.S. Haider et al., Catalytic hydrotreatment of microalgae biocrude from continuous hydrothermal liquefaction: heteroatom removal and their distribution in distillation cuts. Energies **11**(12), 3360 (2018)
43. W. Costanzo et al., Effect of low temperature hydrothermal liquefaction on catalytic hydrodenitrogenation of algae biocrude and model macromolecules. Algal Res.-Biomass Biofuels Bioprod. **13**, 53–68 (2016)
44. S.S. Toor, L. Rosendahl, A. Rudolf, Hydrothermal liquefaction of biomass: a review of subcritical water technologies. Energy **36**(5), 2328–2342 (2011)
45. S. Sun, G. Chung, T.-S. Shin, Volatile compounds of the green alga, capsosiphon fulvescens. J. Appl. Phycol. **24**, 1003–1013 (2012)
46. T.M. Brown, P. Duan, P.E. Savage, Hydrothermal liquefaction and gasification of *Nannochloropsis* sp. Energy Fuels **24**(6), 3639–3646 (2010)
47. P. Duan, P.E. Savage, Hydrothermal liquefaction of a microalga with heterogeneous catalysts. Ind. Eng. Chem. Res. **50**, 52–61 (2011)
48. Y. Guo et al., A review of bio-oil production from hydrothermal liquefaction of algae. Renew. Sustain. Energy Rev. **48**, 776–790 (2015)
49. T. Kabe, A. Ishihara, W.H. Qian, Elucidation of hydrodesulfurization mechanism on molybdenum-based catalysts using 35S radioisotope pulse tracer methods. Catal. Surv. Jpn. **3**(1), 17–25 (1999)

50. H. Tominaga, M. Nagai, Reaction mechanism for hydrodenitrogenation of carbazole on molybdenum nitride based on DFT study. Appl. Catal. A **389**(1–2), 195–204 (2010)
51. V. Schwartz, V.T. da Silva, S.T. Oyama, Push-pull mechanism of hydrodenitrogenation over carbide and sulfide catalysts. J. Mol Catal. A-Chem. **163**(1–2), 251–268 (2000)
52. F. Rota, V.S. Ranade, R. Prins, Stereochemistry of hydrodenitrogenation: the mechanism of elimination of the amino group from cyclohexylamines over sulfided Ni–Mo/γ-Al_2O_3 Catalysts. J. Catal. **200**, 389–399 (2001)
53. R.M. Laine, Comments on the mechanisms of heterogeneous catalysis of the hydrodenitrogenation reaction. Catal. Rev.-Sci. Eng. **25**(3), 459–474 (1983)
54. M. Jian, R. Prins, Mechanism of the hydrodenitrogenation of quinoline over NiMo(P)/Al_2O_3 catalysts. J. Catal. **179**(1), 18–27 (1998)
55. M. Jian, F. Kapteijn, R. Prins, Kinetics of the Hydrodenitrogenation ofortho-Propylaniline over NiMo(P)/Al_2O_3Catalysts. J. Catal. **168**(2), 491–500 (1997)
56. Y. Zhao, J. Czyzniewska, R. Prins, Mechanism of the direct hydrodenitrogenation of naphthylamine on sulfided NiMo/Al_2O_3. Catal. Lett. **88**(3), 155–162 (2003)
57. M.J. Girgis, B.C. Gates, Reactivities, reaction networks, and kinetics in high-pressure catalytic hydroprocessing. Ind. Eng. Chem. Res. **30**(9), 2021–2058 (1991)
58. S.-R. Li et al., Mechanism of C–N bond cleavage in aniline on MoP(001) surface. Acta Phys. Chim. Sin. **32**(2), 465–473 (2016)
59. Y. Guo et al., Ring-opening and hydrodenitrogenation of indole under hydrothermal conditions over Ni, Pt, Ru, and Ni-Ru bimetallic catalysts. Chem. Eng. J. **406**, 126853 (2021)
60. T. Kabe, W. Qian, A. Ishihara, Elucidation of hydrodesulfurization mechanism using 35S radioisotope pulse tracer methods. Catal. Today **39**(1), 3–12 (1997)
61. Y. Chu et al., NiMoNx/γ-Al_2O_3 catalyst for HDN of pyridine. Appl. Catal. A **176**(1), 17–26 (1999)
62. C.N. Satterfield, M. Modell, J.A. Wilkens, Simultaneous catalytic hydrodenitrogenation of pyridine and hydrodesulfurization of thiophene. Ind. Eng. Chem. Process. Des. Dev. **19**(1), 154–160 (1980)
63. V. Schwartz, S.T. Oyama, Reaction network of pyridine hydrodenitrogenation over carbide and sulfide catalysts. J. Mol. Catal. A-Chem. **163**(1–2), 269–282 (2000)
64. M.J. Ledoux, A. Bouassida, R. Benazouz, The use of pyridine and piperidine hdn as probe for activity of molybdenum-based hydrotreatment catalysts. The role of the nickel (part II). Appl. Catal. **9**(1), 41–52 (1984)
65. P. Duan, P.E. Savage, Catalytic hydrothermal hydrodenitrogenation of pyridine. Appl. Catal. B-Environ. **108**(1–2), 54–60 (2011)
66. L. Luo et al., High yield of hydrocarbons from catalytic hydrodenitrogenation of indole under hydrothermal conditions. Energy Fuels **31**(11), 12594–12602 (2017)
67. B.C. Ledesma, M.L. Martinez, A.R. Beltramone, Iridium -supported SBA-15 modified with Ga and Al as a highly active catalyst in the hydrodenitrogenation of quinoline. Catal. Today **349**, 178–190 (2020)
68. M.-T. Nguyen et al., Indole hydrodenitrogenation over alumina and silica alumina-supported sulfide catalysts comparison with quinoline. Ind. Eng. Chem. Res. **56**(39), 11088–11099 (2017)
69. A. Bunch et al., Reaction network of indole hydrodenitrogenation over NiMoS/γ-Al_2O_3 catalysts. Appl. Catal. A **190**(1–2), 51–60 (2000)
70. J.L. Olivé et al., Hydroprocessing of indole and o-ethyl aniline over sulfided CoMo, NiMo and NiW catalysts. Appl. Catal. **19**(1), 165–174 (1985)
71. X. Liu et al., Hydrothermal hydrodeoxygenation of palmitic acid over Pt/C catalyst: mechanism and kinetic modeling. Chem. Eng. J. **407**, 126332 (2021)
72. S.C. Kim, F.E. Massoth, Hydrodenitrogenation activities of methyl-substituted indoles. J. Catal. **189**(1), 70–78 (2000)
73. Y. Guo et al., Hydrothermal reaction of tryptophan over Ni-based bimetallic catalysts. J. Supercrit. Fluids **143**, 336–345 (2019)

74. D. Ferdous, A.K. Dalai, J. Adjaye, Comparison of hydrodenitrogenation of model basic and nonbasic nitrogen species in a trickle bed reactor using commercial NiMo/Al_2O_3 catalyst. Energy Fuels **17**(1), 164–171 (2003)
75. X. Bai et al., Hydrothermal catalytic processing of pretreated algal oil: a catalyst screening study. Fuel **120**, 141–149 (2014)
76. P. Biller, R. Riley, A.B. Ross, Catalytic hydrothermal processing of microalgae: decomposition and upgrading of lipids. Biores. Technol. **102**(7), 4841–4848 (2011)
77. P. Duan, P.E. Savage, Upgrading of crude algal bio-oil in supercritical water. Biores. Technol. **102**(2), 1899–1906 (2011)
78. P.Q. Yuan et al., Catalytic denitrogenation of hydrocarbons through partial oxidation in supercritical water. Fuel **85**(3), 367–373 (2006)
79. P. Duan, P.E. Savage, Catalytic hydrothermal hydrodenitrogenation of pyridine. Appl. Catal. B **108–109**, 54–60 (2011)
80. Y. Xu, P. Duan, B. Wang, Catalytic upgrading of pretreated algal oil with a two-component catalyst mixture in supercritical water. Algal Res. **9**, 186–193 (2015)
81. P. Duan, P.E. Savage, Catalytic hydrotreatment of crude algal bio-oil in supercritical water. Appl. Catal. B **104**(1), 136–143 (2011)
82. Y. Xu, P. Duan, B. Wang, Catalytic upgrading of pretreated algal oil with a two-component catalyst mixture in supercritical water. Algal Res.-Biomass Biofuels Bioprod. **9**, 186–193 (2015)
83. P. Duan et al., Catalytic upgrading of pretreated algal bio-oil over zeolite catalysts in supercritical water. Biochem. Eng. J. **116**, 105–112 (2016)
84. P. Duan et al., Activated carbons for the hydrothermal upgrading of crude duckweed bio-oil. Catal. Today **274**, 73–81 (2016)
85. C. Zhang et al., Catalytic upgrading of duckweed biocrude in subcritical water. Biores. Technol. **166**, 37–44 (2014)
86. D. Xu, P.E. Savage, Supercritical water upgrading of water-insoluble and water-soluble biocrudes from hydrothermal liquefaction of Nannochloropsis microalgae. J. Supercrit. Fluids **133**, 683–689 (2018)
87. H. Topsøe, The role of Co–Mo–S type structures in hydrotreating catalysts. Appl. Catal. A **322**, 3–8 (2007)
88. J.V. Lauritsen et al., Location and coordination of promoter atoms in Co- and Ni-promoted MoS_2-based hydrotreating catalysts. J. Catal. **249**(2), 220–233 (2007)
89. R.R. Chianelli, Fundamental studies of transition-metal sulfide catalytic materials. Abs. Pap. Am. Chem. Soc. **214**, 98-PETR (1997)
90. K.-I. Tanaka, Catalysis controlled by the constitution of active sites, in *Advances in Catalysis*. ed. by D.D. Eley, H. Pines, P.B. Weisz (Academic Press, Elsevier, 1985), pp.99–158
91. M. Salmeron et al., The adsorption and binding of thiophene, butene and H_2S on the basal plane of MoS_2 single crystals. Chem. Phys. Lett. **90**(2), 105–107 (1982)
92. R.R. Chianelli et al., The reactivity of MoS_2 single crystal edge planes. J. Catal. **92**(1), 56–63 (1985)
93. H. Topsøe et al., In situ Mössbauer emission spectroscopy studies of unsupported and supported sulfided Co Mo hydrodesulfurization catalysts: evidence for and nature of a Co Mo S phase. J. Catal. **68**(2), 433–452 (1981)
94. R.G. de Castro et al., Surface-dependent activity of model CoMoS hydrotreating catalysts. J. Catal. **403**, 16–31 (2021)
95. L.L. Qu et al., MAS NMR, TPR, and TEM studies of the interaction of NiMo with alumina and silica-alumina supports. J. Catal. **215**(1), 7–13 (2003)
96. B. Delmon, G.F. Froment, Remote control of catalytic sites by spillover species: a chemical reaction engineering approach. Catal. Rev.-Sci. Eng. **38**(1), 69–100 (1996)
97. N. Nelson, R.B. Levy, The organic chemistry of hydrodenitrogenation. J. Catal. **58**(3), 485–488 (1979)
98. S. Eijsbouts, V.H.J. Debeer, R. Prins, Hydrodenitrogenation of quinoline over carbon-supported transition metal sulfides. J. Catal. **127**(2), 619–630 (1991)

99. T.G. Harvey, T.W. Matheson, Hydroprocessing catalysis by supported ruthenium sulphide. J. Catal. **101**(2), 253–261 (1986)
100. D. Gulkova, Y. Yoshimura, Z. Vit, Mesoporous silica-alumina as support for Pt and Pt-Mo sulfide catalysts: effect of Pt loading on activity and selectivity in HDS and HDN of model compounds. Appl. Catal. B-Environ. **87**(3–4), 171–180 (2009)
101. P. Chen et al., Recent progress of transition metal carbides/nitrides for electrocatalytic water splitting. J. Alloy. Compd. **883**, 160833 (2021)
102. S.T. Oyama, Preparation and catalytic properties of transition metal carbides and nitrides. Catal. Today **15**(2), 179–200 (1992)
103. Z. Mahdavifar, M. Ershadifar, A. Farrokhnia, Electro-optical properties and structural stability perspectives of M_3N and M_2C_2 ($M = Sc, La$) clusters encapsulated in B_{80} fullerene: a density functional theory study. J. Electron. Mater. **47**(1), 550–565 (2018)
104. E. Furimsky, Metal carbides and nitrides as potential catalysts for hydroprocessing. Appl. Catal. A **240**(1–2), 1–28 (2003)
105. M.K. Neylon et al., Catalytic properties of early transition metal nitrides and carbides: n-butane hydrogenolysis, dehydrogenation and isomerization. Appl. Catal. A **183**(2), 253–263 (1999)
106. V. Schwartz, S.T. Oyama, J.G.G. Chen, Supported bimetallic Nb-Mo carbide: synthesis, characterization, and reactivity. J. Phys. Chem. B **104**(37), 8800–8806 (2000)
107. H.A. Al-Megren et al., A comparative study of the catalytic performance of Co–Mo and Co(Ni)-W carbide catalysts in the hydrodenitrogenation (HDN) reaction of pyridine. Appl. Catal. A **329**, 36–45 (2007)
108. T.C. Xiao et al., Preparation of molybdenum carbide-based catalysts for deep HDN. Comptes Rendus De L Academie Des Sci. Serie Ii Fascicule C-Chimie **3**(6), 451–458 (2000)
109. T.C. Xiao et al., Preparation and characterisation of bimetallic cobalt and molybdenum carbides. J. Catal. **202**(1), 100–109 (2001)
110. D.J. Sajkowski, S.T. Oyama, Catalytic hydrotreating by molybdenum carbide and nitride: unsupported Mo_2N and Mo_2C/Al_2O_3. Appl. Catal. A **134**(2), 339–349 (1996)
111. C.W. Colling, L.T. Thompson, The structure and function of supported molybdenum nitride hydrodenitrogenation catalysts. J. Catal. **146**(1), 193–203 (1994)
112. J.-G. Choi et al., Synthesis and characterization of molybdenum nitride hydrodenitrogenation catalysts. Catal. Today **15**(2), 201–222 (1992)
113. S.Z. Li, J.S. Lee, Molybdenum nitride and carbide prepared from heteropolyacid II. Hydrodenitrogenation of indole. J. Catal. **173**(1), 134–144 (1998)
114. I.K. Milad et al., A comparison of bulk metal nitride catalysts for pyridine hydrodenitrogenation. Catal. Lett. **52**(1–2), 113–119 (1998)
115. Y.H. Wang et al., Characterization and catalytic properties of supported nickel molybdenum nitrides for hydrodenitrogenation. Appl. Catal. A **215**(1–2), 39–45 (2001)
116. Y. Li et al., The modification of molybdenum nitrides: the effect of the second metal component. Catal. Lett. **48**(3–4), 239–245 (1997)
117. S.T. Oyama, Novel catalysts for advanced hydroprocessing: transition metal phosphides. J. Catal. **216**(1–2), 343–352 (2003)
118. C. Xia et al., Selenide-based electrocatalysts and scaffolds for water oxidation applications. Adv. Mater. **28**(1), 77–85 (2016)
119. C. Stinner, R. Prins, T. Weber, Formation, structure, and HDN activity of unsupported molybdenum phosphide. J. Catal. **191**(2), 438–444 (2000)
120. P. Clark, W. Li, S.T. Oyama, Synthesis and activity of a new catalyst for hydroprocessing: tungsten phosphide. J. Catal. **200**(1), 140–147 (2001)
121. Z. Deng et al., Hydrodenitrogenation of pyridine on MoP(010): competition between hydrogenation and denitrification. Inorg. Chim. Acta **435**, 30–37 (2015)
122. R.H. Bowker et al., Carbazole hydrodenitrogenation over nickel phosphide and Ni-rich bimetallic phosphide catalysts. Appl. Catal. A **482**, 221–230 (2014)
123. C.A. Badari et al., Catalytic hydrodenitrogenation of propionitrile over supported nickel phosphide catalysts as a model reaction for the transformation of pyrolysis oil obtained from animal by-products. React. Kinet. Mech. Catal. **115**(1), 217–230 (2015)

124. A. Benitez et al., Effect of fluoride on the structure and activity of NiW/Al_2O_3 catalysts for HDS of thiophene and HDN of pyridine. Appl. Catal. A **144**(1–2), 343–364 (1996)
125. A.P. Ferreira et al., CH4-SCR of NO over Co and Pd ferrierite catalysts: effect of preparation on catalytic performance. Catal. Today **119**(1–4), 156–165 (2007)
126. R.D.C. Colman, M.A. Scaramelo Baldanza, M. Schmal, Influence of Mo species on the Pt/NaY catalyst and the effect of sulfur content on the hydrodenitrogenation reaction. J. Phys. Chem. C. **114**(43): 18501–18508 (2010)
127. Z. Sun et al., The effect of CeO_2 on the hydrodenitrogenation performance of bulk Ni_2P. Top. Catal. **55**(14–15), 1010–1021 (2012)
128. Y. Guo et al., Catalytic hydrodenitrogenation of pyridine under hydrothermal conditions: a comprehensive study. Acs Sustain. Chem. Eng. **9**(1), 362–374 (2021)
129. A. Infantes-Molina et al., HDS and HDN on SBA-supported RuS_2 catalysts promoted by Pt and Ir. J. Catal. **305**, 101–117 (2013)
130. Y. Guo et al., Catalytic gasification of indole in supercritical water. Appl. Catal. B **166–167**, 202–210 (2015)
131. Q. Guan et al., Catalytic gasification of phenol in supercritical water over bimetallic Co-Ni/AC catalyst. Environ. Technol. **40**(16), 2182–2190 (2019)
132. L. Jia et al., Noble-metal-free bimetallic alloy nanoparticle-catalytic gasification of phenol in supercritical water. J. Supercrit. Fluids **126**, 79–88 (2017)
133. G. Centeno et al., Effect of different heavy feedstocks on the deactivation of a commercial hydrotreating catalyst. Fuel **100**, 73–79 (2012)
134. Construction and property investigation of transition-metal complexes modified octamolybdate hybrid materials based on V-shaped organic ligands. CrystEngComm **12**: 434–445 (2010)
135. G. Wen et al., Production of hydrogen by aqueous-phase reforming of glycerol. Int. J. Hydrogen Energy **33**(22), 6657–6666 (2008)
136. T.V. Choudhary, S. Parrott, B. Johnson, Understanding the hydrodenitrogenation chemistry of heavy oils. Catal. Commun. **9**(9), 1853–1857 (2008)
137. A. Baiker, D. Monti, Y.S. Fan, Deactivation of copper, nickel, and cobalt catalysts by interaction with aliphatic amines. J. Catal. **88**(1), 81–88 (1984)
138. L. Hegedus, T. Máthé, Hydrogenation of pyrrole derivatives: part V: poisoning effect of nitrogen on precious metal on carbon catalysts. Appl. Catal. A **226**(1–2), 319–322 (2002)
139. X. Chen et al., Structure and composition changes of nitrogen compounds during the catalytic cracking process and their deactivating effect on catalysts. Energy Fuels **31**(4), 3659–3668 (2017)
140. M. Breysse et al., Hydrogen activation by transition metal sulfides. Catal. Rev. **44**, 651–735 (2002)
141. B.M. Vogelaar et al., Hydroprocessing catalyst deactivation in commercial practice. Catal. Today **154**(3–4), 256–263 (2010)
142. J.N. Jocz, P.E. Savage, L.T. Thompson, Heterogeneous catalyst stability during hydrodenitrogenation in supercritical water. Catal. Today **371**, 171–178 (2021)
143. M. Mapiour et al., Effects of the operating variables on hydrotreating of heavy gas oil: experimental, modeling, and kinetic studies. Fuel **89**(9), 2536–2543 (2010)
144. Z. Zhang, Y. Zhao, T. Wang, Spirulina hydrothermal carbonization: effect on hydrochar properties and sulfur transformation. Bioresour. Technol. **306** (2020)
145. S.G. Roussis, R. Cranford, N. Sytkovetskiy, Thermal treatment of crude algae oils prepared under hydrothermal extraction conditions. Energy Fuels **26**(8), 5294–5299 (2012)
146. M. Osada et al., Subcritical water regeneration of supported ruthenium catalyst poisoned by sulfur. Energy Fuels **22**(2), 845–849 (2008)
147. M. Osada et al., Effect of sulfur on catalytic gasification of lignin in supercritical water. Energy Fuels **21**(3), 1400–1405 (2007)
148. M. Dreher et al., Catalysis in supercritical water: pathway of the methanation reaction and sulfur poisoning over a Ru/C catalyst during the reforming of biomolecules. J. Catal. **301**, 38–45 (2013)

149. M. Dreher et al., On-stream regeneration of a sulfur-poisoned ruthenium-carbon catalyst under hydrothermal gasification conditions. ChemCatChem **6**(2), 626–633 (2014)
150. L. Wang, K. Murata, M. Inaba, Highly efficient conversion of gasoline into hydrogen on Al_2O_3-supported Ni-based catalysts: catalyst stability enhancement by modification with W. Appl. Catal. A **358**(2), 264–268 (2009)
151. B. Hwang et al., Density functional theory study for the enhanced sulfur tolerance of Ni catalysts by surface alloying. Appl. Surf. Sci. **429**, 87–94 (2018)
152. E. Laurent, B. Delmon, Influence of water in the deactivation of a sulfided NiMoγ-Al_2O_3 catalyst during hydrodeoxygenation. J. Catal. **146**(1), 281–291 (1994)
153. H. Huang et al., Effect of water on Cu/Zn catalyst for hydrogenation of fatty methyl ester to fatty alcohol. Korean J. Chem. Eng. **26**(6), 1574–1579 (2009)
154. D.L. Bhering, A. Ramírez-Solís, C.J.A. Mota, A density functional theory based approach to extraframework aluminum species in zeolites. J. Phys. Chem. B **107**(18), 4342–4347 (2003)
155. N.C. Concibido et al., Kinetics of deactivation of Pd/C catalyst repeatedly used in the liquid-phase hydrodechlorination of PCE. React. Kinet. Catal. Lett. **90**(1), 127–136 (2007)
156. Q. Guan et al., Catalytic gasification of lignin with Ni/Al_2O_3–SiO_2 in sub/supercritical water. J. Supercrit. Fluids **95**, 413–421 (2014)
157. T. Zhang et al., Enhanced hydrothermal stability of Cu-ZSM-5 catalyst via surface modification in the selective catalytic reduction of NO with NH_3. Appl. Surf. Sci. **375**, 186–195 (2016)
158. T. Usui et al., Improve the hydrothermal stability of Cu-SSZ-13 zeolite catalyst by loading a small amount of Ce. ACS Catal. **8**(10), 9165–9173 (2018)
159. X. Liu, X.T. Bi, Removal of inorganic constituents from pine barks and switchgrass. Fuel Process. Technol. **92**(7), 1273–1279 (2011)
160. H. Li et al., Conversion efficiency and oil quality of low-lipid high-protein and high-lipid low-protein microalgae via hydrothermal liquefaction. Biores. Technol. **154**, 322–329 (2014)
161. K. Anastasakis, A.B. Ross, Hydrothermal liquefaction of the brown macro-alga Laminaria Saccharina: effect of reaction conditions on product distribution and composition. Biores. Technol. **102**(7), 4876–4883 (2011)
162. M. Osada et al., Subcritical water regeneration of supported ruthenium catalyst poisoned by sulfur. Energy Fuels **22**, 845–849 (2008)
163. J. Jiang, P.E. Savage, Metals and other elements in biocrude from fast and isothermal hydrothermal liquefaction of microalgae. Energy Fuels **32**(4), 4118–4126 (2018)
164. J. Jiang, P.E. Savage, Using solvents to reduce the metal content in crude bio-oil from hydrothermal liquefaction of microalgae. Ind. Eng. Chem. Res. **58**(50), 22488–22496 (2019)
165. M.S. Haider, D. Castello, L.A. Rosendahl, The art of smooth continuous hydroprocessing of biocrudes obtained from hydrothermal liquefaction: hydrodemetallization and propensity for coke formation. Energy Fuels **35**(13), 10611–10622 (2021)
166. J. Jiang, P.E. Savage, Influence of process conditions and interventions on metals content in biocrude from hydrothermal liquefaction of microalgae. Algal Res.-Biomass Biofuels Bioprod. **26**, 131–134 (2017)
167. J.G. Reynolds, W.R. Biggs, S.A. Bezman, Reaction sequence of metallopetroporphyrins during heavy residuum upgrading. Prepr.-Am. Chem. Soc. Div. Petrol. Chem. **31** (1986)
168. S. Fukuzumi, T. Honda, T. Kojima, Structures and photoinduced electron transfer of protonated complexes of porphyrins and metallophthalocyanines. Coord. Chem. Rev. **256**(21), 2488–2502 (2012)
169. J.M. Jarvis et al., Impact of iron porphyrin complexes when hydroprocessing algal HTL biocrude. Fuel **182**, 411–418 (2016)
170. M. Gamboa, M. Campos, L. Alfonso Torres, Study of the stability of 5,10,15,20-tetraphenylporphine (TPP) and metalloporphyrins NiTPP, CoTPP, CuTPP, and ZnTPP by differential scanning calorimetry and thermogravimetry. J. Chem. Thermodyn. **42**(5): 666–674 (2010)
171. J. Ryou, S. Hong, G. Kim, Ab initio study of hydrogen binding on Ca-inserted porphyrin. Vacuum **84**(5), 537–539 (2009)

172. Z. Wang et al., Upgrading of hydrothermal liquefaction biocrude from algae grown in municipal wastewater. Fuel Process. Technol. **142**, 147–156 (2016)
173. J. Jiang et al., Fate of iron during hydrothermal liquefaction of hemin. J. Supercrit. Fluids **157**, 104705 (2020)
174. L.A. Hayes et al., Binding of metalloporphyrins to model nitrogen bases: collision-induced dissociation and ion–molecule reaction studies. Int. J. Mass Spectrom. **227**(1), 111–120 (2003)
175. S.A. Stevenson, A. Lisitsyn, H. Knozinger, Adsorption of carbon monoxide on manganese-promoted rhodium/silica catalysts as studied by infrared spectroscopy. J. Phys. Chem. **94**(4), 1576–1581 (1990)
176. M.S. Rana et al., Heavy oil hydroprocessing over supported NiMo sulfided catalyst: an inhibition effect by added H2S. Fuel **86**(9), 1263–1269 (2007)
177. R.L.C. Bonné, P. van Steenderen, J.A. Moulijn, Hydrogenation of nickel and vanadyl tetraphenylporphyrin in absence of a catalyst: a kinetic study. Appl. Catal. A **206**(2), 171–181 (2001)
178. P.C. Mandal et al., Nickel removal from nickel etioporphyrin (Ni-EP) using supercritical water in the absence of catalyst. Fuel Process. Technol. **104**, 67–72 (2012)
179. X.-Q. Tang et al., Demetallization of heavy oils of high metal content through pyrolysis under supercritical water environment. Energy Fuels **34**(3), 2861–2869 (2020)
180. P.C. Mandal, M. Goto, M. Sasaki, Removal of nickel and vanadium from heavy oils using supercritical water. J. Jpn. Petrol. Inst. **57**(1), 18–28 (2014)
181. C.H. Tsai et al., Effect of coal-derived-liquid solvent on the hydrogenation and restrictive diffusion of nickel porphyrins. Fuel Process. Technol. **29**(3), 153–169 (1991)
182. O.N. Fedyaeva, A.V. Shatrova, A.A. Vostrikov, Effect of temperature on bitumen conversion in a supercritical water flow. J. Supercrit. Fluids **95**, 437–443 (2014)
183. Y. Cheng et al., Enhancing the selectivity of the hydrogenation of naphthalene to tetralin by high temperature water. Green Chem. **11**(7), 1061–1065 (2009)
184. K.K. Dailey, T.B. Rauchfuss, Pi-complexes of metalloporphyrins as model intermediates in hydrodemetallation (HDM) catalysis. Polyhedron **16**(18), 3129–3138 (1997)
185. J. Ancheyta, M.S. Rana, E. Furimsky, Hydroprocessing of heavy oil fractions. Catal. Today **109**, 1–2 (2005)
186. M.S. Rana et al., Maya crude hydrodemetallization and hydrodesulfurization catalysts: an effect of TiO_2 incorporation in Al_2O_3. Catal. Today **109**(1–4), 61–68 (2005)
187. F. Vandeneeckoutte et al., Thermal equilibrium between vanadyloctaethylporphyrin and hydrogenated porphyrin during the hydrodemetallation process. React. Kinet. Catal. Lett. **45**, 191–198 (1991)
188. P.W. Tamm, H.F. Harnsberger, A.G. Bridge, Effects of feed metals on catalyst aging in hydroprocessing residuum. Ind. Eng. Chem. Process. Des. Dev. **20**(2), 262–273 (1981)
189. R.A. Ware, J. Wei, Catalytic hydrodemetallation of nickel porphyrins: II. Effects of pyridine and of sulfiding. J. Catal. **93**(1), 100–121 (1985)
190. C. Leyva et al., On the use of acid-base-supported catalysts for hydroprocessing of heavy petroleum. Ind. Eng. Chem. Res. **46**(23), 7448–7466 (2007)
191. R. Agrawal, J. Wei, Hydrodemetalation of nickel and vanadium porphyrins. 1. Intrinsic kinetics. Ind. Eng. Chem. Process Des. Develop. **23**(3), 505–514 (1984)
192. R.A. Ware, J. Wei, Catalytic hydrodemetallation of nickel porphyrins: II. Effects of pyridine and of sulfiding. J. Catal. **93**(1), 122–134 (1985)
193. H.J. Chen, F.E. Massoth, Hydrodemetalation of vanadium and nickel porphyrins over sulfided cobalt-molybdenum/alumina catalyst. Ind. Eng. Chem. Res. **27**(9), 1629–1639 (1988)
194. T. Takahashi, H. Higashi, T. Kai, Development of a new hydrodemetallization catalyst for deep desulfurization of atmospheric residue and the effect of reaction temperature on catalyst deactivation. Catal. Today. Today **104**(1), 76–85 (2005)

MIX
Papier aus verantwortungsvollen Quellen
Paper from responsible sources
FSC® C105338

If you have any concerns about our products,
you can contact us on
ProductSafety@springernature.com

In case Publisher is established outside the EU,
the EU authorized representative is:
Springer Nature Customer Service Center GmbH
Europaplatz 3, 69115 Heidelberg, Germany

Printed by Libri Plureos GmbH
in Hamburg, Germany